Deciphering a Shell Midden

Deciphering a Shell Midden

Edited by

Julie K. Stein

Department of Anthropology and Burke Museum
University of Washington
Seattle, Washington

Academic Press, Inc.

Harcourt Brace Jovanovich, Publishers

San Diego New York Boston London Sydney Tokyo Toronto

Front cover illustration by Timothy D. Hunt

This book is printed on acid-free paper. ♾

Academic Press, Inc.
1250 Sixth Avenue, San Diego, California 92101-4311

United Kingdom Edition published by
Academic Press Limited
24–28 Oval Road, London NW1 7DX

Library of Congress Cataloging-in-Publication Data

Deciphering a shell midden / [edited by] Julie K. Stein.
p. cm.
Includes bibliographical references and index.
ISBN 0-12-664730-5
1. Kitchen-middens. 2. Coastal archaeology. 3. Great Britain--Antiquities. 4. Indians of North America--Northwest Coast of North America--Antiquities. 5. Northwest Coast of North America--Antiquities. I. Stein, Julie K.
CC77.S5D43 1992
930.1--dc20 91-41639
CIP

PRINTED IN THE UNITED STATES OF AMERICA
92 93 94 95 96 97 BB 9 8 7 6 5 4 3 2 1

Contents

8

Shell Midden Lithic Technology: Analysis of Stone Artifacts from British Camp

Kimberly D. Kornbacher

9

Lithic Manufacturing at British Camp: Evidence from Size Distributions and Microartifacts

Mark E. Madsen

10

An Analysis of Fire-Cracked Rocks: A Sedimentological Approach

Timothy W. Latas

11

Shell Midden Deposits and the Archaeobotanical Record: A Case from the Northwest Coast

Margaret A. Nelson

12

Effects of Recovery Techniques and Postdepositional Environment on Archaeological Wood Charcoal Assemblages

Diana M. Greenlee

14

Is Bone Safe in a Shell Midden?

Angela R. Linse

15

Burned Archaeological Bone

Patrick T. McCutcheon

Contributors

Numbers in parentheses indicate the pages on which the authors' contributions begin.

Rinita A. Dalan (43), Center for Ancient Studies and Institute for Rock Magnetism, University of Minnesota, Minneapolis, Minnesota 55455

Pamela J. Ford (283), Mt. San Antonio College, Walnut, California 97189

Diana M. Greenlee (261), Department of Anthropology, University of Washington, Seattle, Washington 98195

Kimberly D. Kornbacher (95, 163), Department of Anthropology, University of Washington, Seattle, Washington 98195

Timothy W. Latas (211), Kleinfelder, Inc., San Diego, California 92123

Angela R. Linse (327), Department of Anthropology, University of Washington, Seattle, Washington 98195

Mark E. Madsen (193), Department of Anthropology, University of Washington, Seattle, Washington 98195

Patrick T. McCutcheon (347), Department of Anthropology, University of Washington, Seattle, Washington 98195

John M. Musser, Jr. (43), Geo-Recon International, Ltd., Seattle, Washington 98155

Margaret A. Nelson (239), Department of Anthropology, University of Washington, Seattle, Washington 98195

Julie K. Stein (1, 25, 43, 71, 95, 135), Department of Anthropology and Burke Museum, University of Washington, Seattle, Washington 98195

Bryn H. Thomas (61), Archaeological and Historical Services, Eastern Washington University, Cheney, Washington 99004

James W. Thomson (61), Pacific Northwest Regional Office, National Park Service, Seattle, Washington 98104

Jason L. Tyler (95), Department of Anthropology, University of Washington, Seattle, Washington 98195

Fran H. Whittaker (25), Department of Anthropology, University of Washington, Seattle, Washington 98195

Preface

The reason why some archaeologists are drawn specifically to shell middens remains a mystery. Shell middens are one of the most stratigraphically complex types of sites in the world. Many archaeologists suggest that the "challenge" is the magnet that draws us. However, I suspect of equal importance is the proximity of shell middens to coasts with magnificent views. Whatever the reasons, there is a core group of archaeologists who turn consistently toward research questions answered only through the analysis of shell middens. This book is written especially for this group, with the hope that the burdens associated with shell middens are lessened.

The chapters in this book are roughly divided into an initial section where the location, environment, and stratigraphy of the site are described (Chapters 1–5), followed by a second section (Chapters 6–15) of specific analyses, organized by material type (stratigraphy, sediments, lithics, floral remains, and faunal remains).

In Chapter 1, "The Analysis of Shell Middens," the research objectives and site location are introduced. In Chapter 2, "Shell Midden Boundaries in Relation to Past and Present Shorelines," the location around British Camp is described, the area before prehistoric occupation is reconstructed, and the history of sea level in the region and its effect on the British Camp shell midden is discussed. In Chapter 3, "Geophysical Exploration of the Shell Midden," an additional reconstruction of the site is provided, using geophysical exploration. This reconstruction provides information on the deeply buried portion of the site and the sediment below the shell midden. In Chapter 4, "Historic Treatment of a Prehistoric Landscape," the effects of the historic British occupation are considered. The evidence is derived from descriptions written by British soldiers and historic photographs. In Chapter 5, "Interpreting Stratification of a Shell Midden," archaeological and geological stratigraphy is discussed. This chapter contains the definitions of terms and includes a discussion of archaeological stratigraphic nomenclature. These five chapters provide the background for the analytical results that follow.

The analyses conducted on material removed from excavation units begins with Chapter 6, "British Camp Shell Midden Stratigraphy." In this chapter the stratigraphy of British Camp is summarized, including the methods used to excavate, record, describe, and correlate deposits. In Chapter 7, "Sediment Analysis of the British

Camp Shell Midden," the sediment data used to document the postdepositional alteration by groundwater are described. The next three chapters are concerned with lithics. In Chapter 8, "Shell Midden Lithic Technology: Analysis of Stone Artifacts from British Camp," the chipped stone artifacts and debitage found at British Camp are described. In Chapter 9, "Lithic Manufacturing at British Camp: Evidence from Size Distributions and Microartifacts," the size distribution of stone debitage is examined to explore the effects of postdepositional movements of debitage in the midden. In Chapter 10, "An Analysis of Fire-Cracked Rocks: A Sedimentological Approach," the attributes used to describe fractured rocks are examined and the origin of the fracture investigated.

The remaining chapters report the analyses of biological remains. In Chapter 11, "Shell Midden Deposits and the Archaeobotanical Record: A Case from the Northwest Coast," charcoal and seeds from strata within the site are compared. In Chapter 12, "Effects of Recovery Techniques and Postdepositional Environment on Archaeological Wood Charcoal Assemblages," the effects of the flotation device on recovery of charred plant remains is examined, including measurements of charcoal's volume and effects of precipitates on the recovery by flotation. In Chapter 13, "Interpreting the Grain Size Distributions of Archaeological Shell," the grain size distribution of each taxa is compared to interpret how many depositional events are recorded by shell from a particular strata. In Chapter 14, "Is Bone Safe in a Shell Midden?," a theoretical dissolution curve for hydroxyapatite is used to investigate whether bone will be preserved in sediments with high alkaline pH. In Chapter 15, "Burned Archaeological Bone," a classification for burned bone is presented to determine the temperature at which bone is burned, as well as the effects of British Camp groundwater on that classification.

Acknowledgments

The San Juan Island Archaeological Project has been supported by many people. The project began in 1983 when the University of Washington, facilitated by the encouragement of Robert Dunnell, agreed to fund an archaeological field school. The National Park Service (NPS) became involved when I asked to examine the British Camp shell midden. Every summer from 1983 to 1989 the project excavated the British Camp shell midden.

The most significant contributions have been made by the staff of the project, who implemented the research design. A rich educational environment was provided by the following staff: (1984) G. Thomas Jones, Pamela Ford, Margaret Nelson, Elizabeth Dennis, James Hale; (1985) G. T. Jones, P. Ford, M. Nelson, Charlotte Beck, Debra Hume; (1986) P. Ford, M. Nelson, Fran Whittaker, Kimberly Kornbacher, Angela Linse, Jana McAnally, D. Hume; (1987) P. Ford, M. Nelson, Patrick McCutcheon, Mary Parr, Walter Bartholomew, Eric Rassmussen, Patricia Rassmussen; (1988) M. Nelson, K. Kornbacher, Kris Wilhelmsen, M. Parr, Stella Spring, W. Bartholomew, E. Rassmussen, P. Rassmussen; (1989) M. Nelson, K. Kornbacher, K. Wilhelmsen, M. Parr, Helen Cullen, E. Rassmussen, P. Rassmussen.

The students contributed as well with their insights and hard work. In 1984 the students were Daniel Alden, Joel Allen, Andrew Ching, David Conca, Shanna Green, Diana Greenlee, Catherine Griffin, Gary Kitchen, Jenny Lemire, Angela Linse, Jana McAnally, Patrick McCutcheon, Leslie Norman, Chris Ottaway, Eric Richardson, Bettina Scheffel, Robin Teas, Sharon Ulsh, Kurt Weihs, and Zeke Wells.

In 1985 the students were Meg Burch, Claire Carlson, John Carson, Kalin Cumbridge, Suzanne Dickerson, Sonja Duke, Laurey Edelstein, Thomas Edney, James Ferrara, Cori Giannuzzi, Maureen Hopper, Dinah Hunter, Alice Keesey, Kathleen Kornbacher, Kimberley Kornbacher, Tim Latas, Charles Lynn, Marguerite McAdams, Julie Mclean, Douglas Miller, Clifford Nicoll, John Picklesimer, Patrick Ryan, Elisabeth Sperling, Matthew Steinkamp, Jason Tyler, Alicia Wise, and Megan Wolf.

In 1986 the students were Walter Bartholomew, Sibel Barut, Vuong Ha, Jenny Holder, Scott King, Quyen Nguyen, Sean Oslin, Mary Parr, Eric Rasmussen, Jonathan Russell, Sarah Sherwood, Mark Smith, Carlyn Smuels, Michelle Spomer, Stella Spring, and Thomas Whitley.

In 1987 the students were Dabney Benjamin, Nancy Bowers, Anthony Cagle, Carol Carr, Amanda Cohn, Melissa Divack, Nolan Feintuch, David Grant, Jerome Hawkins, Siendie Joy, Elonna Lester, Alison MacDougal, Mark Madsen, Linda Quamme, Judy Rosenthal, Genny Stone, and Greg Sullivan.

In 1988 the students were Kari Aunan, Edward Bakewell, Deborah Blehm, Robert Bohus, Karen Carmer, Helen Cullen, Amy Eggler, Dottie Foskin, Kathryn Frost, Kirk Ghio, Dennis Gosser, Montgomery Nelson, Shannon Paulsen, Coral Rassmussen, Celeste Ray, Katherine Russell, Amy Sievers, Heidi Thorsen, Della Valdez, and Annemarie Williams.

In 1989 the students were Daniel Alden, Gregory Bantham, Michael Carmody, Genevieve Carnell, Pat Clabaugh, Dianne Ficarra, Alison Hoff, Shane Kleven, Suzanne Lowery, Christopher Neimeth, Vaughn Rawland, Olive Rieflin, Erika Sakrison, David Sherman, Nicholas Smith, Annette Stillwell, and Barbara Terzian.

I wish to thank them all for their cooperation and for caring about archaeology.

In addition to staff and students, many volunteers helped this project: Betty Ackley, George Black, Wendy Fairfax, Ron Garner, Carol Garner, Judith Greenfield, Susan Harris, Betty Heiman, Edna Knudsen, Vivian Landbeck, Joan May, Susan Meridith, James Meridith, Toni Niemann, Toots Oliver, Mary Jane Pitt, Nancy Todd, George Warner, and Irene Warner. Most of these individuals live on San Juan Island and care deeply about preserving the cultural resources of the island. Their participation was facilitated by the National Park Service Volunteer Program, administered by Detleff Wieck and Steve Gobat, and organized by Irene Warner. Other people who volunteered information and services are Jim Auel and Marshall Sanborn.

This research was possible in a large part because of the support of the National Park Service. First and foremost, I would like to acknowledge the assistance of the Northwest Regional Archaeologist, Jim Thomson. His strong encouragement and financial support helped sustain the research. Also, the Northwest Regional Curator, Kent Bush, taught me the meaning of curation and helped produce a collection that has significant research value. It is completely computerized and curated properly. Perhaps the most significant assistance, however, was given by the superintendents of San Juan Island National Historical Park, Frank Hastings (1983–1985) and Dick Hoffman (1985–1989). They gave us logistical support and permission to be in the park. I wish also to thank NPS staff Steve Gobat, Detleff Wieck, Kathy Wieck, Diane Joy, David Arnold, Wes Callender, Mac Foreman, and Mary Jane Leche, all of whom made our lives easier.

Many colleagues helped this research through the years, giving guest lectures to the field school students, sharing historic information about previous excavations, and giving advice about field strategies. I wish to thank Kenneth Ames, Kitty Bernick, Roy Carlson, Sarah Campbell, Dale Croes, Robert Dunnell, Jon Erlanson, Paul Goldberg, Gar Grabert, Donald Grayson, Robert Greengo, James Haggarty, Phillip Hobler, Richard Inglis, Stephen Kenady, Patrick Kirch, R. G. Matson, Bob Merendorf, Donald Mitchell, Madonna Moss, Astrida Onat, James Plaster, David Pokotylo, Jan Simek, Roderick Sprague, Wayne Suttles, Patrice Teltser, Bryn Thomas, Gail Thompson, Jim Thomson, Gary Wessen, and Rob Whitlam. I would like to thank the

Lummi tribal members, who shared with us their oral history, provided interpretations of archaeological materials, and were interested in our research efforts.

This research was financially supported by the University of Washington Summer Quarter and College of Arts and Sciences; a grant from the University of Washington Graduate Faculty Research Fund; the National Park Service; and private donations by Diane Adkins, Kathy Clayson, Marie Coyle, Michael Ferro, Jeffrey Fintz, Cori Giannuzzi, Peggy Hamernyik, John Hastings, Clarence and Lydia Johnson, Pam Kidd, David Linse, Carole Linse, Nicholas Nelson, Katherine Nelson, Mary Jane Pitt, Vidmantas Raisys, Freya Richardson, Jerry Stein, Eileen Stein, LaVerne Szabo, Barbara Terzian, Nina Watts, William Wolfe, and Molly Wolfe.

I wish to thank the individuals who helped me personally through this project. Mary Parr organized the laboratories in the field and at the University of Washington. As laboratory supervisor, she (with help from Angela Linse, Pam Ford, Joel Allen, and James Hale) organized the entire collection from the moment the objects left the ground until they were curated in their NPS bags. Rick Hilton untangled all our computer problems and helped us back up 60,000 individual catalog entries. Walter Bartholomew cataloged most of the 60,000 objects in this collection. Without his organizational ability the curation effort would have been impossible. Eric Rasmussen built the complicated water recovery system and the screens, our laboratory tables, shelves, and seats, as well as our kitchen in the woods. He spent four summers building magnificent creations with almost no supplies. Patty Rasmussen and Debra Hume fed us magnificently. We could never have completed this research without Patty's homemade pasta and Northern Italian cuisine. Jim Thomson, Bryn Thomas, John Ehrenhard, Eric Rasmussen, Galen Cawlfield, and Dick Hoffman solved our water-shortage problem by rerouting drainage and cleaning out the cistern. Lastly, and most important, my research colleagues Pam Ford, Meg Nelson, Kim Kornbacher, Angela Linse, Mary Parr, and Kris Wilhelmsen have contributed greatly to this finished product. Year after year they participated in planning meetings, interpreted data, and invented new ways to make the excavation strategy better and better. This research is as much theirs as mine. Finally, I wish to thank Stanley Chernicoff who advised me on geological matters, as well as took care of our children and home.

1

The Analysis of Shell Middens

Julie K. Stein

I. Introduction

Shell middens are found in nearly every coastal area of the world, and have been recognized as the remains of prehistoric peoples for a century. Shell heaps observed by naturalists in the eighteenth century were thought to be caused by extraordinary natural phenomena, and the early investigations of shell middens concentrated on establishing their human origins. Once the origin of shell middens was established as cultural, archaeologists shifted away from studying their origin to studying their content. For the last century most research on shell middens focused on the biological remains as indicators of subsistence and the artifactual remains as indicators of culture history. Although subsistence and culture history are crucial aspects of archaeology, the shell middens from which the biological and artifactual remains come have complex origins. The remains on which the reconstructions are based have been affected by postdepositional alterations that change the character of the site. To correct the biases introduced by postdepositional alterations, archaeologists need to examine (as they did a century ago) the origins of shell middens and their diagenesis.

All shell middens have certain properties in common. Shell comes largely from the hard parts of aquatic fauna (freshwater or marine), and sites containing shell are usually located adjacent to aquatic environments (Bailey and Parkington 1988; Sullivan 1980; Seilacher, 1973). Sites that contain shell have: (1) increased porosity, permeability, and alkalinity (Ceci 1984; Sanger 1981); (2) low densities of historically diagnostic artifacts and high densities of shell (Sanger 1985c); and (3) high probabilities of being saturated by the adjacent body of water (Bloom 1983; Stright 1990). These properties must be described and interpreted to decipher the depositional history of a shell midden.

Deciphering a Shell Midden contains results of research conducted at a Northwest Coast shell midden where porosity, alkalinity, low density of artifacts, and saturation by rising sea level have all affected the formation of the site. The research

Deciphering a Shell Midden

focuses on the striking stratigraphic feature of the shell midden: an obvious division of light-colored matrix in the upper portion of the midden and dark-colored matrix in the lower portion. Although this stratigraphy has been observed at other shell middens, it has never been analyzed in detail. This book will demonstrate that the dual stratigraphy stems from the postdepositional saturation of the shell midden by groundwater, which in turn was influenced by the Late Holocene rise in sea level. The groundwater has hydrated the clay and organic matter, darkened the color of the organic matter, leached carbonate from the fine-grained sediment fraction, and produced the characteristic "greasy" feel of the matrix. These processes have been superimposed onto the stratification of subsistence and artifactual material, and need to be separated if appropriate archaeological interpretations are to be made. The properties of the shell midden and its proximity to the shoreline have resulted in drastic alterations of the original stratification of the site.

In this chapter, the general location of the shell midden and history of investigation at this site is described, followed by a detailed discussion of shell midden research. The discussion of research traces the history of shell midden investigations and emphasizes those research projects that use geoarchaeological methods. Last, a comparison is made of the peculiar two-toned stratification discovered at this Northwest Coast shell midden to the stratigraphy of other shell middens in the region, demonstrating that the postdepositional processes discovered here are endemic to the region, and by extension, to most coastal areas of the world.

II. Location and History of Site

The British Camp shell midden, located near the international boundary between the United States and Canada in the state of Washington (Figures 1 and 2), is part of the San Juan Island National Historic Park-British Camp (45SJ24). In 1966 the San Juan Island National Historical Park was established by Congress to preserve the sites of American and British Camps, which were occupied from 1859 to 1872 during the "Pig War" (Thompson 1972). The British Camp shell midden was originally called English Camp after the name of the park. However, the name of the park was changed by the National Park Service in 1985 because the soldiers serving at the encampment were not all English. The prehistoric site is refered to in archaeological literature as English Camp, but to avoid confusion with interpretive concerns at the park, the name British Camp is used in this research.

The site is a large shell midden, over 4 m deep and 300 m long, located along the shores of Garrison Bay on San Juan Island (Figure 3). The British built British Camp on the "site of an old indian village. . . . as usual at such localities there were immense quantities of clam shells on the shore" (Thompson 1972:111). Although the British inhabited British Camp for a very short time, Northwest Coast peoples had inhabited the site for millennia. Fortunately, the establishment of the park preserved the important prehistoric and historic sites from encroaching urban development.

The British Camp locale was inhabited by prehistoric and historic peoples whose artifactual remains are most closely tied to those found in sites within the Gulf of

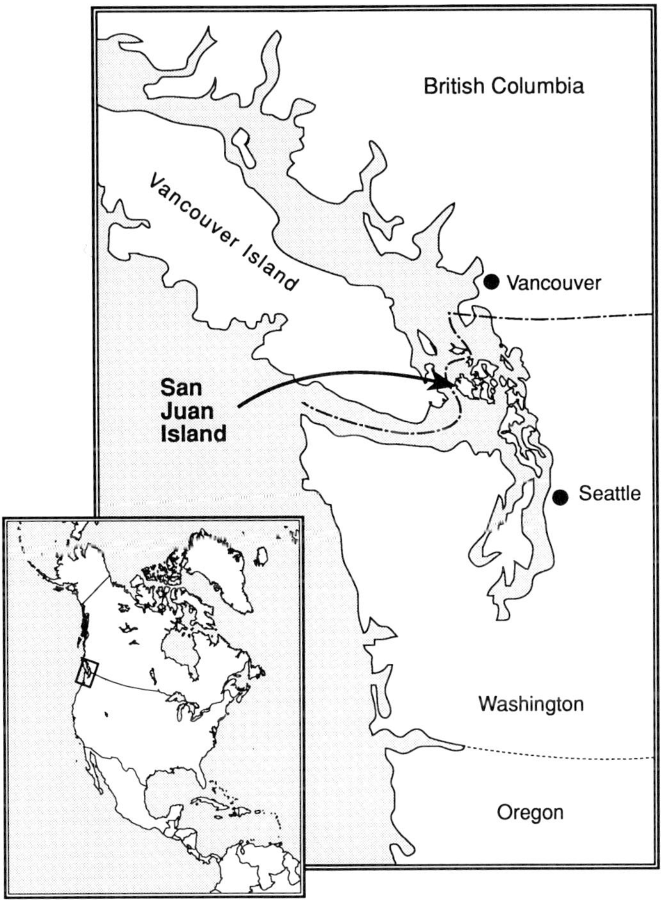

Figure 1 Location of San Juan Island, Washington, near the international border of Canada and the United States.

Georgia area (Carlson 1960; Mitchell 1971, 1990). The Gulf of Georgia area is bounded on the west by the Coast Mountains of Vancouver Island and on the east by the Cascade Mountains. The northern boundary is the entrance to Queen Charlotte Strait and the southern boundary the Strait of Juan de Fuca and the entrance to Puget Sound. Ethnographically this area was inhabited by Central Coast Salish-speaking peoples, who focused their subsistence on marine resources, especially migrating salmon and herring, as well as shellfish, birds, deer, elk, berries, camas, and other plants and animals (Suttles 1951, 1968, 1983, 1987, 1990; Suttles and Elmendorf 1963).

The San Juan Island Archaeological Project (responsible for the research reported in this book) is not the first archaeological excavation of the British Camp midden. In 1950 A. E. Treganza excavated a portion of British Camp as a University of Washington Field School. Roy Carlson was a student of that field school and analyzed

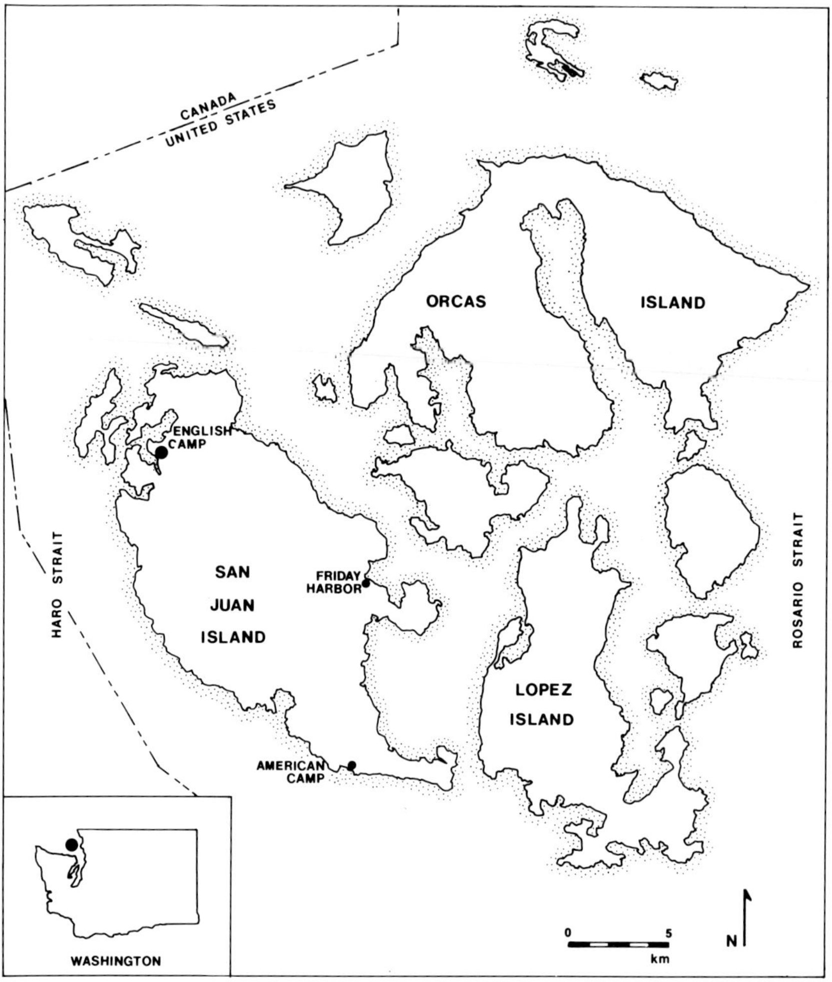

Figure 2 Map of San Juan Island, one of the larger islands in the group called the San Juan Islands. The British Camp shell midden (here labeled English Camp) is a large shell midden, located along the shores of Garrison Bay in the northwestern end of the island. A second park, San Juan Island National Historic Park-American Camp, is located on the southern end of the island. Friday Harbor is the nearest town and the location of the ferry terminal. Drafted by M. Nelson.

some of the Treganza artifacts as part of his master's thesis (Carlson 1954). In 1951 Carlson excavated the Garrison site (45SJ25) adjacent to the British Camp shell midden, as well as several other sites on the island (Carlson 1960). Carlson notes that 45SJ25 and 45SJ24 are continuations of one large site (1960:563).

The next excavations were sponsored by the National Park Service from 1970 to 1978 and directed toward stabilizing and interpreting the historic structures within the park (Sprague 1983). At the same time Sprague directed the historic archaeological

Figure 3 Garrison Bay is in the center of the photo with the sailboats moored in its protected waters. The British Camp shell midden is near the shore of the grassy area on the right. On the far horizon is Vancouver Island, looking across Haro Strait.

investigations (Sprague 1973, 1976, 1983), Stephen M. Kenady, a student at the University of Washington, supervised the excavations of the prehistoric portion of the site. Kenady summarized some preliminary results of excavations conducted in 1970, 1971, and 1972 (Kenady 1971, 1972, 1973). He opened small units at regular intervals along the shoreline, and a larger unit near the commissary (Figure 4). The dark and light stratification is especially clear in the profiles of this large unit (Figure 5). Kenady was the first to investigate the dark and light layers at British Camp (1972), and many of his observations and descriptions contributed significantly to the results reported in this volume.

In the spring of 1983, Pamela Ford and Julie Stein presented a research proposal for excavations at the British Camp site to Frank Hastings (superintendent of San Juan Island National Historic Park), and Jim Thomson (Archaeologist for the Pacific Northwest Region, National Park Service). In the summer of 1983 a small test pit was excavated by Pamela Ford, Margaret Nelson, and Julie Stein. Information from this excavation was used to design the large-scale excavation started there in 1984, called Operation A (see Whittaker and Stein, Ch. 2, this volume).

During the period from 1983 to 1989, other excavations took place at British Camp. Jim Thomson, Cathy Gilbert, and Bryn Thomas (see Thomas and Thomson, Ch. 4, this volume), conducted excavations to reconstruct the historic landscape, stabilize historic features, and locate historic buildings not previously mapped. Following the closing of Operation A in 1989, a new excavation was begun in Operation D. The

Figure 4 In 1972 Stephen Kenady directed the excavation of the large unit in the center of the photo. The commissary is on the left, the block house is near the water on the right. The shell midden extends back from the shore almost to the trees. See Figure 5 in Ch. 2, this volume for a comparison of the location of this unit and Operation A. Photo by S. Kenady.

location of this new investigation is near Treganza's original excavation at the boundary of 45SJ24 and 45SJ25. The results of this research will form the body of a dissertation conducted by Kimberly D. Kornbacher at the University of Washington.

III. History of Shell Midden Research

The term shell midden has a long tradition in archaeology (Christenson 1985; Trigger 1986) and carries connotations that are inaccurate (Claassen 1991a). The word "midden" has its roots in the Scandinavian languages, meaning an accumulation of refuse about a dwelling place. Because most shell probably did not accumulate about a dwelling per se, a more accurate term for sites with remains of shellfish is "shell-bearing site" (Widmer 1989), a term which does not restrict the archaeological deposits to only middens. Although the term shell-bearing site is more accurate, it is also cumbersome and problematic (Claassen 1991a:252). Because the term "shell midden" has a long tradition in archaeology it is used here to describe sites that contain shell, yet I recognize that all the deposits may not necessarily be middens.

Although research conducted at shell middens focuses most frequently on artifacts and fauna (Ambrose 1967), at the British Camp shell midden a new focus dominates the research. Ambrose (1967) summarizes clearly the many research questions addressed by examining shell middens, dividing them into four categories (Table 1): (1) looking at faunal remains in terms of available food supply (e.g., reconstructing meat weights and calories); (2) plotting variations in shellfish species acquired from column samples to infer changing ecological conditions in nearby aqueous habitats (e.g., noting shifts from abundant rocky-bottom bay shellfish species to abundant soft-bottom bay species); (3) delineating ancient shoreline locations by plotting geomorphic positions of shell middens; and (4) ignoring shellfish remains and constructing cultural historical sequences by analyzing the artifacts only. The research at the British Camp shell midden represents a fifth category: (5) examining the shell midden in terms of its depositional and postdepositional processes.

Most of the earliest investigations of shell middens fall into the first and fourth categories of Table 1, those of analyzing faunal remains in terms of available food supply, and for purposes of constructing cultural histories. Some of the earliest shell midden analyses in the western United States are those conducted in California (Nelson 1909; Gifford 1916), where a tradition began in which shell midden size became an estimate of site age, prehistoric diet, and population (Ascher 1959; Cook 1946, 1950; Cook and Heizer 1951; Cook and Treganza 1947, 1950; Greengo 1951; Greenwood 1961; Treganza and Cook 1948). These studies became the models for research conducted in the following decades, especially on the west coast of North America. In the eastern United States shell middens were excavated by the Tennessee

Figure 5 In 1972 Stephen Kenady excavated this large unit, exposing the dual stratigraphy of shell midden with light-colored matrix above shell midden with dark-colored matrix. This view is of the southern profile of the unit observed in Figure 3. Photo by S. Kenady.

Table 1 Categories of Shell Midden Research

Category[a]	Description
1	Examining shellfish remains in terms of available food supply (e.g., reconstructing meat weights and calories)
2	Plotting variations in shellfish species, as measured from column samples in shell middens, to infer changes in ecological conditions within nearby aqueous habitats (e.g., noting shifts from abundant rocky-bottom bay shellfish species to abundant soft-bottom bay species)
3	Delineating ancient shoreline locations by plotting geomorphic positions of shell middens, and inferring that the location of the shell midden indicates the location of the paleoshoreline
4	Ignoring the shellfish remains in the site and constructing cultural historical sequences by analyzing artifacts only
5	Examining the shell midden in terms of its depositional and postdepositional processes

[a]Categories 1–4 are described in Ambrose (1967).

Valley Authority (TVA) and Works Progress Administration (WPA) (Quimby 1979) to conserve archaeological material and information that was being inundated by flooding (Webb 1938), and to arrange them within a chronological framework (e.g., Webb 1946; Webb and DeJarnette 1942; Webb and Haag 1939). The examination of these shell middens led to the concept of a Shell Mound Archaic in the eastern Woodlands (Funkhouser and Webb 1932; Rolingson 1967; see discussion in Claassen 1991b). Unlike the California archaeologists, the WPA/TVA researchers never analyzed the shell as dietary evidence. Instead they were treated as sediment encasing the objects of historical significance (Table 1, category 4). During this same time period shell middens were also examined in other parts of the world, primarily for purposes of creating culture histories and dietary studies (see summaries in Christenson 1985; Trigger 1986; Waselkov 1987).

In the latter half of the twentieth century the traditions begun in California and the eastern United States continued, emphasizing the shell in terms of food supply (Table 1, category 1) (e.g., Akazawa 1980; Barber 1982; Claassen 1982; Clarke and Clarke 1980; Coutts 1975; Deith 1986; Erlandson 1988; Ham 1976, 1982; Klippel and Morey 1986; Koloseike 1969; Lightfoot and Cerrato 1989; Lubell *et al.* 1975; Lyman 1991; McKillop 1984; Noli 1988; Osborn 1977; Parmalee and Klippel 1974; Pozorski 1979; Spenneman 1987; Voigt 1975; Waselkov 1987; Wessen 1988; Yesner 1980). This research assumes that if the shell is in the site, then it contributed significantly to the diet of the prehistoric people. It also assumes that the season of death of the organism is the season in which the people ate the shellfish.

Claassen (1991a) has evaluated the use of shellfish remains as indicators of diet and seasonality. Her work, as well as others (Bailey 1975, 1978; Brennan 1977; Ford 1989; Mackay and White 1987; Monks 1981; Pallant 1990; Rollins *et al.* 1990; Sanger 1981; Wing and Quitmyer 1985) questions the reliability of the assumption behind the dietary and seasonality interpretations. Shellfish found in sites may have been used for more than just protein and dietary considerations. Shellfish have been used as bait (Claassen 1991a), as construction material (Blukis Onat 1985), or as raw material for tool manufacture (Lima *et al.* 1986). Ethnographic observations concerning the methods of collecting, cooking, and storing of shellfish suggest considerable variability in cultural choices and practices (Botkin 1980; Claassen 1986; Ford 1989; Lawrence 1988; Waselkov 1987). The seasonality assignments are problematic due to local climatic conditions and lack of adequate sampling (Claassen 1991a; Monks 1981; Rollins *et al.* 1990). Finally, sampling biases of the archaeological record and of archaeologists themselves influence the results of faunal analyses (Claassen 1991a, 1991b; Grayson 1984; Koike 1979; Rollins *et al.* 1990; Thackeray 1988). All of these considerations have lead to skepticism concerning the assumption that the presence of shellfish remains in sites means that the shellfish were eaten. Even with these reservations, archaeologists still include in most excavation reports of shell middens a list of species present, along with the implicit assumption that the prehistoric people consumed the shellfish.

The second and third categories of Ambrose (see Table 1) have been a major focus of research in only the last two decades. The new emphasis reflects an increased interest in environmental reconstruction (Willey and Sabloff 1980) and geoarchaeology (Watson *et al.* 1984). Archaeologists use shells that are discovered in shell middens to reconstruct changes in the habitats in which the shellfish lived before death (Erlandson 1988; Fairbridge 1976; Lightfoot 1985; Shackleton and van Andel 1986; Thompson 1978; Woodroffe *et al.* 1988). They also use them to reconstruct the location of ancient shorelines (Bailey 1983; Beaton 1985; Bickel 1978; Craig and Psuty 1971; DePratter and Howard 1980; Fladmark 1975; Ruppé 1980; Sandweiss *et al.* 1983; Shackleton 1988). To infer that shells, which are extracted from shell middens, are indicators of shifts in nearby aquatic habitats or shifts in shoreline locations requires accepting the assumption that the shellfish species found in shell middens are adequate reflections of environmental conditions in adjacent habitats. One must assume that people were selecting shellfish randomly, depositing a random sample of the species inhabiting the bay.

The assumption that shellfish in middens are indicators of aquatic habitat rather than indicators of shifts in cultural preferences has also been questioned. One of the most serious concerns is that not only are cultural choices biasing the sample, but also a host of factors are affecting the habitat of shellfish, including water temperature, salinity, and coastal ecology (Ceci 1984; McIntire 1971; Rollins *et al.* 1990; Thomas 1987). If the shellfish in the site change, one does not necessarily know the causal factor for the change. People or one of many habitat characteristics may be responsible. The assumption is further complicated by the fact that the modern environment around the site has been changed drastically by sea level rise, coastal subsidence, and erosion (Little and Andrews 1986; Sanger and Kellogg 1989). Finally, as stated before, sampling biases in the archaeological record, of the investigator, and in the geological record will influence results (Claassen 1991a, 1991b; Grayson 1984; Koike 1979; Rollins *et al.* 1990; Thackeray 1988).

The difficulties with the assumption that shellfish from shell middens can be used alone to interpret environmental conditions demand that other data be collected to support any sort of environmental reconstruction. For example, Sanger has not reconstructed the coastal environments of Maine from the shells found in shell middens alone. His interdisciplinary research at Passamaquoddy Bay (Sanger 1985a, 1986), Damariscotta River (Sanger 1985b; Sanger and Sanger 1986), and Penobscot Bay (Sanger 1988) has used offshore cores, marine geology (Belknap and Hine 1983), and shells from shell middens, to understand the complex interplay between environment and cultural preferences. In Boston Harbor, Jones and Fisher (1990) supplemented the shell midden data with environmental indicators from geomorphology and pollen evidence. These independent tests allowed these scholars to evaluated separately the environmental reconstructions and cultural preferences.

The fifth category (Table 1), proposed as examining shell middens in terms of their depositional and postdepositional processes, has been the focus of research in only the last decade. Although these processes have not been popular in the majority of recent shell midden investigations, some excellent examples exist.

In Australia, Sullivan (1981) investigated shell middens in New South Wales. She used the results to interpret effects of European settlement, which differentially impact sites of various locations. At the site of Pambula Lake, Sullivan (1984) analyzed shell middens using a sedimentological approach, concluding that the shell midden was deposited by repeated visits of small bands of people, rather than by large numbers of people (suggested by ethnohistory). Sullivan and Sassoon (1987) analyzed a shell midden on Loloata Island, Papua New Guinea, reconstructing the depositional processes from sedimentological analyses. Stone (1989) examined shell mounds in northern Australia using paleoecological, ornithological, archaeological, and ethnographic observations. Stone concluded that the mounds are created by scrub fowl. These birds scrape aboriginal shell midden, and other materials found on the landscape, into large mounds. These mounds are then used by aboriginal peoples as locations for huts and to camp, and by the fowl as nests. The reconstruction of the depositional history of these unusual "shell middens" allowed their complex history to be unraveled.

Carter (1990) described the manner in which landsnail lenses are altered by earthworms in England. The landsnail deposits are not technically shell middens, but rather are the systematic burial of large, durable fragment of shell by earthworms. However, the processes described are of interest to excavators of shell middens. Carter plots the grain size and depth of shell fragments, comparing them to the grain size distribution of mineral clasts in the soil profile. He concludes that the surface-casting earthworms are systematically burying the shell fragments of a certain large landsnail species, and mixing the remains of other species with less durable shells. The shell lenses are therefore a function of earthworm ecology and shell durability, and are not appropriate as environmental indicators. In South Africa, Burgess and Jacobson (1984) examine the texture of sediment under a recent shell midden to determine if the depositional history of the site accounts for the lack of earlier sites. The shell midden is located on a dune in the delta of the Kuiseb River. The reason proposed for the lack of earlier sites is the disturbance by the river and the aeolian process that altered earlier sites.

In South America, Salemme (*et al.* 1989) notes shell fauna, radiocarbon dates, and archaeological material at a shell midden in Rio de la Plata littoral, Argentina. The results indicate that the archaeological material was deposited after the shell ridge was created by coastal processes, and that the archaeological materials were added to the surface and mixed postdepositionally. Therefore the shell is not associated with the occupation, and the site is not technically a shell midden.

In the eastern United States, Stein (1980, 1982) used grain size and chemical analyses to reconstruct the depositional history of a Green River shell mound. She determined that the entire mound was transported to that location by people and that the river contributed only minor amounts of silt and clay to the deposition of the mound. She determined that a shell-free layer blanketing the surface of the mound was not the result of leaching, but rather a later accumulation of material from people who did not utilize shell. She also identified that subsurface-casting earthworms had obliterated stratigraphic markers within the matrix (Stein 1983), but that the orientation and location of the shell was not affected (Gorski 1981).

On the Northwest Coast of North America, Moss (1984) used phosphate analysis to detect prehistoric sites on Admiralty Island, Alaska. She compared samples from shell middens, from sites where shell had either been leached or never deposited, from locations merely "suspected" as sites, and from control areas. This approach allowed Moss to design a surveying and testing strategy for sites covered by dense vegetation. Crozier and Amos (1982) conducted extensive textural and chemical analyses of the matrix of Shoemaker Bay shell midden on Vancouver Island, British Columbia, Canada. Unfortunately, the results were not used for any interpretations other than to add quantifiable data to the field descriptions of stratigraphy. Campbell (1981) used soil descriptions to evaluate four depositional units at the Duwamish site, Washington. Textural differences were primarily used to detect alluvial sources versus cultural sources of sediment. The use of coring was especially important in interpreting this site's depositional history. Ham (1982) used textural and chemical analyses to evaluate preservation of the Crescent Beach site near Vancouver, B.C. He found that pH increased with depth, suggesting that although leaching may have affected the upper 60 cm of the profile, it has not been extensive throughout the site. McDowell (1989) conducted geomorphological, textural, and chemical analyses of samples from shell middens at North Yaquina Head, Oregon. Her data are used to reconstruct the depositional history of the four strata observed during excavation. The lowest stratum, which is high in phosphorus and low in calcium, is thought to be a land surface occupied prehistorically before the shell midden deposition occurred (1989:17).

IV. Stratigraphy of Shell Middens

A. *Stratigraphy of British Camp Shell Midden*

The British Camp research reported here is another example of a site where depositional and postdepositional factors have been investigated. The shell midden is stratigraphically complex, with abundant layering of greater or lesser amounts of shell and darker and lighter matrix. This stratification can be lumped into two large contrasting layers: a dark-colored shell layer at the bottom of the excavation profile (resting on glacial drift or beach deposits) and an overlying lighter colored shell layer that appears to contain a greater amount of shell than does the lower darker layer. The difference in shell density between the two layers may be the effect of the dark "greasy" matrix in the lower layer. The matrix sticks to the shell making it more difficult to see.

The difference in strata can be explained in two ways: either (1) the two strata represent the deposition of two different kinds of material (representing two different kinds of cultural activities); or (2) the deposition of only one kind of material has been altered after its deposition so as to appear as two strata instead of one. These two explanations can be tested by examining the contents of the shell midden. If the stratification is caused by depositional (cultural) events, then the kinds of artifacts, fauna, flora, and sediments within the two strata should change at the boundary of the two strata. If the stratification is caused by postdepositional events, then artifacts should remain the same across the boundary and evidence of weathering should be found in one of the two strata observed.

At the British Camp site some artifactual material did change over time, but surprisingly the shellfish, plants, tools, and chipped-stone debitage did not always change at the same boundary as that of the light and dark strata. In many cases the changes occurred in locations independent from the boundary defining the dual stratification, and were sometimes above the boundary and sometimes below. These results suggest that the stratification observed in the field by noting light and dark colors of the matrix was not caused by depositional (cultural) events, but rather by postdepositional processes. Those postdepositional processes, if allowed to weather the shell for millennia, would completely leach the carbonate from all the shells leaving a shell-free layer at the base of the midden.

At most sites a strongly contrasting physical stratigraphy, such as observed at British Camp, would be used to aggregate artifact assemblages. The artifacts from one layer would be grouped and counted, then compared to the counts of artifacts from another layer. If the light/dark strata, or a shell-free layer below a shell layer, were used as the grouping devise, then some layers that were really older would be mixed with others that are much younger, or the shell-free layer would be misinterpreted as a deposit made by people who did not collect shell. Such misinterpretations may have been responsible for some of the confusion in Northwest Coast culture histories (Abbott 1972; Burley 1980). The same confusion was not perpetuated at British Camp because the procedures described in this volume were followed.

B. *Stratigraphy of Other Northwest Coast Shell Middens*

The stratification observed at San Juan Island has been noted by archaeologists at other shell middens. Although other sites differ in ages, they all are located near the coastline, suggesting that the relationship with the shoreline and sea level may be crucial. Many of the Northwest Coast sites are located in the straits between Vancouver Island and the mainland of British Columbia (Gulf of Georgia area), however, that distribution may reflect the greater concentration of archaeological activity in that area. Although most archaeologists recognize that the saturation by seawater of the deeper portions of sites has affected the stratification, all but a few use the physical light/dark stratification to divide the artifacts in the sites into "cultural components," even though this stratigraphy may be postdepositionally produced and may have nothing to do with the deposition of the site and artifacts.

At the Montague Harbour site (DfRu13), on Galiano Island (an island located just northwest of San Juan Island) a dual stratification was discovered that is similar to that seen at British Camp (Mitchell 1971:83–88). Zone A (at the surface) contained numerous lenses and layers of shell with ash, charcoal, fire-cracked rock, and light matrix. Below that was Zone B, which resembled Zone A, except that it was black when wet and dried to dark grey. Zone B was subdivided into Zone B-1 and B-2, with B-2 characterized by an almost total absence of shell. When this lowest shell-free zone was first encountered Mitchell thought it represented very early deposits. Artifacts, however, were similar throughout Zone B-1 and B-2 suggesting to Mitchell that "Zone B-2 may be shell-free not because of great age but because, while submerged, the calcareous materials were leached more rapidly from the deposit." (1971:88). Mitchell was one of the first archaeologists in the region to notice that water had altered the lowest portions of the site. He recognized that the shell-free layer may have once had shell. Mitchell did use the stratigraphic boundary between the light- and dark-colored matrix to separate the cultural units, however, he considered the physical stratigraphy to be produced postdepositionally.

At the Georgeson Bay site (DfRu24), also on Galiano Island, Haggarty and Sendey (1976) also noted a lower shell-free layer described as "slightly compacted black sandy soil with traces of shell evident only along its upper margin" (1976:16). This shell-free layer differs from the above shell midden in soil color, quantities of fire-cracked rock, and decreasing percentages of shellfish remains. The lower stratigraphic unit is considered a separate cultural component. Although the authors do not consider the possibility, the site may have been leached of carbonate and the shell-free layer is not as culturally significant as it once appeared.

Other sites with this stratigraphy exist close to British Camp. On Sucia Island (just northeast of San Juan Island) Kidd (1969:40–41) excavated the Fossil Bay site (45SJ105), describing it as having a dark layer at its base (see especially Plate V, p. 41). Kidd evidently did not encounter any shell-free layers at the base. At the Helen Point Site (DfRu8) on Mayne Island, Carlson (1970:114) divided the shell midden into three layers. The deepest layer is described as "compact, greasy, black deposit which contains charcoal, bone, fire-cracked rock . . . , but only a minute amount of highly fragmented shell." Carlson uses the three physical stratigraphic units to define the three cultural phases at the site. On Gabriola Island, Burley (1989) suggests that the False Narrows midden (DgRw4) "was subject to a 3 m drop in sea level during the Late Marpole time period of circa 1600 to 1800 years ago" (1989:10). Because Burley wrote the report long after the excavation was completed (and did not witness the excavation), the description of stratigraphy is not detailed. However, a drop in sea level is contrary to evidence reported at other sites (Whittaker and Stein, Ch. 2, this volume).

Farther north from San Juan Island, near Courtenay, B.C. (on the east central coast of Vancouver Island), shell middens have been found that contain a dark layer at the base of the midden. For example, Capes (1977:62) describes the stratigraphy at the Millard Creek site (DkSf2). In one area of the site (area 2D) she describes a

Layer 3 as having dark soil underlying Layer 2 and 1. Nearby at the Puddleduck site (DkSf26), Mitchell (1988:2) reports that part of the site, where deeper deposits are found, contains a dual stratification of upper shell-rich Zone I, underlain by a dark layer with fine fragments of shell. Both these sites suggest that shell middens in the area of Courtenay, B.C. may have been subjected to saturation by rising sea level.

Near Port Hardy, B.C. (on the northeast coast of Vancouver Island), other sites have a similar dual stratigraphy. At the O'Connor site (EeSu5), Chapman (1982:76) reports finding three strata, a dual stratification within the shell midden, underlain by a shell-free dark layer. The lowest shell-free strata did not contain bone tools similar to the ones found above, in the shell midden. However, Chapman does not believe that preservation differences are responsible for the lack of bone tools in the shell-free stratum because she found other bone "negating an argument that poor preservation was responsible for . . . the absence of a developed bone tool industry in these shell-less and often wet deposits" (1982:123). The shell midden overlying the shell-free stratum is divided into a lower zone with a dark, greasy matrix, highly fragmented shell, and often decomposed fish remains, above which is a zone with high concentrations of shell. Although Chapman does not think that preservation is responsible for the lack of the bone tool tradition in the lower levels of the shell midden, the decrease in shell with depth, the observation of wetness, and the difference in bone preservation from shell preservation (Linse, Ch. 14, this volume) suggests that saturation may indeed be affecting the preservation of carbonate and perhaps bone in the site.

At the Bear Cove site (EeSu8), also near Port Hardy, B.C., C. Carlson (1979) found three layers that contain artifacts. The lowest was a banded beach gravel zone, above which was a black, greasy loam containing no shell, followed by an upper layer of shell midden. The radiocarbon ages on charcoal from the shell-free zone cluster around 4300 years ago (1979:183). Although the author suggests that the lower shell-free layer is the result of early deposition by people who did not utilize shell, I suggest that the Bear Cove site may have been affected over the last 5000 years by weathering induced by rising sea level, which leached the lower layer to make it appear shell-free. The Echo Bay site (EeSo1) on Gilford Island (across from Port Hardy near the mainland of B.C.) has shell midden deposits almost 6 m thick, the lower portions of which are under water (Mitchell 1981:106).

In central B.C. (farther to the north and across from Queen Charlotte Island), sites near Prince Rupert have been excavated that also exhibit stratification indicative of inundation. The McNaughton Site (ElTb10) is reported to have an early black, greasy deposit in its lowest portion (Carlson 1976), and Namu (ElSx1) has stratification including a dark, shell-free layer at its base (R. L. Carlson 1979, 1990; Hester and Nelson 1978). The lowest deposits at Namu are 9000 years old (Luebbers 1978:26; Carlson 1990:66), and may have experienced thousands of years of leaching. If shell was there originally, soil thin-section (Courty *et al.* 1989) could be used to discern their outlines in soil voids, even if the remains have been leached away.

Shell middens are also found along the major rivers that empty into the coastal waters of the Northwest. Along the Fraser River the Glenrose Cannery site (DgRr6) had lower wet layers, some of which did contain shell. These layers were described as tan, however, and were controlled by events of the Fraser River and its delta rather than directly by sea level (Matson 1976). The shell middens located on the

Fraser may be primarily affected by fluvial/deltaic interactions (Calvert 1970:54–55), although shell middens located closer to the coast and within the delta may be affected by rising sea level (Borden 1970; Ham 1982; Matson *et al.* 1980).

Further south on the mainland (in the state of Washington) on the Skagit River delta, Bryan (1963:6) described sites as "often being found underwater." Although Bryan does not describe the stratification found in the sites he excavated, he notes that sites along the western shores of Washington and on the islands off the coast show evidence of saturation by the sea (see also Fladmark 1975, 1983; Grabert and Larsen 1973; Larsen 1971, 1972; Mitchell 1971; Thompson 1978).

These examples demonstrate that some sites on the Northwest Coast have stratification suggestive of inundation, weathering, and leaching processes similar to those described for British Camp. Not all Northwest Coast sites were inundated. For example, on San Juan Island the Cattle Point site (45SJ1) may have been uplifted, or at least deposited on elevated, Early Holocene marine terraces (King 1950). At Yaquina Head shell midden in Oregon the midden is found on old marine terraces, also elevated above the beach (McDowell 1989). However, the many examples described here suggest that the relationship between sea level and shell middens must be considered, not only on the Northwest Coast but on every coast where sea level has fluctuated.

V. Summary

Any shell midden could be affected by processes similar to those at British Camp as long as shell, organic matter, and a body of water are present. When the local water table rises saturation and chemical alterations of the middens will occur. Thus, the stratification of shell middens will result both from initial deposition and from post-depositional effects. As our predecessors in the 18th century knew, the formation processes of shell middens are complex. To decipher them requires unique excavation and analytic procedures. Procedures used at British Camp and results reported here allow shell middens located anywhere in the world to be deciphered.

Acknowledgments

The following people may not know the importance of their contribution to this chapter, but I would like to thank them. David Sanger encouraged me to explore shell midden research when he asked me to coorganize a symposium on shell middens at the Society for American Archaeology in 1985, and Cheryl Claassen continued the effort by asking me to participate in a symposium on shell middens at the Society for American Archaeology in 1989. Both of these individuals have been inspirational to me, and although they did not read this manuscript in this form, much of this chapter is the result of my exposure to their scholarly work, as well as their helpful comments over the years.

References

Abbott, D. N.

1972 The utility of the concept of phase in the archaeology of the southern Northwest Coast. *Syesis* **5**, 267–278.

Akazawa, T.
1980 Fishing adaptation of prehistoric hunter-gatherers at the Nittano Site, Japan. *Journal of Archaeological Science* **7**, 324–344.
Ambrose, W. R.
1967 Archaeology and shell middens. *Archaeology and Physical Anthropology in Oceania* **2**, 169–187.
Ambrosiani, B.
1977 Comments on units of archaeological stratification. *Norwegian Archaeological Review* **10**, 95–97.
Ascher, R.
1959 A prehistoric population estimate using midden analysis and two population models. *Southwestern Journal of Anthropology* **15**, 168–178.
Bailey, G. N.
1975 The role of molluscs in coastal economies: the results of analysis in Australia. *Journal of Archaeological Science* **2**, 45–62.
1978 Shell middens as indicators of postglacial economies: a territorial perspective. In *The early postglacial settlement of northern Europe*, edited by P. Mellars. London: Duckworth. Pp.38–63.
1983 Problems of site formation and the interpretation of spatial and temporal discontinuities in the distribution of coastal middens. In *Quaternary coastlines and marine archaeology*, edited by P. M. Masters and N. C. Flemming. London: Academic Press. Pp.559–582.
Bailey, G. N., and J. Parkington
1988 The archaeology of prehistoric coastlines: an introduction. In *The archaeology of prehistoric coastlines*, edited by G. N. Bailey and J. Parkington. Cambridge: Cambridge University Press. Pp.1–10.
Barber, R.
1982 The Wheeler's Site: a specialized shellfish processing station on the Merrimack River. Peabody Museum Monograph, **7**. Cambridge, Massachusetts.
Beaton, J. M.
1985 Evidence for a coastal occupation time-lag at Princess Charlotte Bay (North Queensland) and implications for coastal colonization and population growth theories for Aboriginal Australia. *Archaeology in Oceania* **20**, 1–20.
Belknap, D. F., and A. C. Hine,
1983 Evidence for sea level lowstand between 4500 and 2400 B.P. on the southeast coast of the United States: Discussion. *Journal of Sedimentary Petrology* **53**, 680–682.
Bickel, P. M.
1978 Changing sea levels along the California coast: anthropological implications. *Journal of California Anthropology* **5**, 6–20.
Bloom, A. L.
1983 Sea level and coastal changes. In *Late-Quaternary Environments of the United States. The Holocene, Vol. 2*, edited by H. E. Wright, Jr. Minneapolis: University of Minnesota Press. Pp.42–51.
Blukis Onat, A. R.
1985 The multifunctional use of shellfish remains: from garbage to community engineering. *Northwest Anthropological Research Notes* **19**, 201–207.
Borden, C. E.
1970 Cultural history of the Fraser-Delta region: An outline. *B. C. Studies* No. 6–7, 95–112.
Botkin, S.
1980 Effects of human exploitation on shellfish populations at Malibut Creek, California. In *Modeling change in prehistoric subsistence economies*, edited by T. K. Earle and A. L. Christenson. New York: Academic Press. Pp.121–139.
Brennan, L. A.
1977 The midden is the message. *Archaeology of Eastern North America* **5**, 122–137.
Bryan, A. L.
1963 An archaeological survey of northern Puget Sound. Idaho State University Museum, Occasional Papers No. 11. Pocatello, Idaho.

Burgess, R. L., and L. Jacobson
1984 Archaeological sediments from a shell midden near Wortel Dam, Walvis Bay, South Africa. *Palaeoecology of Africa* **16**, 429–435.

Burley, D. V.
1980 Marpole: Anthropological reconstruction of a prehistoric Northwest Coast culture type. Simon Fraser University, Department of Archaeology Publication No. 8. Burnaby, British Columbia.
1989 Senewelets: Culture history of the Nanaimo Coast Salish and the False Narrows midden. Royal British Columbia Museum, Memoir No. 2. Victoria, British Columbia.

Calvert, S. G.
1970 The St. Mungo cannery site: A preliminary report. In *Archaeology in British Columbia, new discoveries*, edited by R. L. Carlson. *B.C. Studies*, Special Issue. Pp.54–76.

Campbell, S. K.
1981 The Duwamish No. 1 site: A lower Puget Sound shell midden. University of Washington, Office of Public Archaeology. Institute for Environmental Studies Research Reports 1. Seattle, Washington.

Capes, K. H.
1977 Archaeological investigations of the Millard Creek site, Vancouver Island, British Columbia. *Syesis* **10**, 57–84.

Carlson, C.
1979 The early component at Bear Cove. *Canadian Journal of Archaeology* **3**, 177–194

Carlson, R. L.
1954 *Archaeological investigations in the San Juan Islands.* M. A. Thesis. University of Washington, Seattle.
1960 Chronology and culture change in the San Juan Islands, Washington. *American Antiquity* **25**, 562–586.
1970 Excavations at Helen Point on Mayne Island. *B.C. Studies* No. 6–7, 113–125.
1976 The 1974 excavations at McNaughton Island. In *Current research reports*, edited by R. L. Carlson. Simon Fraser University, Department of Archaeology. Burnaby, Canada.
1979 Early period on the central coast of British Columbia. *Canadian Journal of Archaeology* **3**, 211–228.
1990 Cultural antecedents. In *Handbook of North American Indians, Northwest Coast: Volume 7*, edited by W. Suttles. Washington D.C.: Smithsonian Institution. Pp.60–69.

Carter, S. P.
1990 The stratification and taphonomy of shells in calcareous soils: implications for land snail analysis in archaeology. *Journal of Archaeological Science* **17**, 495–507.

Ceci, L.
1984 Shell midden deposits as coastal resources. *World Archaeology* **16**, 62–74.

Chapman, M. W.
1982 Archaeological investigations at the O'Connor site, Port Hardy. In *Papers on central coast archaeology*, edited by P. M. Hobler. Simon Fraser University, Department of Archaeology, Publication No. 10. Pp.65—132. Burnaby, British Columbia.

Christenson, A. L.
1985 The identification and study of Indian shell middens in Eastern North America: 1643–1861. *North American Archaeologist* **6**, 227–243.

Claassen, C.
1982 Shellfishing patterns: An analytical study of prehistoric shell from North Carolina coastal middens. Ph.D. dissertation, Department of Anthropology, Harvard, Boston. University Microfilms, Ann Arbor, Michigan.
1986 Temporal patterns in marine shellfish-species use along the Atlantic Coast in the Southeastern United States. *Southeastern Archaeology* **5**, 120–137.
1991a Normative thinking and shell-bearing sites. In *Archaeological method and theory, Vol. 3*, edited by M. B. Schiffer. Tucson: University of Arizona Press. Pp.249–298.
1991b Gender, shellfishing, and the Shell Mound Archaic. In *Engendering archaeology: women and prehistory*, edited by M. W. Conkey and J. M. Gero. Cambridge: Basil Blackwell. Pp.276–300.

Clarke, L. R., and A. H. Clarke
1980 Zooarchaeological analysis of mollusc remains from Yuquot, B. C. In *The Yuquot project, vol. 2*, edited by W. J. Folan and J. Dewhirst. History and Archaeology 43, National Historic Parks and Sites Branch, Ottawa, Canada.
Cook, S. F.
1946 A reconsideration of shellmounds in respect to population and nutrition. *American Antiquity* **12**, 50–53.
1950 Physical analysis as a method for investigating prehistoric habitation sites. *Reports of the University of California Archaeological Survey* **7**, 2–5.
Cook, S. F., and R. F. Heizer
1951 The physical analysis of nine indian mounds of the Lower Sacramento Valley. *University of California Publications in American Archaeology and Ethnology* **40**, 281–312.
Cook, S. F., and A. E. Treganza
1947 The quantitative investigation of aboriginal sites: Comparative analysis of two California Indian mounds. *American Antiquity* **13**, 135–141.
1950 The quantitative investigations of indian mounds. *University of California Publications in American Archaeology and Ethnology* **40**, 223–261.
Courty, M. A., P. Goldberg, and R. Macphail
1989 *Soils and micromorphology in archaeology.* New York: Cambridge University Press.
Coutts, P. J. F.
1975 *The seasonal perspective of marine-oriented prehistoric hunter-gatherers. In* Growth rhythms and the history of the earth's rotation, edited by G. D. Rosenberg and S. K. Runcorn. London: John Wiley and Sons. Pp.243–252.
Craig, A. K., and N. P. Psuty
1971 Paleoecology of shell mounds at Otuma, Peru. *Geographical Review* **61**, 125–132.
Crozier, S. N., and J. Amos
1982 Archaeological matrix analysis, Site DhSe 2, Port Alberni, British Columbia. Appendix III. In *Alberni prehistory: Archaeological and ethnographic investigations on western Vancouver Island*, edited by A. D. McMillan and D. E. S. Claire. Penicton, British Columbia: Theytus Books. Pp.150–173.
Deith, M. R.
1986 Subsistence strategies at a Mesolithic camp site: Evidence of stable isotope analyses of shells. *Journal of Archaeological Science* **13**, 61–78.
DePratter, C. B., and J. D. Howard
1980 Indian occupation and geologic history of the Georgia coast: a 5,000 year summary. In *Excursions in southeastern geology*, edited by J. D. Howard, C. B. DePratter, and R. W. Frey. Geological Society of America Annual Meeting Guidebook 20, Boulder, Colorado. Pp.1–65.
Erlandson, J.
1988 The role of shellfish in prehistoric economies: a protein perspective. *American Antiquity* **53**, 102–109.
Fairbridge, R. W.
1976 Shellfish-eating preceramic Indians in coastal Brazil. *Science* **192**, 353–359.
Fladmark, K. R.
1975 A paleoecological model for Northwest Coast prehistory. Archaeological Survey of Canada Mercury Series. No. 43. National Museum of Man, Ottawa.
1983 Comparison of sea-levels and prehistoric cultural developments on the east and west coasts of Canada. In *The evolution of maritime culture on the Northeast and Northwest Coasts of America*, edited by R. J. Nash. Pp.65–76. Simon Fraser University, Department of Archaeology, Publication No. ll. Burnaby, British Columbia.
Ford, P. J.
1989 Archaeological and ethnographic correlates of seasonality: problems and solutions on the Northwest Coast. *Canadian Journal of Archaeology* **13**, 133–150.
Funkhouser, W. D., and W. S. Webb
1932 *An archaeological survey of Kentucky.* University of Kentucky, Reports in Archaeology and Anthropology, Vol 2. Lexington, Kentucky.

Gifford, E. W.
1916 Composition of California shellmounds. *University of California Publications in American Archaeology and Ethnology* **12**, 1–29.
Gorski, L.
1981 Microstratigraphic analysis at the Carlston Annis site, 15BT5, Butler County, Kentucky. Unpublished Master's thesis, Department of Anthropology, University of Missouri, Columbia.
Grabert, G. F., and C. E. Larsen
1973 Marine transgressions and cultural adaptation: preliminary tests of an environmental model. In *Prehistoric maritime adaptations of the circumpolar zone*, edited by W. Fitzhugh. The Hague: Mouton. Pp.229–251.
Grayson, D. K.
1984 *Quantitative zooarchaeology*. New York: Academic Press.
Greengo, R. E.
1951 Molluscan species in California shell middens. *University of California Archaeological Survey* Report No. 13.
Greenwood, R. S.
1961 Quantitative analysis of shells from a site in Goleta, California. *American Antiquity* **26**, 416–420.
Haggarty, J. C., and J. H. W. Sendey
1976 *Test excavation at Georgeson Bay, B. C.* British Columbia Provincial Museum, No. 19, Occasional Paper Series, Victoria, British Columbia.
Ham, L. C.
1976 Analysis of shell samples from Glenrose. In *The Glenrose cannery site*, edited by R. G. Matson. National Museum of Man Mercury Series, Archaeological Survey of Canada Paper 52, Ottawa, Canada.
1982 Seasonality, shell midden layers, and coast Salish subsistence activities at the Crescent Beach Site, DgRr 1. Unpublished Ph.D. dissertation, Department of Anthropology, University of British Columbia, Vancouver.
Hester, J. J., and S. M. Nelson
1978 *Studies in Bella Bella prehistory*. Department of Archaeology, Simon Fraser University, Burnaby, British Columbia, Canada.
Jones, J. R., and J. J. Fisher
1990 Environmental factors affecting prehistoric shellfish utilization, Grape Island, Boston Harbor, Massachusetts. In *Archaeological geology of North America*, edited by N. P. Lasca and J. Donahue. Geological Society of America, Centennial Special Volume 4, Boulder, Colorado. Pp.137–146.
Kenady, S. M.
1971 An interim report on the prehistoric excavations at English Camp. Unpublished manuscript, Burke Museum, University of Washington, Seattle.
1972 Research design for the 1972 season at Garrison Bay. Unpublished manuscript, Burke Museum, University of Washington, Seattle.
1973 Environmental and functional change in Garrison Bay. In *Miscellaneous San Juan Island reports 1970–1972*, edited by S. M. Kenady, S. A. Saastamo, and R. Sprague. University of Idaho Anthropological Research Manuscript Series, No. 7, Moscow. Pp.39–83.
Kidd, R. S.
1969 The archaeology of the Fossil Bay site, Suçia Island, northwestern Washington state, in relation to the Fraser Delta Sequence. *National Museums of Canada Bulletin 232, Contributions to Anthropology VII: Archaeology, Paper No. 2*. 32–67.
King, A. R.
1950 Cattle Point: A stratified site in the southern Northwest Coast region. *American Antiquity* Memoir 7, supplement to vol. 15, 1–94.
Klippel, W., and D. Morey
1986 Contextual and nutritional analysis of freshwater gastropods from Middle Archaic deposits at the Hays Site, Middle Tennessee. *American Antiquity* **51**, 799–813.
Koike, H.
1979 Seasonal dating and the valve-pairing technique in shell-midden analysis. *Journal of Archaeological Science* **6**, 63–74.

Koloseike, A.
1969 On calculating the prehistoric food resource value of molluscs. University of California, Archaeological Survey Annual Report 1969, Los Angeles.

Larsen, C. E.
1971 An investigation into the relationship of change in relative sea level to social change in the prehistory of Birch Bay, Washington. Unpublished Master's thesis, Department of Anthropology, Western Washington University, Bellingham, Washington.
1972 The relationship of relative sea level to the archaeology of the Fraser River Delta, British Columbia. Paper presented at the 1972 Annual Meeting of the Geological Society of America, Minneapolis, Minnesota.

Lawrence, D. R.
1988 Oysters as geoarchaeological objects. *Geoarchaeology* **3**, 267–274.

Lightfoot, K.
1985 Shell midden diversity: a case example from coastal New York. *North American Archaeologist* **6**, 289–324.

Lightfoot, K., and R. Cerrato
1989 Regional patterns of clam harvesting along the Atlantic Coast of North America. *Archaeology of Eastern North America* **17**, 31–46.

Lima, T. A., E. M. Botelho de Mello, and R. C. Pinheiro da Silva
1986 Analysis of Molluscan remains from the Ilha de Santana Site, Macaé, Brazil. *Journal of Field Archaeology* **13**, 83–97.

Little, E. A., and J. C. Andrews
1986 Prehistoric shellfish harvesting at Nantucket Island. *Bulletin of the Massachusetts Archaeological Society* **47**, 18–27.

Lubell, D., F. A. Hassan, A. Gautier, and J. Bailles
1975 The Capsian Escargotieres. *Libyca* **23**, 43–121.

Luebbers, R.
1978 Excavations: Stratigraphy and artifacts. In *Studies in Bella Bella prehistory*, edited by J. J. Hester and S. M. Nelson. Department of Archaeology, Simon Fraser University, Publication No. 5, Burnaby, British Columbia. Pp.11–66.

Lyman, R. L.
1991 *Prehistory of the Oregon Coast: The effects of excavation strategies and assemblage size on archaeological inquiry*. San Diego, California: Academic Press.

McDowell, P. F.
1989 Geomorphology and soils. In *Archaeology of the North Yaquina Head shell midden, central Oregon coast*, edited by R. Minor. U.S. Dept. of the Interior, BLM, Cultural Resource Series No. 3, Portland, Oregon. Pp.6–18.

McIntire, W. G.
1971 Methods of correlating cultural remains with stages of coastal development. In *Introduction to coastal development*, edited by J. A. Sears. Massachusetts Institute of Technology Press, Cambridge. Pp.188–203.

Mackay, R., and J. P. White
1987 Musseling in on the NSW coast. *Archaeology in Oceania* **22**, 107–111.

McKillop, H.
1984 Prehistoric Maya reliance on marine resources: Analysis of a midden from Moho Cay, Belize. *Journal of Field Archaeology* **11**, 25–35.

Matson, R. G.
1976 The Glenrose cannery site. Archaeological Survey of Canada, Mercury Series No. 52. National Museum of Man, Ottawa.

Matson, R. G., D. Ludowicz, and W. Boyd
1980 Excavations at Beach Grove (DgRs 1) in 1980. Unpublished manuscript, Laboratory of Archaeology, University of British Columbia.

Mitchell, D. H.
1971 Archaeology of the Gulf of Georgia area, a natural region and its culture type. *Syesis* **4**, suppl. 1, 1–228.

1981 Test excavations at randomly selected sites in Eastern Queen Charlotte Strait. *B.C. Studies* No. 48, 103–123.
1988 The J. Puddleduck site: a northern Strait of Georgia Locarno Beach component and its predecessor. Contributions to Human History, No. 1, Royal British Columbia Museum, Victoria, British Columbia.
1990 Prehistory of the coasts of southern British Columbia and northern Washington. In *Handbook of North American Indians, volume 7 Northwest Coast*, edited by W. Suttles. Washington, D. C.: Smithsonian Institution. Pp.340–358.

Monks, G. G.
1981 Seasonality studies. In *Advances in archaeological method and theory, vol. 4*, edited by M. B. Schiffer. New York: Academic Press. Pp.177–240.

Moss, M. L.
1984 Phosphate analysis of archaeological sites, Admiralty Island, southeast Alaska. *Syesis* **17**, 95–100.

Nelson, N. C.
1909 Shellmounds of the San Francisco Bay region. *University of California Publications in American Archaeology and Ethnology* **7**, 309–356.

Noli, D.
1988 Results of the 1986 excavation at Hailstone Midden (HSM), Eland's Bay, Western Cape Province. *South African Archaeological Bulletin* **43**, 43–48.

Osborn, A. J.
1977 Strandloopers, mermaids, and other fairy tales: Ecological determinants of marine resource utilization, the Peruvian case. In *For theory building in archaeology*, edited by L. R. Binford. New York: Academic Press. Pp.157–206.

Pallant, E.
1990 Applications of molluscan microgrowth analysis to geoarchaeology: A case study from Costa Rica. In *Archaeological geology of North America*, edited by N. Lasca and J. Donahue. Geological Society of America, Centennial Special Volume 4, Boulder, Colorado. Pp.421–430.

Parmalee, P., and W. Klippel
1974 Freshwater mussels as a prehistoric food resource. *American Antiquity* **39**, 421–434.

Pozorski, S. G.
1979 Prehistoric subsistence and diet of the Moche Valley, Peru. *World Archaeology* **11**, 163–184.

Quimby, G. I.
1979 A brief history of WPA archaeology. In *The uses of anthropology*, edited by W. Goldschmidt. American Anthropological Association, Washington, D. C. Pp.110–123.

Rolingson, M. A.
1967 Temporal perspective on the archaic cultures of the Middle Green River region, Kentucky. Ph.D. dissertation, Department of Anthropology, University of Michigan, University Microfilms, Ann Arbor.

Rollins, H. B., D. H. Sandweiss, and J. C. Rollins
1990 Mollusks and coastal archaeology: A review. In *Archaeological geology of North America*, edited by N. Lasca and J. Donahue. Geological Society of America, Centennial Special Volume 4, Boulder, Colorado. Pp.467–478.

Ruppé, R. J.
1980 The archaeology of drowned terrestrial sites: a preliminary report. *Florida Bureau of Historic Sites and Properties Bulletin (Tallahassee)* **6**, 35–45.

Salemme, M., L. Miotti, and M. Aquirre
1989 Holocene settlement in the Rio de La Plata Littoral (Argentina): a methodological approach. *Geoarchaeology* **4**, 69–80.

Sandweiss, D. H., H. B. Rollins, and J. B. Richardson, III.
1983 Landscape alteration and prehistoric human occupation on the north coast of Peru. *Annals of Carnegie Museum* **52**, 277–298.

Sanger, D.
1981 Unscrambling messages in the midden. *Archaeology of Eastern North America* **9**, 37–42.
1985a Sea-level rise and archaeology in Passamaquoddy Bay: Archaeology and sediment cores. Maine Geological Survey, Open-file Report 85-73, Augusta.

1985b Sea-level rise and archaeology in Damariscotta River. Maine Geological Survey, Open-file Report 85-74, Augusta.

1985c Seashore Archaeology in New England. *Quarterly Review of Archaeology* **6**, 3–4.

1986 An introduction to the prehistory of the Passamaquoddy Bay region. *American Review of Canadian Studies* **16**, 139–159.

1988 Maritime adaptations in the Gulf of Maine. *Archaeology of Eastern North America* **16**, 81–99.

Sanger, D., and D. C. Kellogg

1989 Prehistoric archaeology and evidence of coastal subsidence on the coast of Maine. In *Neotectonics in Maine*, edited by W. A. Anderson and J. H.W. Borns. Maine Geological Survey Bulletin No. 40, Augusta. Pp.107–126.

Sanger, D., and M. J. Sanger

1986 Boom and bust on the river, the story of the Damariscotta Oyster Shell Heaps. *Archaeology of Eastern North America* **14**, 65–78.

Seilacher, A.

1973 Biostratinomy: the sedimentology of biologically standardized particles. In *Evolving concepts in sedimentology*, edited by R. N. Ginsburg. John Hopkins University Press, Baltimore, Maryland. Pp.159–177.

Shackleton, J. C.

1988 Reconstructing past shorelines as an approach to determining factors affecting shellfish collecting in the prehistoric past. In *The archaeology of prehistoric coastlines*, edited by G. Bailey and J. Parkington. Cambridge University Press, Cambridge. Pp.11–21.

Shackleton, J. J., and T. H. van Andel

1986 Prehistoric shore environments, shellfish availability, and shellfish gathering at Franchthi, Greece. *Geoarchaeology* **1**, 127–143.

Spenneman, D. H. R.

1987 Availability of shellfish resources on prehistoric Tongatapu, Tonga: Effects of human predation and changing environment. *Archaeology in Oceania* **22**, 81–96.

Sprague, R.

1973 Location of the Pig Incident, San Juan Island. In *Miscellaneous San Juan Island reports 1970–1972*, edited by S. M. Kenady, S. A. Saastamo, and R. Sprague. University of Idaho Anthropological Research Manuscript Series, No. 7, Moscow. Pp.17–38.

1976 The submerged finds from the prehistoric components, English Camp, San Juan Island, Washington. In *The excavation of water-saturated archaeological sites (wet sites) on the Northwest Coast of North America*, edited by D. R. Croes. National Museum of Man, Mercury Series, Archaeological Survey of Canada, Paper No. 50, Ottawa. Pp.78–85.

1983 San Juan archaeology, vol. 1 and 2. Laboratory of Anthropology, University of Idaho, Moscow.

Stein, J. K.

1980 Geoarchaeology of the Green River shell mounds, Kentucky. Ph.D. dissertation, Department of Anthropology, University of Minnesota, University Microfilms, Ann Arbor.

1982 Geologic analysis of the Green River shell middens. *Southeastern Archaeology* **1**, 22–39.

1983 Earthworm activity: A source of potential disturbance of archaeological sediments. *American Antiquity* **48**, 277–289.

Stone, T.

1989 Origins and environmental significance of shell and earth mounds in Northern Australia. *Archaeology in Oceania* **24**, 59–64.

Stright, M. J.

1990 Archaeological sites on the North American continental shelf. In *Archaeological geology of North America*, edited by N. Lasca and J. Donahue. Geological Society of America, Centennial Special Volume 4, Boulder, Colorado. Pp.439–465.

Sullivan, M.

1980 Specific generalizations. In *Holier than thou*, edited by I. Johnson. Department of Prehistory, Research School of Pacific Studies, The Australian National University, Canberra. Pp.143–146.

1981 Ninety years later: a re-survey of shell middens on Wagonga Inlet and Pambula Lake, N.S.W. *Archaeology in Oceania* **16**, 81–86.
1984 A shell midden excavation at Pambula Lake on the far south coast of New South Wales. *Archaeology of Oceania* **19**, 1–15.
Sullivan, M. E., and M. Sassoon
1987 Prehistoric occupation of Loloata Island Papua New Guinea. *Australian Archaeology* **24**, 1–9.
Suttles, W.
1951 The economic life of the coast Salish of Haro and Rosario Straits. Ph.D. dissertation, Department of Anthropology, University of Washington. (Published in 1974 by Garland Press, New York.)
1968 Coping with abundance: Subsistence on the Northwest Coast. In *Man the Hunter*, edited by R. B. Lee and I. DeVore. Chicago, Illinois: Aldine. Pp.56–68.
1983 Productivity and its constraints: A coast Salish case. In *Indian art traditions of the Northwest Coast*, edited by R. L. Carlson. Burnaby, British Columbia: Archaeology Press. Pp.67–87.
1987 *Coast Salish essays*. Seattle, Washington: University of Washington Press.
1990 Central Coast Salish. In *Handbook of North American Indians, volume 7 Northwest Coast*, edited by W. Suttles. Washington, D.C.: Smithsonian Institution. Pp.453–475.
Suttles, W., and W. W. Elmendorf
1963 Linguistic evidence for Salish prehistory. In *Symposium on language and culture*, edited by V. E. Garfield. Proceedings of the 1962 Annual Spring Meeting of the American Ethnological Society, Seattle, Washington. Pp.41–52.
Thackeray, J. F.
1988 Molluscan fauna from Klasies River, South Africa. *South African Archaeological Bulletin* **43**, 27–32.
Thomas, K. D.
1987 Prehistoric coastal ecologies: a view from outside Franchthi Cave, Greece. *Geoarchaeology* **2**, 231–240.
Thompson, E. N.
1972 Historic resource study: San Juan Island National Historical Park, Washington. Denver Service Center, National Park Service, U.S. Department of the Interior, Denver, Colorado.
Thompson, G.
1978 Prehistoric settlement changes in the southern Northwest Coast: A functional approach. University of Washington, Department of Anthropology, Reports in Archaeology No. 5, Seattle.
Treganza, A. E., and S. F. Cook
1948 The quantitative investigation of aboriginal sites: Complete excavation with physical and archaeological analysis of a single mound. *American Antiquity* **13**, 287–297.
Trigger, B. G.
1986 *Native shell mounds of North America*. New York: Garland.
Voigt, E. A.
1975 Studies of marine mollusca from archaeological sites: Dietary preferences, environmental reconstructions and ethnological parallels. In *Archaeozoological studies*, edited by A. T. Clason. Amsterdam: North Holland. Pp.87–98.
Waselkov, G. A.
1987 Shellfish gathering and shell midden archaeology. In *Advances in archaeological method and theory, vol. 10*, edited by M. B. Schiffer. Orlando, Florida: Academic Press. Pp.93–210.
Watson, P. J., S. A. LeBlanc, and C. L. Redman
1984 *Archeological explanation: The scientific method in archeology.* New York: Columbia University Press.
Webb, W. S.
1938 An archaeological survey of the Norris Basin in eastern Tennessee. Bureau of American Ethnology, Bulletin No. 118, Washington, D.C.
1946 Indian Knoll, Site Oh 2, Ohio County, Kentucky. University of Kentucky, Reports in Archaeology and Anthropology 4(3), Lexington, Kentucky.
Webb, W. S., and D. L. DeJarnette
1942 An archaeological survey of Pickwick Basin in the adjacent portions of the states of Alabama, Mississippi and Tennessee. Bureau of American Ethnology, Bulletin No. 129, Washington D.C.

Webb, W. S., and W. G. Haag
1939 The Chiggerville site, Site 1, Ohio County, Kentucky. University of Kentucky, Reports in Archaeology and Anthropology 4(1), Lexington.
Wessen, G. C.
1988 *The use of shellfish resources on the Northwest Coast: the view from Ozette. In* Prehistoric economies of the Pacific Northwest coast, edited by B. L. Isaac. Research in Economic Anthropology, Supplement 3. JA1 Press, Greenwich, Connecticut. Pp.179–210.
Widmer, R.
1989 Archaeological research strategies in the investigation of shell-bearing sites, a Florida perspective. Paper presented at Society for American Archaeology, Atlanta.
Willey, G. R., and J. A. Sabloff
1980 *A history of American archaeology. 2nd edition.* San Francisco: W. H. Freeman.
Wing, E., and I. Quitmyer
1985 Screen size for optimal data recovery: a case study. In *Aboriginal subsistence and settlement archaeology of the Kings Bay locality, vol. 2*, edited by W. Adams. University of Florida, Department of Anthropology, Reports of Investigation, 2, Gainesville. Pp.49–59.
Woodroffe, C. D., J. Chappell, and B. G. Thom
1988 Shell middens in the context of estuarine development, South Alligator River, Northern Territory. *Archaeology in Oceania* **23**, 95–103.
Yesner, D. R.
1980 Maritime hunter-gatherers: ecology and prehistory. *Current Anthropology* **21**, 727–750.

2

Shell Midden Boundaries in Relation to Past and Present Shorelines

Fran H. Whittaker
Julie K. Stein

I. Introduction

Deposition of shell middens in coastal areas drastically alters the configuration of the past and present shoreline. The deposition of hundreds of cubic meters of shell, sediment, artifacts, and organic matter change the processes operating at that location. Deep embayments are filled to form shallow shelves. Undulating coastlines are straightened. And gravel-sized shells blanket the shore as they erode from the midden. Add to these physical changes the effects of sea level fluctuations and tectonic events, and the extent of coastal disruption becomes clear. Investigating shell middens requires that the topography before the occupation, the sea level and tectonic history, and the historic relationship of shoreline and shell midden all be reconstructed. To accomplish these reconstructions the boundaries of the shell midden must be found, the configuration of the shell midden on the paleosurface reconstructed, and the sea level history for the region be determined.

In this chapter we attempt to provide these reconstructions. First, we describe the augering and mapping method used to discover the boundaries of the shell midden. Then the topography of the modern surface is described, followed by the description of the topography of the surface below the shell midden. The difference between these two surfaces allows us to reconstruct some crucial information: the location of the prehistoric shoreline, the place of the oldest occupation, and the amount and age of sea level changes affecting Garrison Bay. These investigations support interpretations reported in this book that are based on other types of data, that is, that sea level changes and processes of shell midden deposition are responsible for some of the stratification observed at British Camp.

II. Augering at British Camp

The augering of the British Camp site on San Juan Island began during the 1985 field season. This project was undertaken to determine the depth and extent of the shell midden. While archaeological material containing shell was naturally exposed along most of the local shoreline, its extent away from the shore was unclear. An excavation of one (1 × 1 m) unit near the shore was opened in 1983. At this location the shell midden was observed to extend to or below the water table. Augering was the logical method to determine the depth of the midden over the full extent of the site.

Coring and augering (along with remote sensing, see Dalan *et al.*, Ch. 3, this volume) permit fast, cost-efficient, and resource-conservative investigation of archaeological deposits. A site's horizontal and vertical extent or the relationship of cultural and noncultural deposits can easily be determined, and a large area can be examined with a minimum of disturbance (Stein 1986, 1991). Coring removes a cutting of sediment using a hollow cylinder; augers cut through sediments using a helical motion, disturbing the context of the material. An augering tool is used in situations where numerous large objects prevent the cutting of a continuous column of sediment. At the British Camp site, large durable shells, pebbles, and cobbles of igneous and metamorphic rocks required the use of a 4-inch bucket auger. The auger allowed us to penetrate the midden and determine its thickness.

Once the extent of the midden was determined, the second goal of the augering project was to reconstruct the location of Garrison Bay shoreline prior to the formation of the shell midden. We used augering to reconstruct the location of a wave-cut bank and the shoreline before occupation. This reconstruction allowed an assessment of the extent of cultural deposits at the site as well as information about land and sea level fluctuations.

Augering and coring have received little use at southern Northwest Coast archaeological sites. A notable exception is the work done at the Duwamish Site in Washington. There Campbell (1981) described her use, and previous uses of, a manual screw-type auger and posthole digger, which determined that the site represented a single stratigraphic component with considerable horizontal variation. With the addition to the literature of the British Camp study, the potential of minimally destructive augering techniques in the southern Northwest Coast will be fully demonstrated.

III. Methods

The modern topographic surface of the study area was mapped using a transit and stadia rod in a controlling-point survey method. All readings are related to a primary reference, Datum A (grid coordinate 294/306), located near the large excavation unit (Operation A)(Figure 1). Datum A provides both horizontal and vertical control for the mapping project. Datum A is set in cement, and four other permanent datums are installed to facilitate the mapping of a larger area. Surface elevations are taken for each augering location individually.

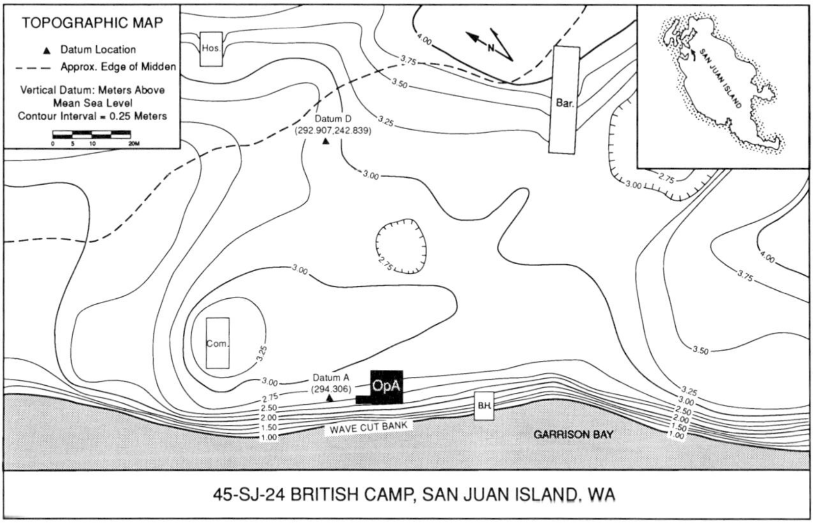

Figure 1 Modern topographic map of British Camp. Rectangles are outlines of historic buildings (B.H. is Block House; Bar. is Barracks; Com. is Commissary; Hos. is Hospital). The shaded area labeled OpA is the location of the 1983–1989 University of Washington archaeological excavation Operation A. The grid coordinate system of the archaeological excavation is used to locate Datum A and Datum D. The grid is expressed in meters. The contours are expressed as meters above mean sea level, using the horizontal datum "National Geodetic Vertical Datum of 1929" (as surveyed through NPS contract. Map drawn by Gary Hickman, 1989. Map available through NPS Northwest Region office.)

The orientation of all project maps corresponds to the grid system used by the project. This arbitrary grid is specifically designed to be parallel to the wave-cut bank and allow for the exposure of stratigraphy perpendicular to the shoreline. This strategy is designed to provide the excavators with stratigraphic exposures of the shoreline progradation process in cross section. On the topographic map (Figure 1), directions are oriented according to the arbitrary project grid. Both true and magnetic north are indicated on the map. Directions specified in the text refer to grid directions rather than relative to magnetic or true north.

The augering procedure involves cutting away a small circular section of sod, inserting the bucket auger, and twisting it into the ground. As the blades of the auger cut downward (Figure 2), sediment rises into the bucket. The bucket auger is attached to extension rods of various lengths (Figure 3). Most augering is done using 2 m extension rods. The maximum depth of midden reached is 4 m. Often large rocks are encountered that are too large to fit through the mouth of the bucket. On these occasions, the auger is moved a few centimeters away and a new hole begun. Occasionally, several new starts are required at one location before the base of the midden is reached.

After the auger bucket fills with sediment its contents are examined and the depth of the hole measured. Major changes in stratigraphy and the approximate depth of

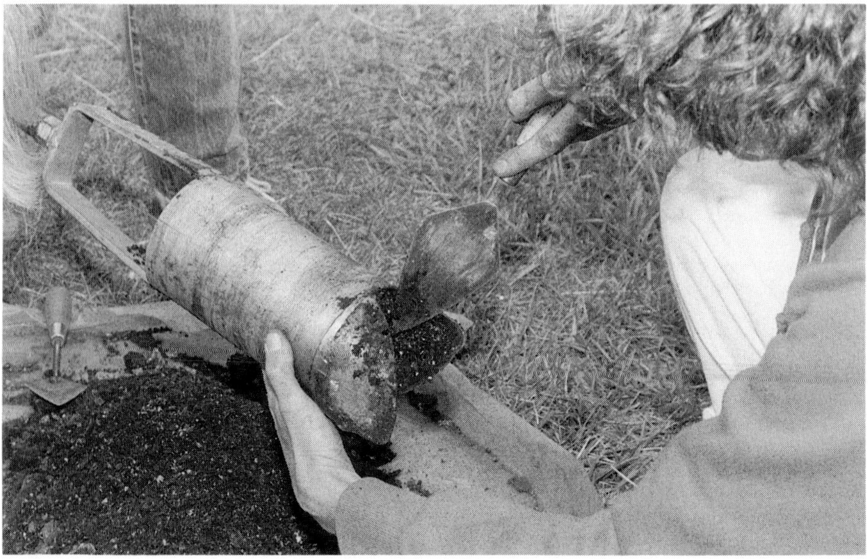

Figure 2 The 4-inch diameter bucket auger. Shell midden is being extracted from the bit with a trowel. Photo by M. Levin.

the boundaries are also recorded during augering. The moment the auger bit passes below the midden, the depth is measured on the extension rods of the auger. When glacial drift, which is devoid of cultural material, is observed in the bucket, the depth is measured and recorded. Sometimes several more buckets are removed to confirm that the base of the midden has actually been reached. The glacial drift beneath the midden deposits is distinct from the shell midden in both color and texture. It is lighter in color, friable, and sandy while the midden is dark, rocky, and feels crunchy while the auger cuts through it.

In the deeper parts of the midden near the shoreline, groundwater is encountered prior to the base of the shell midden. In a few places, the base of the midden is deeper than the length of the extension rods (greater than 4 m in depth). In those locations the depth of the midden is simply recorded as greater than 4 m.

IV. Modern Topography

The British Camp site is located on Garrison Bay in the northwestern portion of San Juan Island. The bay itself is protected from large waves by peninsular portions of the island lying to the west (Figure 4). Most terrain beside Garrison Bay slopes steeply toward the water. The area including the parade grounds at historic British Camp, shown on the modern topographic map and in Figure 5, is exceptional. This acreage is one of the largest tracts of land that is relatively level and treeless. To the north, east, and southeast the slopes rise to 20 m in elevation within 75 m of the site.

Figure 3 Bucket auger with two extension rods attached to bit head, allowing penetration to a depth of 4 m. Photo by Ken Alford.

Presumably this level land pictured in Figure 5 attracted settlers among prehistoric peoples as it did the Europeans who chose this location (Thompson 1972).

The topography of the study area detailed in Figure 1 is a gently rolling, nearly level surface above the shoreline. The edge of Garrison Bay is marked today by the presence of a 2.5-m high erosional landform called a wave-cut bank. At the top of the bank a level, slightly undulating surface begins. Near the western edge of the map, the course of an ephemeral drainage originating on the slope to the north is prominent. Winter flooding has been observed in this western drainage near the

Figure 4 Aerial photograph of Garrison Bay at high tide looking east. British Camp parade grounds (and shell midden) is the treeless, grassy area in the upper left portion of the photo. Guss Island is in the center of Garrison Bay. Areas examined by Treganza and Carlson (1960) are located along the shoreline left (west) of British Camp.

shoreline where the wave-cut bank is not steep. British accounts from the mid-1800s describe this area to the west of the hospital and commissary (marked Hos. and Com. on the topographic map, Figure 1) as marshy. The shell midden disappears in this section of the site. However, a 1972 excavation uncovered preserved prehistoric basketry close to the shore, west of the commissary (Sprague 1976).

The area shown on the modern topographic map was a focal point of the British occupation in the last century, and there is no doubt that the British changed the shape of the land surface (see Thomas and Thomson, Ch. 4, this volume). The configuration of the midden as seen today is the product of both historic and prehistoric landscape modification (e.g., Onat 1985). An exploration by William I. Warren of the U.S. Boundary Commission in 1860 described a deep bay or inlet with a swamp at the northern end (Thompson 1972:111). Thompson speculates that this may be Westcott Bay, but we believe the description to be of Garrison Bay. This particular description is especially interesting, since Warren describes camping near the remains of an old native lodge on the bay. During the encampment of the British, notes on the work performed by the soldiers stationed there mention the dismantling of an Indian house. They also mention the heavy amount of physical work required of the men for the clearing of timber (Thompson 1972:202). So one can assume that the surrounding area was wooded prior to the British occupation. This limited amount of information reminds one of the difficulties encountered when trying to envision landscapes of just a century before.

The locations of park buildings and reconstructed nineteenth century buildings have been included on the maps and are useful as reference points. The commissary is about 20 m from shore at the highest point of an oblong knob referred to as the supratidal platform. This modern platform rises a half meter above the surrounding topography. East of this, the locations of reference Datum A and the Operation A (Op. A) excavations are shown. The blockhouse (B.H.) is located directly on the wave-cut bank where its base extends well out into the tidal zone. Other historic park structures, the barracks and hospital, are located farther north. North of these structures the grade steepens upward, and a slope leads up to the remains of a historic homestead (the Crook House) at the crest of the hill. A dashed line on Figure 1 shows the approximate inland extent of the shell midden deposits, the boundary of which was determined through augering.

Thus, the majority of the area shown on the map represents archaeological deposit. The shell midden continues beyond the edges of the map to the east where it abuts a bedrock slope (see Figure 4). To the west it curves along the shore for at least several hundred meters. The heavily forested area in Figure 4 is the location of Treganza's excavation (1950), Carlson's excavation (1960), and an excavation by the University of Washington in 1988, 1990, and 1991.

Figure 5 Aerial view of British Camp parade grounds. University of Washington excavation (Operation A) is in right center of photo. Area and buildings are roughly the same area as depicted in the topographic map (Figure 1). Photo by G. T. Jones.

V. Prehistoric Topography

The paleotopographic reconstruction is presented in Figure 6. This map depicts the landscape as it existed prior to its substantial alteration by human occupation with the deposition of the large shell midden. We mapped the topography of the land surface under the midden by subtracting the elevation of the base of the midden from the surface elevation measurement for each auger hole. These measurements provide the information required to construct a contour map of the paleotopography of the preoccupation surface.

The paleoshoreline is reconstructed by noting the location of the wave-cut bank in the paleotopograohy. The wave-cut bank today is the location of the highest tide, the base of which is defined by the 1-m coutour interval on the modern topographic map (Figure 1). The wave-cut bank is expressed as a slope of approximately 60°. Such a slope was discovered in plotting the paleotopographic map and is assumed to be the location of the paleoshoreline. The base of the paleo wave-cut bank is defined by the 0-m coutour interval. If this paleo wave-cut bank represents the location of the paleoshoreline (as the modern wave-cut bank represents the modern shoreline), then mean sea level must have been 1 m lower than it is today.

The most striking alterations caused by the addition of the shell midden are the marked changes in the shape and location of the coastline. To the east of the Operation A excavation area, deposition of the midden into the bay caused the shoreline to prograde more than 25 m. This progradation changed the shoreline from wavy,

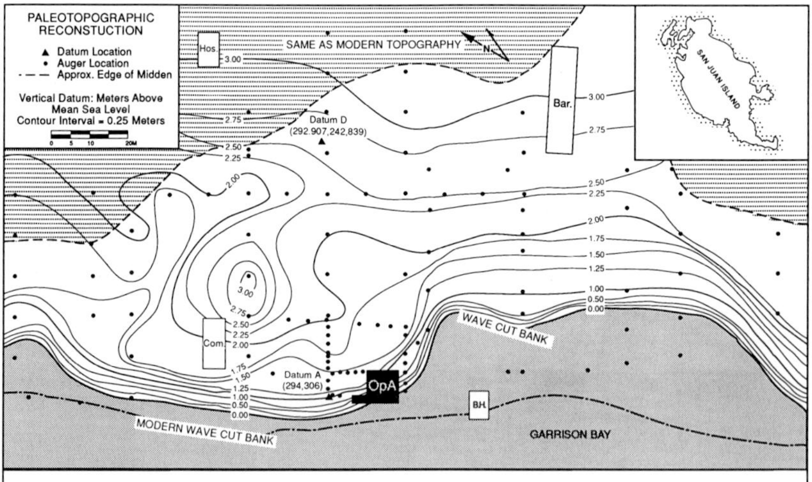

Figure 6 Paleotopographic reconstruction of British Camp area. The surface depicted here is an approximation of the landscape before the prehistoric inhabitants deposited any shell midden. The paleoelevations are calculated by subtracting the depth of the shell midden from the surface elevation. Outlines of buildings (as shown on the topographic map, Figure 1) are shown to allow easy comparison of the modern topography and paleotopography. All locations augered are marked as black dots.

with a promontory near the commissary, to straight. The entire section beneath the block house was filled by prehistoric dumping.

Another aspect of the paleoshoreline is that the elevation of the base of the wave-cut bank is one meter *lower* in elevation than the base of the wave-cut bank today (Figure 1 vs. Figure 6). Further, the elevation of the surface of the prehistoric intertidal platform is more than a meter lower at its top than its elevation today. The location and elevation of the wave-cut banks are known to be controlled by changes in relative positions of both the land and the sea. The changes in the horizontal location of the wave-cut bank is obviously due to the addition of cultural deposits, while the change in elevation is less straightforward. The elevation of the high water line may have risen in the last few millennia by either the sea level rising or the land subsiding through tectonism, or by a combination of both.

VI. Discussion

The major differences between the paleotopography and modern topography of the British Camp site are: (1) the change in the location of the wave-cut bank, especially in the eastern half of the map; (2) the accretion of midden on a supratidal platform immediately east of the commissary; and (3) the change in the elevation of the wave-cut bank before and after the occupation.

A. *Wave-Cut Bank Location*

The location of the wave-cut bank changed as prehistoric people deposited sediment (their waste) into the intertidal zone and erosion removed material from other areas of the site. A shoreline that is now relatively straight was more crooked in the past. The deposition of cultural material into the bay resulted in a 25-m progradation of the shoreline in the eastern part of the site. The wave-cut bank bordering the reconstructed embayment in the eastern part of the site would have been open water before occupation began. As people dumped material over this bank, prograding the land seaward, the configuration of the wave-cut bank changed. As the waves undercut the wave-cut bank in the central portion of the site, any projecting portion of the bank would have been removed by erosion. The amount of progradation of the central and western part of the shoreline is less clear due to limited augering in that area. But modern erosion all along the shoreline has definitely removed some of the midden. The former extent of the midden out in the bay could be reconstructed by drilling into the substrate of Garrison Bay.

B. *Accretion of Midden*

Using the modern and ancient topography at British Camp, we constructed a speculative model of the depositional sequence. The first accumulations of the shell midden on the supratidal platform occurred near the commissary and in the nearby embayment. The supratidal platform seems a favorable location for prehistoric habitation. Across its surface 0.5 to 1.0 m of shell midden has been deposited horizontally on the platform. This platform is believed to be the original location for settlement or other depositional activity, with the first dumping into the intertidal zone occurring near here. The elevation of the intertidal zone was raised by this deposition, and eventually these peripheral dumping areas also became suitable for occupation. Later habitation spread onto the newly created land. Prehistoric people actually increased the size of the supratidal land. If this model is correct, then the oldest deposits should be found on, and immediately adjacent to, the supratidal platform (near the commissary) and the youngest deposits near the present shoreline in the east.

Both stratigraphic and radiocarbon data tend to support this picture of the depositional history. In the stratification of Operation A, a marked change in the orientation of layers from the west end to the east is observed. In the west (unit 306/300) the facies are oriented horizontally, while in the eastern end (unit 310/304) they dip to the southeast. The location of Operation A seems to have encountered the boundary between the area where deposits were laid down on a horizontal surface (perhaps on the original supratidal platform) and on an inclined surface (perhaps the edge of the original wave-cut bank). The inclined strata seen in the profile during the excavation of Operation A (unit 310/304) are pictured in Figure 7.

The radiocarbon dates (see Stein *et al.*, Ch. 6, this volume for a discussion of all dates) further support our model. Deposits of greater age are found in the western portion of Operation A when compared to facies of equivalent depths in the eastern section. Samples of charcoal submitted for radiocarbon analysis from an eastern unit (310/304, facies 1D03, feature 2, 1D, 1B) date younger than samples from facies of equivalent depth in a western unit (306/300 facies 1F, 1L, 1W, 2E)(Table 1). For example, dates obtained on samples from two facies, each found about 80 cm below the present (nearly horizontal) surface, were compared from the two areas of Operation A. In the west (306/300 facies 2E; 85 cm below surface) the calibrated range is 930–670 B.P. using Stuiver and Reimer (1986). In the east (310/304 facies 1D 03; 78 cm below surface) the calibrated range is 528–489 B.P.(see Table 1). The calibrated ranges of these dates do not overlap (Figure 8), tending to support the idea that the oldest deposits at the site are found nearest to and on top of the supratidal platform. This trend is corroborated by the rest of the samples dated from these two units (Table 1, Figure 8). In every case samples from similar depths are younger in the eastern unit with the inclined strata than in the western unit with the horizontal strata.

Previous findings at other parts of the British Camp site also provide general support for our model. Kenady (1973) excavated a large unit between our Operation A and the commissary. His photographs of the stratigraphy of this unit (Figure 9) indicate horizontal layering of the 1 m of cultural deposits and numerous pits excavated prehistorically into the glacial drift. These features indicate only that people inhabited the area, not the age of that habitation. In another unit excavated under the present location of the commissary, Kenady found preserved planks of cedar near the surface, which he interpreted as the remains of a protohistoric plankhouse. The British reportedly dismantled the remains of an "indian house" located somewhere

Figure 7 The upper portions of the east profile of unit 310/304 have shell lenses dipping to the east and south. Photo by K. Kornbacher.

Table 1 Radiocarbon Measurements from Units 306/300 and 310/304

Unit	Facies	Depth – cm below surface	Lab	Age B.P.	Standard deviation 1σ	Calib. date B.P.[a]	Calib. range B.P.[b]
306/300	1F	40	QL-4153	430	40	509	520–482
306/300	1L	60	QL-4154	810	80	727	792–672
306/300	1W	82	QL-4155	1000	40	924	971–796
306/300	2E	85	QL-4156	830	70	735	930–670
310/304	1B	35	WSU-3514	160	60	271/12	290–64
310/304	1D	43	WSU-3515	370	70	470	508–314
310/304	1D 03	78	WSU-3516	450	50	510	528–489
310/304	feature 2	90	WSU-3153	355	50	461	494–315

[a]All dates are on charcoal. Calibrations were calculated using the PC program CALIB & DISPLAY Version 2.1.
[b]After Stuiver and Reimer (1986).

on the parade grounds (Thompson 1972). Kenady suggests that the cedar planks he discovered are the remains of this house, and he believes that they pinpoint the location of habitation during the protohistoric period.

C. *Elevation of Wave-Cut Bank*

The low elevation of the base of a former wave-cut bank indicates that the relative position of land and sea has changed since (or during the time that) the shell midden was deposited, causing an approximate rise of 1 m in the relative position of shoreline and sea level. Wave-cut banks are erosional features, whose location is

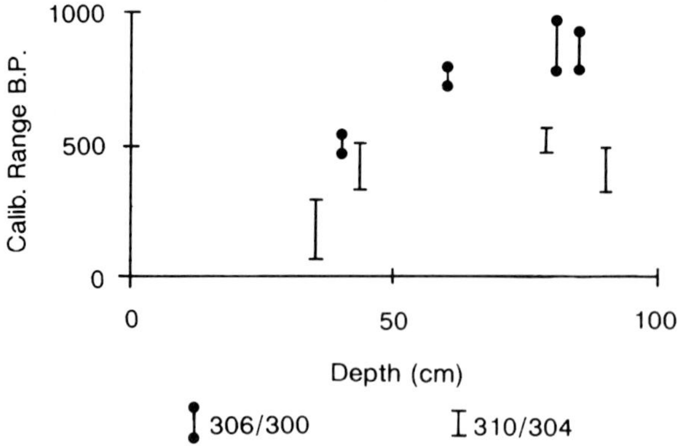

Figure 8 Plot of radiocarbon dates from Table 1. This visual representation of the data shows the consistently younger ages assigned to unit 310/304 where, according to our model of deposition, early dumping into the intertidal zone would have occurred. Drafted by T. Hunt.

Figure 9 Unit excavated by Steve Kenady in 1972. Note the horizontal layering exposed in the profile. The pits being excavated by the archaeologists had been dug prehistorically into the underlying glacial drift. Photo used by permission of S. Kenady.

dictated by the elevation of the highest high tides. The location of the old wave-cut bank and its elevation in the paleotopographic map indicates that before the prehistoric people deposited any midden in this part of the site, the elevation of this high water line was about 1 m lower than it is today. Whether the rise in the water (or lowering of the land) occurred before deposition of the shell midden cannot be addressed given the manner in which these data were collected.

The 1-m rise in sea level has resulted in the inundation of cultural deposits originally laid down on dry land. Brackish groundwater now saturates the lowest portion of most of the midden. The height of the groundwater parallels the modern topography and the rise and fall of the tide, as well as seasonal fluctuations of the local water table. Groundwater is now affecting all shell midden below 2.0 m in elevation. In some portions of the site the cultural deposits are deeper than that elevation, and that portion of the shell midden is saturated.

D. *Regional Sea Level Chronology*

Most of the research on vertical land movements and changing relative sea level pertinent to this study is from southwestern British Columbia, supported by the Geological Survey of Canada and from Washington, supported by the Washington Department of Ecology, Shorelines and Coastal Zone Management Program. The postglacial sea level reconstructions are divided into the inner coast zone, middle coast zone, and outer coast zone (Clague *et al.* 1982:600), with San Juan Island (and eastern Vancouver Island) falling into the middle coast zone.

The sea level history of the middle coast zone begins with melting of the glaciers at the end of the Pleistocene around 13,000 ^{14}C years B.P. (All radiocarbon ages cited in this section are uncalibrated. See Stuiver *et al.* 1991 for current calibration information.) At that time relative sea level was higher than its present position, perhaps as high as 75 to 175 m in elevation (Clague *et al.*, 1982:605–606, Figure 6). Rapid glacio-isostatic rebound accompanied deglaciation, and the land rose causing relative sea level to drop to a level below its present position for the period 9200 to 5500 ^{14}C years ago. Evidence for this low, early Holocene relative sea level stand comes from buried peat deposits, wave-cut benches extending below the zone of modern wave action, drowned river mouths, and growth and movement of sand dunes (Clague *et al.* 1982; Clague 1989; Clague and Brobowsky, 1990; Mathews *et al.* 1970). These data suggest that between 9200 and 5500 B.P. relative sea level in the "middle coast zone" dropped to a position of -11 m below its present position.

By 5000 ^{14}C years ago sea level rose to within a few meters of its present position. At Island View Beach on eastern Vancouver Island north of Victoria, a fossil stump was found extending through a submerged layer of peat (indicating a time of lower sea level) 1.0 to 1.5 m below high tide and has been dated to about 2000 ^{14}C years B.P. (Clague 1989:235; Mathews *et al.* 1970:697). This evidence suggests that relative sea level at Island View Beach was at least 1 m lower 2000 years ago than it is at present.

At Westcott Bay and Third Lagoon on San Juan Island, salt marsh deposits show that relative sea level rose between 2 and 3 m between 5000 and 3000 ^{14}C B.P. and has not varied by more than 1 m since about 2000 ^{14}C B.P. (Beale 1990). The marsh at Third Lagoon (at the southern end of the island in San Juan Island National Historic Park-American Camp) developed from a tidal flat about 1000 B.P. These data suggest to Beale that northern Puget Sound and the San Juan Islands have not experienced sudden vertical crustal movement greater than 1 m in the past 1000 years (1990:54). Beale agrees with Clague and Bobrowsky (1990) that this region has witnessed only about 1 m of sea level rise since 2000 B.P.

Further evidence of lower sea level in the Late Holocene is derived from stratigraphic relationships at other archaeological sites (Clague *et al.* 1982:599–600; Shipman 1989:12–14). For example, at Montague Harbour on Galiano Island (off the eastern coast of Vancouver Island, north of Victoria) the lower portion of a shell midden is inundated at high tide (Mitchell 1971). This midden has been cored recently (Clague and Bobrowsky 1990:248), with the discovery of peat below the cultural deposits. This peat and the shell midden extend seaward below the modern beach gravel, indicating that a Late Holocene marine transgression occurred, ending about 2000 B.P. At many other shell middens in the Gulf of Georgia and northern Puget Sound, the middens are eroding with their lower portions being inundated with seawater at high tide (Grabert and Larsen 1975; Larsen 1972). The British Camp shell midden is one of these middens. We have watched large sections of the midden fall into the bay as trees are undercut and the rootmasses (with attached midden) are washed away.

In many of the buckets brought up by the auger in the old embayment now filled by midden (the area presently under the block house), granule-size and pebble-size gravel immediately underlies the bottom of the shell midden. This stratigraphic relationship implies deposition of midden on near-shore, wave-sorted beach gravels. The stratigraphic relationship of the gravels to the midden is very difficult to observe. Since these auger buckets are extracted from portions of the midden well below the water table, the sides of the hole tend to collapse after each withdrawal. A continuous core needs to be obtained to resolve the origin of the gravels and to investigate the possibility of finding peat below the shell midden.

E. *Crustal Movements in the Region*

While a gradual change in sea level and land movement probably occurred at eastern Vancouver Island and in the region of San Juan Island, sudden changes in land subsidence occurred along the outer coast during great subduction earthquakes (Atwater 1987). When these large (magnitude of about 8 or 9) earthquakes occur, the coastline (near the zone of subduction) can undergo either uplift or subsidence. Atwater's examination of the outer coast of Washington suggests that subsidence occurred rapidly. During these events of substantial seismic activity, peaty sediments of coastal marshes sank 0.5 to 2.0 m, followed by rapid burial of the peat with intertidal mud 0.5 m in thickness. The intertidal muds eventually came to support a new marsh, and peat deposits once again began to accumulate. At Willapa Bay, Washington a series of six peats, each separated stratigraphically by intertidal muds, implies at least six large earthquakes in the past 3500 years alone.

The idea that crustal deformation through tectonism has occurred in the Straits of Georgia region of southwestern British Columbia is now being investigated (Beale 1990; Clague 1989; Clague *et al.* 1982; Clague and Bobrowsky 1990). The fossil stump growing through the peat at Island View Beach has been reexamined for indications of a rapid transgression of sea level at about 2000 radiocarbon years ago. The peat bed is 1.0 to 1.5 m below modern high tide, and the fossil stump itself has been dated to 2040 ± 130 years B.P. (Mathews *et al.* 1970:697). Above the peat is a mud (of about 0.5 m in thickness) containing brackish water foraminiferal species. The stratigraphic sequence is similar to the sequence observed by Atwater (1987), suggesting that a seismic event may have been responsible for the subsidence of the eastern shore of southern Vancouver Island. Clague and Bobrowsky (1990) now believe that the evidence found at Island View Beach is not the result of a rapid seismic event, rather the result of a gradual rise in sea level after the transgression dating to around 2000 B.P. Likewise Beale (1990) finds no evidence of rapid subsidence on San Juan Island. She interprets the sequence as resulting from a gradual relative rise in sea level.

While researchers do not agree on the nature of the change, the local sea level fluctuations they observe also affected British Camp, which is located on the northwestern portion of San Juan Island directly south of Westcott Bay and across Haro Strait from Island View Beach. The distance from Island View Beach across the strait to British Camp is approximately 15 km, and Westcott Bay is 2 km north of Garrison Bay, sharing the same outlet channel to Haro Strait. Obviously any seismic event proposed for Westcott Bay and Island View Beach would also have struck Garrison Bay. The geology of San Juan Island is different from that of Vancouver Island, and there are bedrock faults under Haro Strait (Brandon *et al.* 1988). The coastal zone of Garrison Bay may have been uplifted relative to sea level, stayed the same elevation relative to sea level, or subsided. However, the evidence at Island View Beach, Westcott Bay, and British Camp suggests a gradual submergence unlike that proposed by Atwater.

VII. Conclusions

The augering project at British Camp was carried out over several years while the excavation of Operation A was under way. In combination with modern topographic data, the augering results were used to reconstruct the paleotopography of the area prior to the deposition of the extensive shell midden.

Comparison of the topographic and paleotopographic maps shows that progradation of the shoreline has taken place in the last 2000 years. Further, the shape of the shoreline has been completely altered. Prior to midden deposition the supratidal platform was the highest and best-drained portion of the landscape. It was probably the location of the earliest cultural deposition, with later deposits moving outward into the bay. A major change in the level of the shoreline is indicated by the presence of a wave-cut bank 1 m lower than the modern one. The relevant geological data for the area indicate that either seismic activity or a rise in sea level accounts for this shift in relationship.

Acknowledgments

The mapping of the British Camp Site was begun by Dr. G. Thomas Jones and completed by Fran Whittaker. The augering was done by Julie Stein and Fran Whittaker with the assistance of many field school students, especially Joel Allen, Angela Linse, Eric Rasmussen, Quyen Nguyen, and Vuong Ha. Fran Whittaker would like to gratefully acknowledge the support of the Dorothy Danforth-Compton Graduate Fellowship during portions of the work represented in this chapter. The authors would like to thank Brian Atwater for his help with the sea level information and for reviewing the completed manuscript.

References

Atwater, B. F.

1987 Evidence for great Holocene earthquakes along the outer coast of Washington State. *Science* **236**, 942–944.

Beale, H.

1990 Relative rise in sea level during the past 5000 years at six salt marshes in northern Puget Sound, Washington. Report prepared for Washington Department of Ecology, Shorelands and Coastal Zone Management Program, Olympia, Washington.

Brandon, M. T., D. S. Cowan, and J. A. Vance

1988 The late Cretaceous San Juan thrust system, San Juan Islands, Washington. Geological Society of America Special Paper 221.

Campbell, S. K.
1981 The Duwanish No. 1 site: a lower Puget Sound shell midden. Research Report No. 1, University of Washington Office of Public Archaeology, Seattle, Washington.
Carlson, R. L.
1960 Chronology and culture change in the San Juan Islands, Washington. *American Antiquity* **25**, 562–586.
Clague, J. J.
1989 Late Quaternary sea level change and crustal deformation, southwestern British Columbia. *Geological Survey of Canada*, paper 89-1E, Pp.233–236.
Clague, J. J., and P. T. Bobrowsky
1990 Holocene sea level change and crustal deformation, southwestern British Columbia. *Geological Survey of Canada*, paper 90-1E, Pp.245–250.
Clague, J. J., J. R. Jarper, R. J. Hebda, and D. E. Howes
1982 Late Quaternary sea levels and crustal movements, coastal British Columbia. *Canadian Journal of Earth Science* **19**, 597–618.
Grabert, G. F., and C. E. Larsen
1975 Marine transgression and cultural adaptation: preliminary tests of an environmental model. In *Prehistoric maritime adaptations of the circumpolar zone*, edited by W. Fitzhugh. The Hague: Mouton. Pp.229–251.
Kenady, S. M.
1973 Environment and functional change in Garrison Bay. In *Miscellaneous San Juan Island reports 1970–1972*, edited by S. M. Kenady, S. A. Saastamo, and R. Sprague. Department of Sociology and Anthropology, University of Idaho, Moscow. Pp.39–83.
Larsen, C. E.
1972 The relationship of relative sea level to the archaeology of the Fraser River Delta, British Columbia. Paper presented at the 1972 Annual Meeting of the Geological Society of America, Minneapolis, Minnesota.
Mathews, W. H., J. G. Fyles, and H. W. Nasmith
1970 Postglacial crustal movements in southwestern British Columbia and adjacent Washington state. *Canadian Journal of Earth Sciences* **7**, 690–702.
Mitchell, D. M.
1971 Archaeology of the Gulf of Georgia area, a natural region and its cultural types. *Syesis* **4**, suppl. 1.
Onat, A. R. B.
1985 The multifunctional use of shellfish remains: from garbage to community engineering. *Northwest Anthropological Research Notes* **19**, 201–207.
Shipman, H.
1989 "Vertical land movements in coastal Washington: implications for relative sea level changes." Report prepared for the Washington Department of Ecology, Shorelands and Coastal Zone Management Program, Olympia, Washington.
Sprague, R.
1976 The submerged finds from the prehistoric component, English Camp, San Juan Island, Washington. In *The excavation of water-saturated archaeologoical sites (wet sites) on the Northwest Coast of North America*, edited by D. R. Croes. Canada National Museum of Man, Mercury Series, Archaeological Survey Papers 50, Ottawa. Pp.78–85.
Stein, J. K.
1986 Coring archaeological sites. *American Antiquity* **51**, 505–527.
1991 Coring in CRM and Archaeology: A Reminder. *American Antiquity* **56**, 138–142.
Stuiver, M., and J. Reimer
1986 User's guide to the programs CALIB and DISPLAY 2.1. Quaternary Isotope Lab, Quaternary Research Center (AK-60), University of Washington, Seattle.
Stuiver, M., T. F. Braziunas, B. Becker, and B. Kromer
1991 Climatic, solar, oceanic, and geomagnetic influences on late-glacial and Holocene atmospheric $^{14}C/^{12}C$ change. *Quaternary Research* **35**, 1–24.

Thompson, E. N.
1972 Historic resource study. San Juan Island National Historical Park, Washington. National Park Service, U.S. Department of the Interior. Denver, Colorado.

Treganza, A. E., and students
1950 Excavations at English Camp San Juan Island, Washington, July-August 1950: field notes, reports, artifact list, photos, and maps. University of Washington Summer Field School. Notebook on file at National Historic Park, Friday Harbor, Washington.

3

Geophysical Exploration of the Shell Midden

Rinita A. Dalan
John M. Musser, Jr.
Julie K. Stein

I. Introduction

Geophysical methods offer archaeologists several options in the identification of subsurface phenomena. First, they often require less time and labor than more traditional archaeological investigative techniques such as coring and excavation. As a result, larger areas may be explored using geophysical methods, yielding a more representative picture of subsurface conditions. Second, innovations in geophysical instrumentation and data processing systems enhance greatly the ability of the archaeologist to interpret and present data relating to either naturally or culturally induced changes in physical properties of subsurface deposits. Finally, geophysical methods involve measurements on the surface and generally do not disturb subsurface deposits. As archaeological sites are being increasingly disturbed and destroyed by development and vandalism, and as the excavation of a site must also be considered a destructive procedure, the ability of geophysical methods to explore without disturbing is perhaps their most important feature. These characteristics of geophysical methods are particularly useful for the investigation of large, stratified shell middens. Overviews of the use of geophysical methods in archaeology are provided in Aitken 1974; Tite 1972; Weymouth 1986; and Wynn 1986.

The archaeological site explored in this study, British Camp, is a large Northwest Coast shell midden site with a complex depositional history. To understand the processes of site formation and postdepositional alteration, the University of Washington's San Juan Island Archaeological project employed an excavation strategy which involved the separate removal and analysis of discrete depositional units, or facies. As this excavation approach was time consuming, only a small portion of the site has been examined. Geophysical methods provided a means of exploring a much larger

Deciphering a Shell Midden

portion of the site without disturbing remaining intact deposits. A necessary adjunct to the successful utilization of geophysical methods was the availability of augering data for the site (Stein 1986). Sufficient augering data exists to both compare against the geophysical results and use as an aid in interpretation of the geophysical data.

Limited geophysical surveys were conducted at British Camp between the summer of 1986 and the spring of 1987. The general purpose of the program was to assess the potential of select geophysical methods for aiding in the interpretation of the formation of the British Camp shell midden. Specifically, the geophysical surveys were designed to provide information on the extent (vertical and horizontal) of the midden, layering or structure within the midden, and landforms on which the midden was deposited.

As different geophysical methods measure changes in different physical properties of subsurface deposits, employing more than a single method in an investigation is often advantageous. A variety of geophysical methods were used at British Camp including electrical resistivity, seismic refraction, and magnetic surveys. The magnetic survey did not produce conclusive results and will not be discussed further. The electrical resistivity and seismic refraction surveys, both of which provided useful data, are the subject of the remainder of this chapter. While each of these methods is summarized briefly below, more detailed discussions may be found in Beck 1981; Dobrin 1976; Griffiths and King 1981; Keller and Frischknecht 1966; Parasnis 1986; and Redpath 1973.

II. The Electrical Resistivity Method

The electrical resistivity method is utilized to determine the configuration of subsurface materials based upon differences in their electrical properties, i.e., differences in their electrical resistivities. The electrical resistivity of a material, in the most basic terms, is a unit measure of the resistance it provides to an electric current made to flow through it (McNeill 1980). As current can be conducted through the moisture-filled pores and passages, the resistivity of most granular soils and rocks is controlled generally more by porosity, water content, and water quality, than by the inherent conductivity of the matrix materials. An exception would be units of high clay content, since clay functions as an ionic conductor.

In an electrical survey, a direct current or a very low-frequency synchronous alternating current is applied to the ground through a pair of nonpolarizing electrodes (A and B), and the resulting potential difference established by this current across a second set of electrodes (M and N) is measured. Figure 1 shows an exaggerated view of this concept in a uniform medium. The ratio of the potential difference to the current (commonly designated as E/I or resistance), multiplied by an array constant (K), will yield a quantity with the dimensions of resistivity (expressed in ohm-meters). This quantity is referred to as the apparent resistivity (ρ_a). Variations in apparent resistivity reflect changes in subsurface conditions and may be analyzed to give quantitative information such as the thicknesses and true resistivities of the subsurface units (Beck 1981).

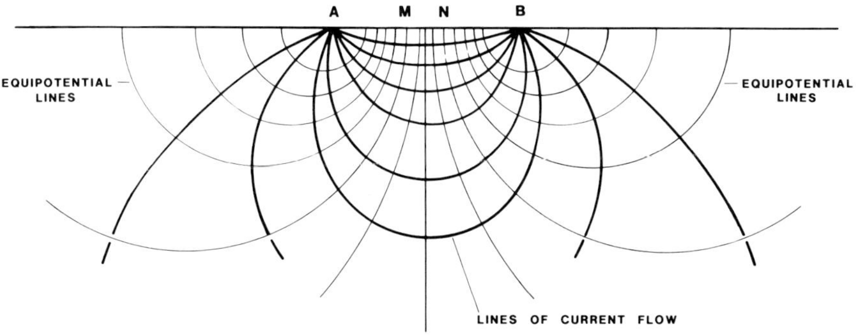

Figure 1 Schematic of an electrical resistivity survey. A and B indicate the locations of the current electrodes and M and N represent the potential electrodes. The distribution of current flow in a uniform medium is shown by the heavy solid lines while the light solid lines are the equipotential lines established by the current.

An ABEM SAS-300 earth resistivity meter was used at British Camp (Figure 2). This system sequentially applies two DC pulses of opposite polarity (160 V at 20 mA) during a 3.6-second interval. The measured potential differences are arithmetically averaged, and the corresponding resistance for the averaged potential is calculated by internal microprocessors and displayed for operator notation at the end of each cycle. Averages of up to 64 cycles may be employed in order to override fluctuating spontaneous potential (noise) generated in the earth's near surface.

Electrical methods can be used either for profiling or depth sounding. Both horizontal electrical profiles (HEP) and vertical electrical soundings (VES) were obtained at British Camp. HEP were used to define lateral midden limits while VES were used primarily to determine vertical midden limits and to study midden structure.

A. *Vertical Electrical Soundings*

Vertical electrical soundings are sometimes referred to as "electric drilling" (Beck 1981) and can be thought of as roughly analogous to a core where information is obtained about materials lying below the surface at a particular location. In order to study the variation of resistivity with depth, the spacing between the various electrodes is gradually increased. The effect of materials at depth becomes more pronounced with the expansion of the electrode array.

Depending upon site conditions and the purpose of the study, several different electrode "arrays" or configurations may be used. For this study, we used the Schlumberger array, where M and N (the potential stakes) are kept constant while A and B (the current stakes) are expanded symmetrically with respect to the midpoint of the array. The four electrodes are spaced so that the ratio of one-half the distance between the outer current electrodes (AB/2) to the distance between the inner potential electrodes (MN) is greater than three. By keeping MN small, the influence of lateral

Figure 2 Resistivity surveying at British Camp.

variations in the subsurface is reduced. Our electrode array was expanded from AB/2 = 1 m to a maximum of AB/2 = 30 m. Maximum depth penetration in our study varied between 15 and 20 m.

The VES data were plotted on a log–log graph at six points per decade and solved for layer thicknesses and resistivities using a modified version of a computer program developed by the U.S. Geological Survey for the inversion of VES curves (Zhody 1975).

B. *Horizontal Electrical Profiling*

Horizontal electrical profiling is sometimes referred to as "electric trenching" (Beck 1981) and is accomplished using techniques similar to those described for the vertical electrical soundings. The four electrodes are moved, maintaining a constant electrode separation, along a traverse line. Apparent resistivities are calculated for the recorded resistance data and plotted as a function of their position along the traverse.

At British Camp, two different AB/2 spacings were utilized. The first HEP profiles were conducted with an AB/2 spacing of 4 m and an MN spacing of 0.25 m. As established from the VES data, this provided a depth penetration of approximately 1.6 m. Subsequent HEP profiles employed an AB/2 spacing of 2 m, which provided a depth penetration of approximately 1.3 m. Readings were taken at 4-m intervals along each traverse line.

III. The Seismic Refraction Method

The seismic refraction method is utilized to determine the configuration of subsurface materials based upon differences in their elastic properties and densities. The elastic properties and density of a medium determine the velocity at which elastic waves travel through it.

In the seismic refraction method, an elastic (compressional) wave is generated at a shallow depth (the shot point) and the resulting particle motion of the ground at in-line points on the surface is recorded. By measuring the travel times through the ground of the generated compressional waves to a number of surface points, the velocities of wave propagation in the ground and the depths of boundaries between layers of differing velocities below each geophone can be calculated. Travel times are detected at the geophones, in direct contact with the earth, by converting the particle motion of the ground into electric signals. These signals are sent to a seismograph which amplifies, filters, records, and displays the data. The raw data, which consists of travel times and distances from each shot point, is converted into a cross section of the different velocity layers which often, though not always, corresponds to the geologic cross section.

The energy source used in this study was a sledgehammer and steel plate. Seismic lines consisted of 24 geophones at 3-m spacings for a total length of 69 m. Each seismic line was composed of two 12-channel (geophone) spreads. Seismic energy was generated at the midpoints, endpoints, and off both ends of each spread. The recording system consisted of an EG&G Nimbus Instruments 1210-F single enhancement, 12-channel seismograph and Mark Products 14-*Hz* geophones. Data interpretation was accomplished using a computerized version of Derecke Palmer's Generalized Reciprocal Method to solve for refractor depths with depth interpretations provided below each geophone. These results were cross checked by manual calculations using the method of differences. At British Camp, the seismic refraction profiles were used to provide information on the depth to bedrock and, to some extent, the preoccupational topography.

IV. Field Survey

Electrical resistivity studies at British Camp consisted of ten vertical electrical soundings and five horizontal electrical profiles (Figure 3). VES 4–7 were located along a line parallel to the wave-cut bank and approximately 3 m behind the block excavation (Op. A), while VES 1–3 and 7–10 were located along the 314 (N/S) grid line, perpendicular to the wave-cut bank. HEP 1 was run parallel to the wave-cut bank, approximately 14 m behind the block excavation (Op. A), while HEPS 2–5 were run perpendicular to the bank at various locations. For HEPS 1 and 2, an AB/2 spacing of 4 m was used. HEPS 3–5 employed an AB/2 spacing of 2 m. HEP 5 was conducted over the HEP 2 location.

Two seismic refraction profiles were accomplished at the site (Figure 3). SL-1 was obtained along a line coincident with HEPS 2 and 5 and VES 1–3 and 7–10. SL-2 was obtained along a line coincident with HEP 1.

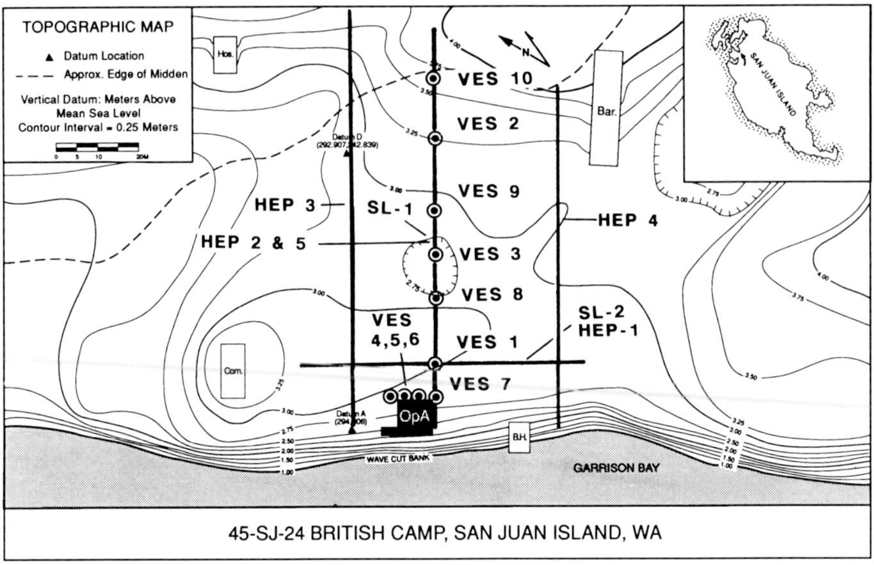

Figure 3 Map of British Camp, San Juan Island, showing the locations of the vertical electrical soundings (VES 1–10), horizontal electrical profiles (HEP 1–5), and seismic refraction lines (SL-1 and SL-2). The shaded box designates the Op. A excavation area while open boxes are historic buildings reconstructed by the National Park Service.

V. Results

A. *Vertical Electrical Soundings*

The VES data are presented in two sections: Section 1 (VES 4–7) includes those soundings located along the 297 (E/W) grid line parallel to the bank; Section 2 (VES 1–3 and 7–10) includes those soundings located along the 314 (N/S) grid line perpendicular to the bank (Figure 3). Calculated layer thicknesses and apparent resistivities are plotted for each sounding (Figures 4 and 5). Vertical lines represent the sounding locations while the dashed horizontal lines connect interpreted equivalent resistivity layers between the VES sounding points.

Section 1 (parallel to the wave-cut bank) shows several distinct resistivity layers (Figure 4). The upper three resistivity layers, with an apparent resistivity averaging around 220 ohm-meters, are interpreted as representing the midden above the water table. The characteristic drop from approximately 200 to 100 ohm-meters at about 1.5 m below the surface probably represents the surface of the water table. As these soundings were conducted in the spring, the water table is fairly high. The water table depth agrees well with data from the nearby block excavation (Op. A), where in 1983 the water table was recorded at 1.8 m below the surface.

The bottom of the midden is most likely represented by the next drop in resistivity, from approximately 100 to 40 ohm-meters. This depth, at about 2 m below the surface, corresponds well with information gained from augering. Below the mid-

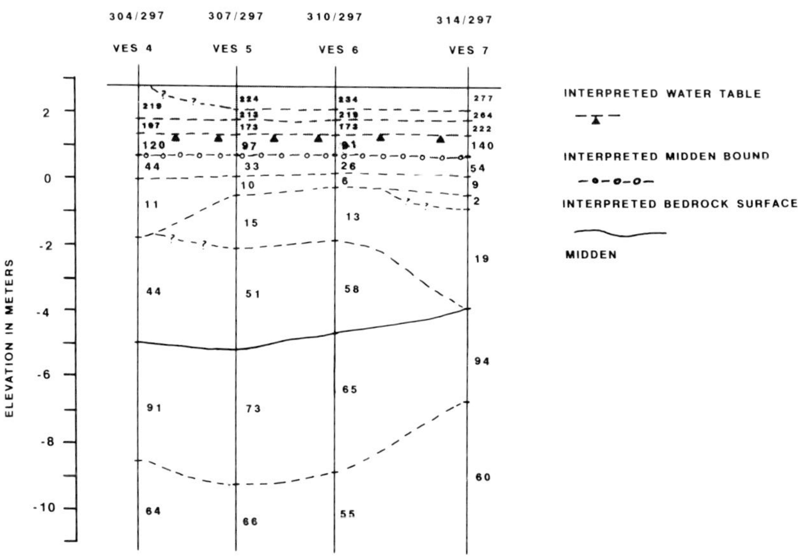

Figure 4 Vertical electrical section across soundings 4–7. Vertical lines represent sounding locations with the appropriate site grid coordinates indicated above. Calculated layer thicknesses and apparent resistivities (in ohm-meters) are presented for each sounding. Interpreted equivalent resistivity layers are connected with horizontal lines. The midden is represented by the top four resistivity layers. The surface of the water table is located near the bottom of the midden. Bedrock is located between 7 and 8 m below the ground surface.

den are fine-grained deposits overlying coarser grained materials (probably glacial drift), with bedrock located between 7 and 8 m below the surface.

Resistivity layers within the midden cannot be interpreted as significant cultural or lithologic units with any certainty. The first drop in resistivity at approximately 0.7 m appears to correspond to the boundary between the light-colored, dry stratum and the black stratum (Stein, Kornbacher, and Tyler, Ch. 6, this volume). As expected, higher resistivities are exhibited by the light-colored stratum. Abundant large shell fragments increase its porosity and decrease the amount of contained water. Due to an increased ability to retain moisture, the underlying dark stratum exhibits lower resistivities.

For the second VES section (perpendicular to the wave-cut bank), only the data from the upper levels is presented (Figure 5). For clarity's sake, the vertical scale has been exaggerated by a factor of 15. VES 1, 2, and 3, with resistivities shown in parentheses, were obtained in the summer when the ground was very dry. The remaining soundings (7, 8, 9, and 10) were obtained in March when the ground was quite wet. As the soundings completed during the different times of the year produced different resistivity values for those layers located above the water table, relative (as opposed to absolute) resistivity changes were used to interpret equivalent resistivity layers.

In Figure 5, similar patterns to those observed in VES Section 1 are evident. As before, there is a change from higher to lower resistivity values within the midden,

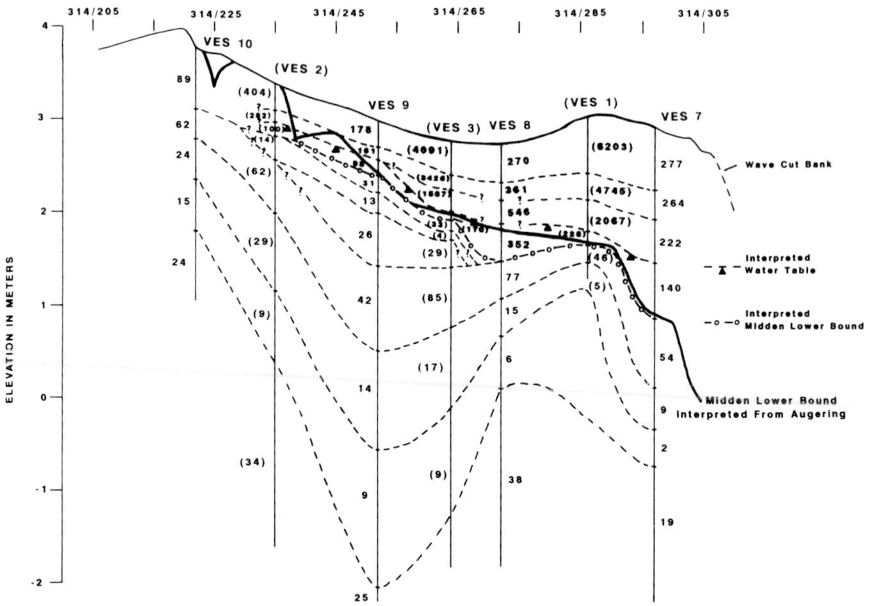

Figure 5 Vertical electrical section across soundings 1–3 and 7–10. Site grid coordinates are indicated above. Parentheses indicate soundings that were accomplished during relatively dry conditions, producing higher resistivity values for those layers located above the water table. As in Figure 4, interpreted equivalent resistivity layers are connected with horizontal lines. The top section shows the midden and the water table sloping gently toward the wave-cut bank. The midden lower bound interpreted from auger data (solid heavy line) agrees well with the midden lower bound interpreted from the VES results (dotted/dashed line). The wedge of sediments underlying the midden on the left half of the figure probably represents an alluvial fan.

a drop in resistivity probably representing the surface of the water table, and another drop interpreted as the bottom of the midden. There is one area (VES 8) where anomalous readings were obtained. At this location, instead of decreasing resistivity values within the midden, a 270-ohm-meter layer overlies layers with higher resistivity values, followed by a drop in resistivity which probably represents the water table. The cause for the dissimilar pattern at VES 8 is unknown at this time, but it may be the result of an indeterminate source of disturbance (either historic or prehistoric), and should be investigated through excavation.

The lower midden boundary as interpreted from the VES data corresponds well with that determined by augering. The only areas where the boundary interpreted from the VES data differs markedly from the augering results is at the VES 8 location in the above noted anomalous area and in the region between VES 2 and VES 9. At the VES 8 location, the lower midden boundary interpreted from the VES data is approximately 35 cm lower than that estimated by augering. The nearest augering location, however, is 4 m away. Likewise, the lower midden boundary in the area between VES 2 and VES 9 was interpolated in a straight line fashion based solely upon the electrical data. The reconstruction of this same section using augering data confirms that the boundaries established at VES 2 and VES 9 are correct, but also shows that the midden surface actually rises and falls between these two points. Thus, differences in estimated midden thicknesses occur only as the result of widely spaced augering and sounding locations.

The sequence interpreted from the second VES section (Figure 5) indicates that bedrock ranges from 4 to 8 m below the surface. The lower portion of the section does not appear on Figure 5 due to the exaggerated vertical scale. On top of the rock surface are various layers of coarse- and fine-grained sediments. The vertical electrical soundings also provide information on lateral midden limits. VES 10 does not display the characteristic high resistivity values and layering associated with the presence of the midden. Augering confirms the lack of midden at this location.

B. *Horizontal Electrical Profiling*

As augering and VES data are available for comparison with HEP 2, 3, and 5, these profiles will be discussed in the most detail. Results from HEP 1 and 4 will be summarized only briefly.

Figure 6 presents the results of HEP 3 (conducted perpendicular to the wave-cut bank along the 294 N/S grid line). The upper portion of the figure shows the plot of apparent resistivity versus the position along the traverse. The wave-cut bank is at the right and the uplands are located to the left. The corresponding midden profile, reconstructed from augering data, is shown below with dots signifying augering locations. Augering locations are widely spaced inland and narrowly spaced near the wave-cut bank.

The depth penetration for HEP 3 was approximately 1.3 m. As indicated by the VES data, the midden has a higher resistivity than the materials directly below it. Therefore, when the midden is thick in relation to the profiling depth, as evident on the right near the wave-cut bank, most of the current flow is through the higher resistivity midden and the resultant HEP values are correspondingly high. When the midden thins and becomes much shallower than the depth penetration of the array, which occurs at approximately 45 m from the landward end of the transect (at about grid coordinates 294 N/245 E), most of the current flow is through the underlying lower resistivity materials and the measured resistivity values are correspondingly low. Thus, the HEP 3 plot can be used to define the general lateral limits of the midden as well as to provide a rough indication of relative midden thickness. As shown in Figure 6, the HEP 3 results agree with the data provided by augering.

Data from HEP 2 and 5, obtained along the previously discussed second VES section (perpendicular to the wave-cut bank along the 314 N/S grid line), also compare favorably with augering data and with the VES results (Figure 7). HEP 5, run with a shorter electrode spacing and thus a shallower penetration depth (1.3 as opposed to 1.6 m) is more precise in defining the lateral limits of the midden. Peaks in measured resistivities occur on both profiles in the same general location as the previously discussed anomalous area discovered in the interpretation of VES 8 and perhaps signify either a thickening of the midden and/or increased resistivities due to disturbance or coarsening of the deposits.

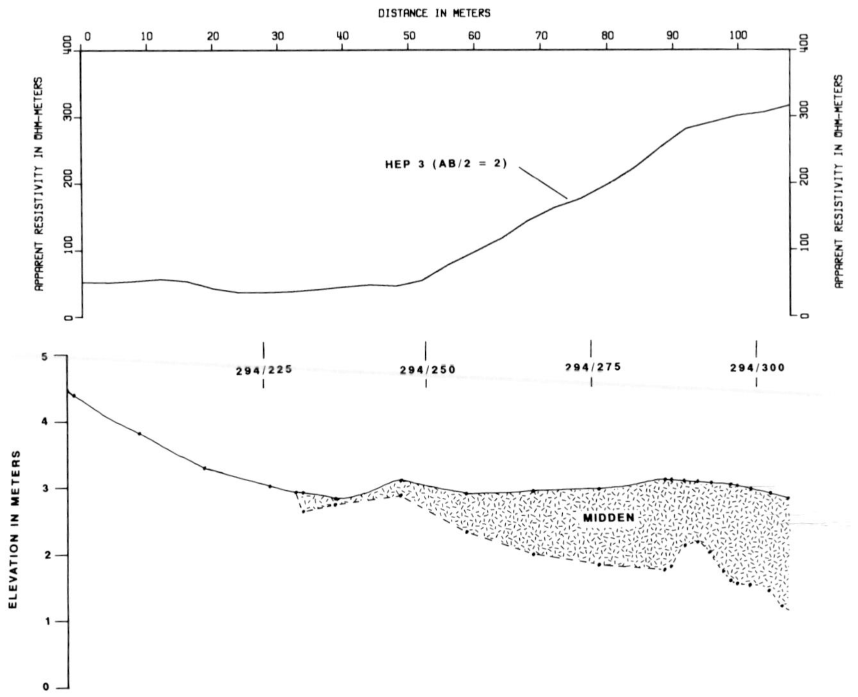

Figure 6 Horizontal electrical profile 3 and the corresponding midden profile obtained by augering. The upper portion of the figure plots apparent resistivity versus distance along the profile. An AB/2 spacing of 2 m yielded a profiling depth of approximately 1.3 m below the surface. The midden profile reconstructed from augering along this same line is shown below. Augering locations (represented by dots) are widely spaced inland (*left*) and more narrowly spaced approaching the wave-cut bank (*right*). Site grid coordinates are indicated between the two plots. The horizontal electrical profile is basically a reverse image of the midden profile, i.e., as the midden lower boundary rises, the measured apparent resistivity decreases.

HEP 4 (Figure 8), located perpendicular to the wave-cut bank (along the 344 N/S grid line) and approximately 30 m to the southeast of HEP 5, provides similar results to those obtained by HEP 5. Although some variation is evident, HEP 4 indicates a relatively thick midden near the wave-cut bank. A marked drop to lower resistivity values is observed at about 15 m from the landward end of the transect (approximately 65 m from the bank). This suggests that the edge of the midden, or at least a very thin midden, is located at this point.

HEP 1, located parallel to the wave-cut bank, indicates generally thicker midden deposits to the southeast and a relatively shallow midden to the northwest (Figure 9). This profile was useful for showing the dipping midden surface, which was not apparent on VES Section 1 due to its limited lateral extent. HEP 1, like HEP 2, was conducted with a larger electrode spacing and hence provided a deeper profiling depth. For this reason, the low resistivity values probably indicate that the midden is thinning to depths of 0.5 m or less rather than the absence of midden at the left

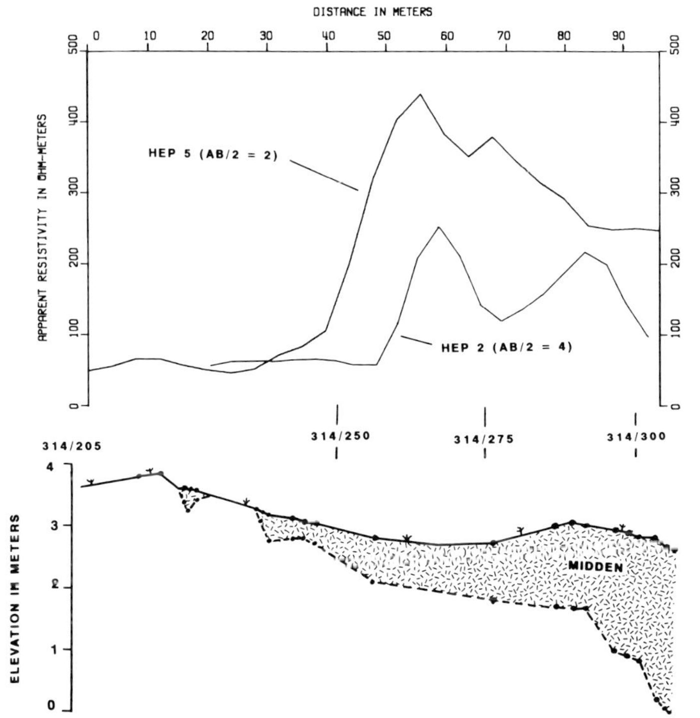

Figure 7 HEP 2 and 5 and the corresponding midden profile obtained by augering. The upper portion of the figure shows the two horizontal electrical profiles obtained along the 314 N/S grid line. Auger data along this same line yielded the midden profile shown below. Site grid coordinates are indicated between the two plots. The uplands are located to the left and the wave-cut bank is located to the right. As in Figure 6, apparent resistivity values generally decrease as the midden thins. The relatively low resistivity, flat portions of the resistivity profiles (*left*) correspond to the regions where the midden is thin or absent. Here, most of the current flow is through the underlying lower resistivity materials. The shorter array spacing for HEP 5 (AB/2 = 2 m as opposed to AB/2 = 4 m for HEP 2) and the corresponding shallower penetration depth (1.3 m as opposed to 1.6 m for HEP 2), allowed the midden to be detected further to the northeast (*left*).

end of the profile. The presence of a thick midden on the right side of the profile agrees with data provided by augering. Augering indicated that the midden is over 4 m thick in this area and further to the southeast.

C. *Seismic Refraction*

Profiles of the two seismic refraction lines are presented in Figures 10 and 11. Seismic Line 1 was run perpendicular to the existing wave-cut bank (along the 314 N/S grid line). Seismic Line 2 was run parallel to the wave-cut bank (perpendicular to Seismic Line 1 along the 286 E/W grid line). The two seismic lines intersect at the location of VES 1 (314 N/286 E).

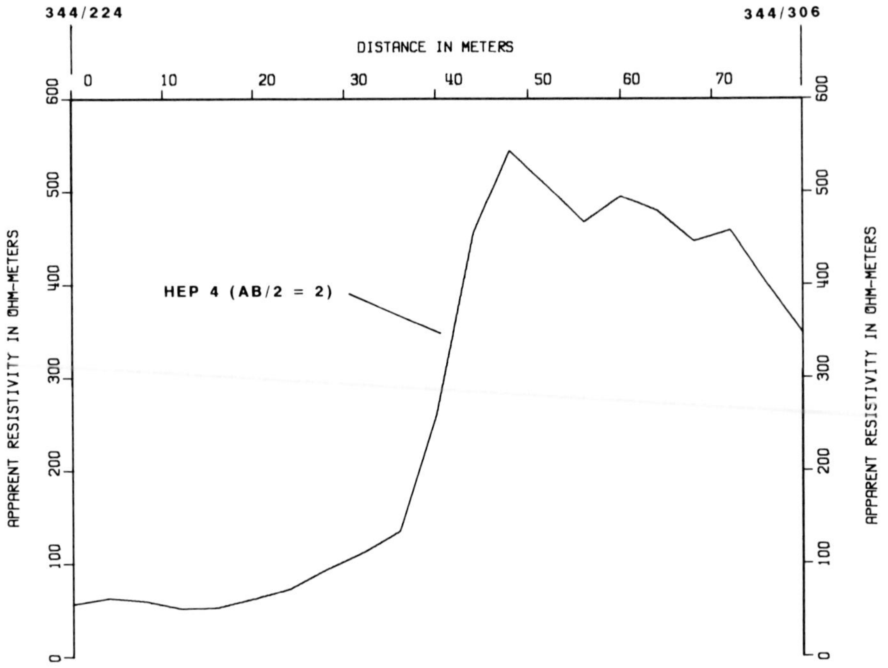

Figure 8 Horizontal electrical profile along the 344 N/S grid line (HEP 4). Approximate grid coordinates are shown above. High apparent resistivities indicate a relatively thick midden near the wave-cut bank (*right*). The sharp drop in apparent resistivities on the left to values averaging around 100 ohm-meters indicates that the midden becomes very thin or is absent in this area.

The seismic profiles depict the velocity layering present along each of the two seismic lines. The seismic velocities of each layer are presented in meters per second (m/s). The calculated depths of boundaries between layers of differing velocities are shown as horizontal lines. The sinuous vertical lines represent locations where lateral velocity changes were first noted within a layer.

Along seismic line 1 (Figure 10), three velocity layers are present. The surficial velocity layer, comprised of the midden above the water table, is represented by a velocity of 550 m/s. The region where the 550-m/s velocity layer is not present (approximately between grid points 260 and 240 E) corresponds to the area where VES and augering data indicate the midden and the water table both began to rise sharply (Figure 5). In this region, there is generally an insufficient volume of unsaturated midden to allow its discernment as a separate velocity layer.

The velocity of the second layer (1550 m/s) is probably more an expression of the water table than the second layer materials themselves. As the water table velocity dominated the material velocity, little information was provided about these deposits. Their characteristics will have to be inferred from augering and the other geophysical results. Resistivity data (Figure 5) suggest that various fine- and coarse-grained materials comprise this seismic layer.

The higher velocities of the third layer (2600 to 3450 m/s) probably represent bedrock. This velocity range is typical for schistose rocks (Latas, Ch. 10, this volume). The calculated bedrock surface is seen to dip gently toward the wave-cut bank, from a high of approximately 4 m below the surface to a low of approximately 10 m below the surface. This agrees with the data from VES Section 2, which indicates the presence of the bedrock surface from between 4 and 8 m below the surface.

On the profile obtained along seismic line 2 (Figure 11), the three velocity layers correspond to those found along seismic line 1. The surficial midden layer, with velocities of 550 to 800 m/s, thickens to the southeast, which is in agreement with the slope of the midden profile interpreted from HEP 1 (Figure 9). The localized increase in first layer thickness (from 6 to 15 m from the left end of the profile, approximately between 287 and 296 N) and the higher velocities in this region may represent historic disturbance or previously excavated archaeological material (Kenady 1973).

The third velocity layer, representing bedrock, also deepens to the southeast. The bedrock surface ranges from approximately 4.5 m in the northwest to approximately 14 m in the southeast. An earlier arm of Garrison Bay may have occupied the area of the deeper bedrock (see Whittaker and Stein, Ch. 2, this volume). The change in

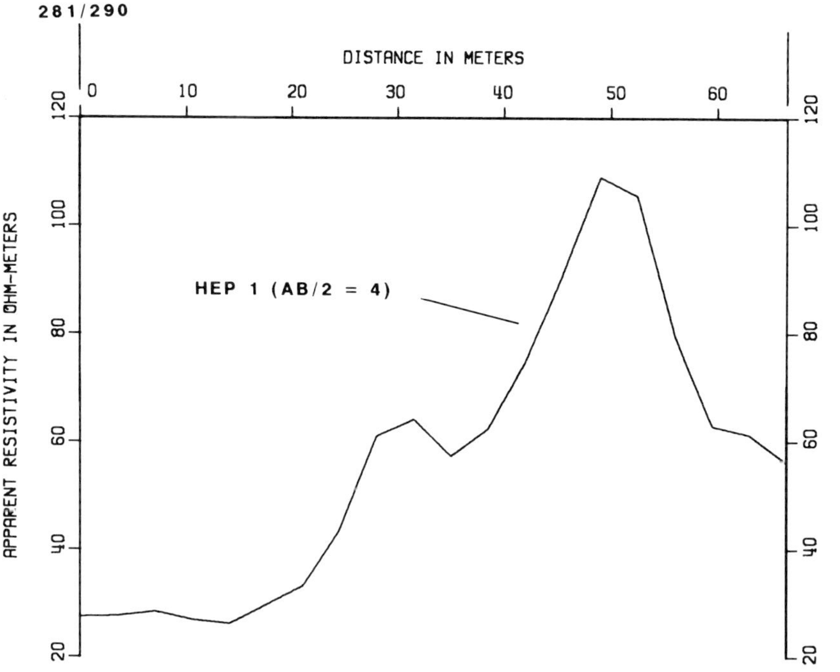

Figure 9 Horizontal electrical profile 1. Calculated apparent resistivities indicate generally thicker midden deposits southeast (*right*) and relatively shallow midden deposits northwest (*left*).

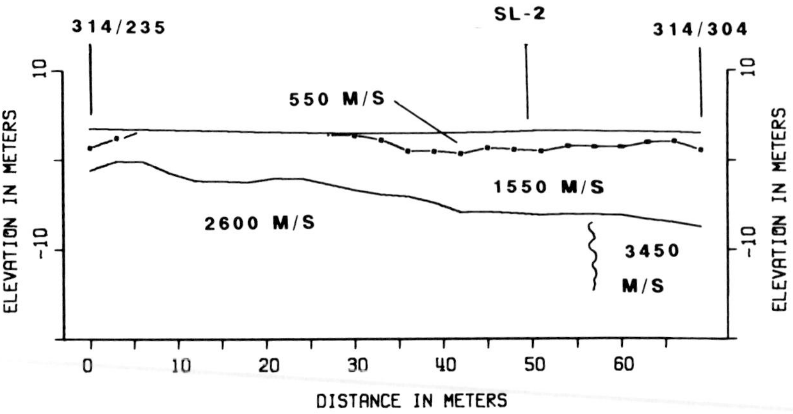

Figure 10 Velocity layers along seismic line 1. The 550-m/s velocity layer represents the midden above the water table. The 1550-m/s velocity layer is comprised of unconsolidated sediments below the water table. The third layer (with velocities ranging from 2600 m/s to 3450 m/s) represents bedrock. Both the midden and the bedrock surface slope gently toward the southwest (*right*). The wave-cut bank is located just beyond the right-hand side of the profile.

rock velocity (from 5800 to 2600 m/s) at approximately 28 m from the northwest end of the line (at approximately 310 N) indicates a change in rock type. The 2600-m/s velocity material, as in seismic line 1, is interpreted as representing a schistose formation. The 5800-m/s velocity represents a much harder, highly erosion-resistant material.

VI. Discussion

The geophysical data, in conjunction with augering results, provide the following information regarding the preoccupational topography and the formation of the midden in the area covered by the geophysical surveys. From the two vertical electrical sections, the surface upon which the midden was deposited is reconstructed. This surface is a platform comprised of both coarse-grained and fine-grained sediments. The deeper, predominately coarse-grained materials are probably Late Wisconsin glacial drift that mantles the bedrock throughout the area. Shallow, fine-grained materials are probably representative of an alluvial fan, much like the one that is observable today extending from the base of the slope at the northern edge of the site. The buried fan, deposited on the glacial drift, was created from sediments eroded from this same bedrock-controlled slope. Occupation and the deposition of the midden resulted in the burial of the toe of this fan. More recent fan deposits interfinger with historic deposits as well.

The resistivity data suggest that the midden accreted vertically in the gently sloping platform area discussed above. Along the edges of the platform, midden also prograded seaward from the shoreline. The prograded midden is found predominantly

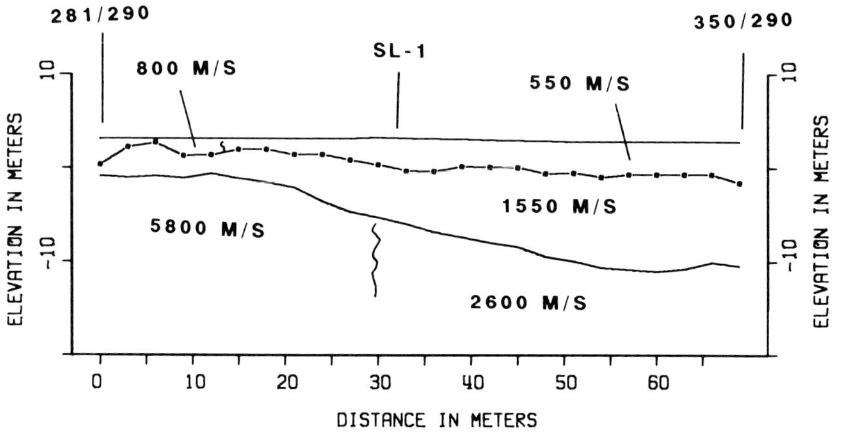

Figure 11 Velocity layers along seismic line 2. The surficial velocity layer (with velocities of 550 to 800 m/s) represents the midden above the water table. The 1550-m/s layer is made up of various fine- and coarse-grained sediments. The third velocity layer represents bedrock with the 5800-m/s material being harder than the 2600-m/s material. Both the lower midden and upper bedrock surfaces deepen to the southeast (*right*).

in the south portion of the geophysical survey area (near the intersection of HEP 1 and HEP 4 and next to the wave-cut bank). Here, as documented by augering, the midden attains thicknesses of greater than 4 m.

The seismic data support the above interpretation and provide additional information on the bedrock surface and its character. Within the area investigated, the bedrock surface ranges from approximately 4 to 14 m below the present ground surface. In general, bedrock appears to slope toward the south and east. Although the bedrock surface is undulatory, reflecting the events associated with the glacial history of San Juan Island, there is no location where bedrock is in contact with the midden. The prehistoric inhabitants selected a gently sloping surface for occupation composed of unconsolidated sediments.

VII. Conclusions

The geophysical results demonstrate the value of using select methods for understanding the British Camp shell midden and its depositional history. The resistivity of the midden, due to its coarse texture and porosity, was sufficiently different from that of the more silty glacial drift and alluvium on which it was deposited. The differences in measured resistivities were particularly apparent in soundings obtained during the dry summer months. Similar results have been obtained in other archaeological studies conducted during varying weather conditions (Al Chalabi and Rees 1962; Carr 1982).

Due to the differences in electrical properties of the British Camp deposits, electrical resistivity surveys provided a rapid, inexpensive, and reliable method for determining the vertical and horizontal limits of the midden. Horizontal electrical profiles

100 m in length were accomplished in 30 minutes or less. After determining the general boundaries of the midden by this method, augering could then be employed efficiently only at locations near the edge of the midden. Vertical electrical soundings to depths of 15 to 20 m typically took 20 minutes per sounding.

In addition to determining horizontal and vertical midden limits, geophysical methods proved useful for providing information on the structure of the midden. Vertical electrical soundings indicated the depth and presence of the water table and some suggestions as to the layering within the midden. Augering and further exposure by excavation, together with analyses of controlled sediment samples, permit an interpretation of the specific processes responsible for this layering. Interpretation of the VES sections also allowed a reconstruction, pertaining to a limited area of the site, of buried landforms such as the bedrock surface, the glacial deposits, the alluvial fan, and the changing shoreline. Seismic refraction surveys complemented the electrical results, proving particularly useful in defining the preoccupational topography, especially the relative positions of the bedrock surface.

The results of this research also suggest that a coordinated approach, using several geophysical methods together with augering or some other direct means of exploration, can be considered when attempting to interpret the depositional history of a site. The application of geophysical methods in archaeology can be extended beyond the location and definition of cultural features and site limits to a description and interpretation of site geomorphology. Where significant contrasts in certain physical properties exist, a coordinated approach such as this one can be used profitably to provide several types of information necessary for an understanding of depositional and postdepositional environments and processes.

Acknowledgments

Equipment and software for data interpretation were generously provided by Geo-Recon International, Seattle, Washington. In addition, the authors would like to thank Clyde A. Ringstad and Wolf M. Krieger, both of Geo-Recon International, for their assistance with the field survey and their helpful suggestions and comments on the content of this chapter. We are also grateful to the National Park Service and Superintendent Dick Hoffman for providing us with the opportunity to conduct this research. Kennan Haynes graciously typed the final manuscript.

References

Aitken, M. J.
1974 *Physics and archaeology* (second edition). Oxford: Clarendon Press.

Al Chalabi, M. M., and A. I. Rees
1962 An experiment on the effect of rainfall on electrical resistivity anomalies in the near surface. *Bonner Jahrbucher* **162**, 226–271.

Beck, A. E.
1981 *Physical principles of exploration methods*. New York: John Wiley & Sons.

Carr, C.
1982 *Handbook on soil resistivity surveying*. Evanston, Illinois: Center for American Archaeology Press.

Dobrin, M. B.
1976 *Introduction to geophysical prospecting* (third edition). New York: McGraw-Hill.

Griffiths, D. H., and R. F. King
1981 *Applied geophysics for geologists and engineers* (second edition). Oxford: Pergamon Press.
Keller, G. V., and F. C. Frischknecht
1966 *Electrical methods in geophysical prospecting*. Oxford: Pergamon Press.
Kenady, S. M.
1973 Environmental and functional change in Garrison Bay. In *Miscellaneous San Juan Island reports, 1970–1972*, edited by S. M. Kenady, S. A. Saastamo, and R. Sprague. Department of Sociology/Anthropology, University of Idaho, Moscow. Pp.39–83.
McNeill, J. D.
1980 Electrical conductivity of soils and rocks. Geonics Limited, Technical Note TN-5. Mississauga, Ontario, Canada.
Parasnis, D. S.
1986 *Principles of applied geophysics* (fourth edition). New York: Chapman and Hall.
Redpath, R. B.
1973 Seismic refraction exploration for engineering site investigations. Technical Report E-73-4. U.S. Army Engineering Waterway Experiment Station, Explosive Excavation Research Laboratory, Livermore, California.
Stein, J. K.
1986 Coring archaeological sites. *American Antiquity* **51**, 505–527.
Tite, M. S.
1972 *Methods of physical examination in archaeology*. London and New York: Seminar Press.
Weymouth, J. W.
1986 Geophysical methods of archaeological site surveying. In *Advances in archaeological method and theory*, vol. 9, edited by M. B. Schiffer. New York: Academic Press. Pp.311–395.
Wynn, J. C.
1986 A review of geophysical methods used in archaeology. *Geoarchaeology* **1**(3), 245–257.
Zhody, A. A. R.
1975 Automatic interpretation of Schlumberger sounding curves using modified Dar Zarrouk functions. U.S. Geological Survey Bulletin 1313E.

4

Historic Treatment of a Prehistoric Landscape

Bryn H. Thomas
James W. Thomson

I. Introduction

Archaeologists who excavate shell-bearing sites have realized recently that postdepositional processes severely affect their interpretations of prehistoric events (Waselkov 1987:146–150). These postdepositional processes include compaction (Muckle 1985), inundation (Rollins *et al.* 1990), translocation of small-sized objects (Sanger 1981) and bioturbation (Stein 1983). One important process not considered explicitly in this recent emphasis is the effect of reoccupation and land disturbance during the historic period. In the last 100 years many shell-bearing sites have had their topography altered, vegetation cleared, prehistoric material (especially shell) mined, and their coastal portions eroded or artificially stabilized. These historic manipulations can be substantial and affect interpretations of prehistoric events.

The purpose of this chapter is to examine the effect of the British and subsequent occupation on the prehistoric shell midden, (45SJ24) at Garrison Bay. To accomplish this task, historical photographs beginning with the British occupation will be reviewed to obtained data that document activities affecting the site. Of the many historical resources available, photographs were chosen for this study due to their inclusiveness and objectivity. Historic activities and changes in the landscape are often clearly identified in the photographs even when the intent of a historic photographer may have been to record something else.

A. *Historical Background*

In 1846 the United States and Great Britain settled upon a boundary for their respective claims in the Pacific Northwest, the 49th north parallel. The treaty further identified a line extending from the mainland to the Pacific between the continent and Vancouver Island. Vancouver Island, which extended south of the parallel, was to be retained by the British. Unfortunately, the treaty did not specify which of two channels, Rosario or Haro, located between the continent and Vancouver Island, was the true boundary. Hence, ownership of the San Juan Islands was in dispute. The Hudson's Bay Company claimed these islands in 1845. Three years later the Oregon Territory was established and territorial jurisdiction of the United States was extended to include the islands.

Dispute over ownership did not arise, however, until the Hudson's Bay Company and American settlers were occupying San Juan Island jointly. The Company had established fishing stations on the south end in the early 1850s and by 1853 had built an agricultural station called Bellevue Farm. By 1859, 25 American settlers, many of whom came north for the Fraser River gold rush, were living on the island. Several settled around Bellevue Farm. None of these early settlements were in the vicinity of the site at Garrison Bay.

Tensions existed between the Company and the settlers, but the event that brought the dispute to a confrontation occurred when an American, Lyman Cutlar, shot one of the Company's pigs. Disagreement over replacement costs and exaggerated accounts of the incident (e.g., the British threat to jail Cutlar) aggravated the situation. Shortly after the shooting, settlers petitioned General Harney for U.S. military protection. In July, 1859, American troops landed at Griffin Bay along the south shore of San Juan Island and established a camp. Within four months, a joint occupation agreement was signed and the British sent Captain Prevost of the Royal Navy to survey the island for a British encampment. The British examined six areas and chose one on the north end of San Juan Island along Garrison Bay. The area they chose included the shell midden site that is the subject of this study.

Ethnographic data indicates that the San Juan Islands were occupied during the eighteenth and early nineteenth centuries by the Central Coast Salish (Suttles 1990:453–475). The tribal dialect spoken by the inhabitants at Garrison Bay is confusing. According to Suttles (1951:36) Sannich, Songish, and Lummi informants associated their origins with this area, but the latter identified a winter village site on Garrison Bay. Such a village could have consisted of one house, several houses, or rows of houses parallel to the shore. Winter dwellings (Suttles 1990:462) were 20–60 feet wide and at least twice as long, although a house could extend indefinitely if the terrain permitted. In any event, no evidence has been found that the British observed remains of an American Indian village upon arrival of the Royal Marines. Thus, the site may have been abandoned since the early to mid-nineteenth century. It is possible that village remnants left after abandonment were not substantial enough to attract the troop's attention or were not judged significant enough to be recorded.

British and American troops continued to occupy San Juan Island for the next twelve years. Resolution of the boundary dispute was delayed by both sides due to the American Civil War and more pressing British diplomatic matters elsewhere. In the interim, relations between the two posts were amicable and the most demanding disagreements arose over jurisdiction of civil matters.

The dispute ended with the signing of the Treaty of Washington when the boundary issue was submitted for arbitration and award to the Emperor of Germany. On October 1872 the boundary was decreed to be Haro Channel and one month later the British Ambassador informed the United States Secretary of State that the Royal Marines had evacuated San Juan Island.

Soon after the British departure, an American settler, William Crook, purchased the former camp from the federal government. The Crook family maintained several of the old military buildings, including the blockhouse and a barracks, but many deteriorated and were removed. In the mid-1950s, in an effort to preserve the camp site, the state of Washington purchased acreage from the Crooks for a park and locus for interpreting the history of the boundary conflict. However, in order to assure long-term preservation of the park's resources, the state pursued federal ownership of the property. In 1966, Congress established San Juan National Historic Park, and since then several historical and archaeological investigations have been conducted.

II. Disturbances to the Prehistoric Record by the Historic Period Occupations

The appearance of the shell midden at the onset of British occupation is not entirely clear from historic record. In 1859 Garrison Bay and environs were described by Lieutenant Roche as being,

> About three quarters of a mile in a SSE direction there is a large patch of water, half lake, half swamp, on the northern shore of which is a situation admirably adapted for an encampment. It slopes gently to the S.W., is well sheltered, has a good supply of water and grass, and is capable of affording manoeuvering ground for any number of men that are likely to be required in that locality, there being a large extent of Prairie land, interspersed with some very fine oak Timber.
>
> (Thompson 1972:199)

From this account, it seems that the American Indian site was abandoned and partially overgrown, since clearing trees was one of the first tasks of the marines upon landing (Thompson 1972:202). No historic accounts have been found that describe shell deposits or surface features that might have existed at the time of the British encampment. However, the surface of the shell midden at British Camp is very flat, unlike the surface of other middens in the area. Wessen (1988:62–65) in a survey of the archaeological resources of the San Juan Islands described other shell middens of the area as having "earthworks" characterized by "circular depressions," "semi-circular trenches," and "mounds." Similar surface topography may have existed in the British encampment area and been leveled by the marines. At Garrison Bay itself, in a wooden area northwest of the former military parade ground, a series of topographic ridges of shell are still present. Investigated by Treganza (1950) and the University of Washington's San Juan Island Archaeological Project in 1990 and 1991, these ridges form an open-ended rectangle and are thought to outline the remains of a prehistoric dwelling (see Chapter 1 by Stein, this volume). Prior to the historic occupation, similar features may have characterized the portion of the site subsequently occupied by the British. Undoubtedly these topographic irregularities would have been disturbed by historic activities of the British marines and the Crook farm. It is the identification of such historic disturbances that we discuss below.

The effect of the British occupation on the shell midden was immediate. The earliest (ca 1860) known photograph of British Camp illustrates the tents and a few buildings associated with the initial encampment (Figure 1). More importantly, the photograph shows marines with spades and hoes near a garden surrounded by a sapling fence. Disturbances that ensued in the first months are seen in the planting of a garden, the placement of a sapling fence, and the felling of trees like the one seen near the tents in the background. Moreover, a mound can clearly be seen behind the level garden plot. This mound may have resulted from British excavation into the midden; alternatively, it may be an as yet undisturbed ridge of shell forming one wall of a rectangular feature as seen in the woods to the northeast. The other sides of such a rectangular feature are not clear in this photograph if indeed that is the ridge's origins.

A ca 1865 photograph of the waterfront parade ground illustrates some of the cutting and filling that had taken place during the initial five years of British occupation (Figure 2). The parade ground seems to be flat with the only mounds being

Figure 1 View of British Camp to the north taken about 1860. This photograph illustrates marines with spade and hoes near the garden along the shore of Garrison Bay. The garden which has been leveled for planting is seen in the foreground and surrounded by a sapling fence. A ridge, which is probably a remnant of the prehistoric landscape, is seen to the north behind the garden fence. Catalogue number HP12720; British Columbia Archives and Records Service, Victoria, B.C.

Figure 2 A view to the southeast from Garrison Bay taken about 1865. The terrace in the foreground has been leveled. The white area along the beach front is the wave-eroded exposures of the shell midden. The midden is also exposed in the road parallel to the shoreline. The shell midden in the ramp extending into the bay on the left and the path from the lower terrace to the officers' quarters on the right hillside is shell that has been transported for construction material. Remnants of a shell midden ridge are visible under the buildings left of the flagpole and in the distance. Catalogue number HP95343; British Columbia Archives and Records Service, Victoria, B.C.

visible in the distance near the hillside. Roads or paths have been cut into the shell midden between buildings. The spoils from these cuts may have been used to construct the boat ramp seen perpendicular to the shoreline in the center of the photograph. In other instances the midden shell may have been deliberately utilized for decoration and paving as seen in the paths in the formal flower garden depicted in a ca 1865 photograph (Figure 3).

Along with digging up the midden and moving it elsewhere there was certainly substantial compaction of the sediments. A photograph taken about 1869 gives an idea of the density of buildings on the parade ground and the intensity of pedestrian traffic to which the midden was subjected (Figure 4).

Thompson (1972:212–213), citing sources from 1872–1875, compiled a list of 35 buildings and structures associated with the main encampment. Those affiliated with the officers' quarters are seen in several photographs. One taken in the early 1860s shows the winding path leading from the parade ground to both the newly constructed officers' quarters and rock wall-faced terraces (Figure 5). A considerable amount of landscape manipulation was required to accommodate and situate these officers' quarters. The marines made large cuts into the hillside, and sediment from the excavations was used as fill material, thereby expanding the building area of the housing terraces. After expanding the surface area of each, the marines constructed stone retaining walls to stabilize and reinforce each terrace. By the end of the British

Figure 3 A view north of the British Camp waterfront, ca 1865. The photograph illustrates the shell midden exposures in the road parallel to the beach, around the flag pole, and the flower garden in the right foreground. The shell midden around the flag pole and in the garden has been disturbed and displaced by the construction of these features. Catalogue number HP35261; British Columbia Archives and Records Service, Victoria, B.C.

occupation five of these terraces and retaining walls had been constructed. Surveys and test excavations conducted by the National Park Service of one terrace and of the stairway indicate that shell midden was transported by the British to the hillside and utilized as building material (Thomas 1987, 1990).

Impacts to the prehistoric shell midden did not end, of course, with the British withdrawal from the site. As mentioned earlier, the Crook family operated a farm at British Camp for over 75 years, from 1872 to the mid-1950s. During this period the family planted an orchard, herded sheep, and operated machinery. An early twentieth century photograph illustrates the Crooks' orchard and sheep grazing in the pasture which was formerly the military parade ground (Figure 6). The amount of disturbance to the shell midden by farming is unknown, but it is assumed that some of the farming activities would be similar to those practiced by the British.

III. Conclusions

This review of the historic photographs does not provide an exhaustive listing of the historic impacts to the shell midden site at Garrison Bay. It does, however, catego-

Figure 4 View to the southeast, ca 1869, showing the density of buildings and intensity of pedestrian traffic to which the midden was subjected during the British occupation. Catalogue number HP12719; British Columbia Archives and Records Service, Victoria, B.C.

Figure 5 View to the east in the early 1860s of the newly constructed officers' quarters. The winding path leading from the lower terrace is visible on the lower left of the photograph. Likewise, a rock wall-faced house terrace, which appears white in the photograph, is seen slightly left of center in front of one of the houses. Shell utilized for the path and terrace construction was transported from the prehistoric midden. Catalogue number HP12903; British Columbia Archives and Records Service, Victoria, B.C.

Figure 6 A photograph taken about 1915 shows the Crook family sheep and orchard on the former military parade ground and Garrison Bay site. National Park Service, San Juan Historic Park photo file.

rize many of the more obvious ones. Further, it suggests that the shell midden topography at the onset of British occupation was probably irregular and characterized by the mound and depression features found elsewhere at sites. The apparent absence of these prehistoric features, except for those in the forested area at the northern side of the parade ground, is due to historic activities including vegetation clearing, cut and fill episodes, compaction, extracting midden for building material, and agricultural development. The knowledge and identity of these activities provides the prehistorian with an estimate of postdepositional impacts to the prehistoric record.

Acknowledgments

The authors would like to thank the following: the superintendent and staff of San Juan National Historical Park for their support, encouragement, and sharing of historical information pertinent to the British impact upon the Garrison Bay site. The park staff was always available for assistance during several testing investigations of the British remains at Garrison Bay. Julie K. Stein, Curator of Archaeology, University of Washington, and Director of the San Juan Island Archaeological Project has shared data and insights into the preparation of this chapter. Roderick Sprague, Director of the Alfred E. Bowers Laboratory of Anthropology, University of Idaho, conducted historic archaeological investigations at Garrison Bay during the 1970s and has shared his knowledge of the site with the authors. We thank Cathryn Gilbert, Historical Landscape Architect, National Park Service, Seattle, Washington, for her initial work in interpreting the historical landscape at Garrison Bay and for sharing her field data and expertise with the authors. The British Columbia Archives and Record Service, Victoria, British Columbia, has kindly given permission to use many of the historical photographs used in this chapter.

References

Muckle, R. J.

1985 Archaeological considerations of bivalve shell fish taphonomy. Master's thesis, Department of Archaeology, Simon Fraser University, Burnaby, British Columbia.

Rollins, H. B., Sandweiss, and J. C. Rollins

1990 Mollusks and coastal archaeology; A review. In *Archaeological geology of North America*, Centennial special vol. 4, edited by N. P. Lasca and Donahue. Geological Society of America, Boulder, Colorado. Pp.467–478.

Sanger, David

1981 Unscrambling messages in the midden. *Archaeology of Eastern North America*, Vol. 9, 37–42.

Stein, J. K.

1983 Earthworm activity: A source of potential disturbance of archaeological sediments. *American Antiquity* **48**, 277–289.

Suttles, W. P.

1951 Economic life of the coast Salish of Haro and Rosario Straits. Unpublished Ph.D. dissertation, Department of Anthropology, University of Washington, Seattle.

1990 Central coast Salish. In *Handbook of North American Indians, volume 7*. Washington: Smithsonian Institution.

Thomas, B. H.

1987 Report of the test excavations of the rock features at the second terrace on the officers' quarters hillside, British Camp, San Juan Island National Historical Park. In *Historic landscape report: American Camp and British Camp 1987*. U.S. National Park Service, Northwest Regional Office, Seattle.

1990 Letter report to Jim Thomson on 1990 test excavations of officers' stairway at British Camp, San Juan Island Historical Park, Washington. U.S. National Park Service, Northwest Regional Office, Seattle.

Thompson, E. N.

1972 Historic resource study: San Juan Island National Historic Park, Washington. Denver Service Center, U.S. National Park Service.

Treganza, A. S.

1950 Profile drawings and field summary. Unpublished materials, U.S. National Park Service. Northwest Regional Office, Seattle.

Waselkov, G. A.

1987 Shellfish gathering and shell midden archaeology. In *Advances in archaeological method and theory, vol. 10*. Orlando: Academic Press. Pp.93–210.

Wessen, G. C.

1988 A technical overview of the prehistoric archaeology of the San Juan Islands region. A report prepared for the U.S. National Park Service, Wessen & Associates, Seahurst, Washington.

5

Interpreting Stratification of a Shell Midden

Julie K. Stein

I. Introduction

Stratigraphy and archaeology became intimately linked in the eighteenth century when the debate concerning human antiquity intensified. In the following centuries that intimacy has continued and is reflected in the insistence of investigators to include descriptions of "site stratigraphy" in every site report published. Although the relationship between stratigraphy and archaeology has been maintained, in many ways the two disciplines have grown apart in the last century. Stratigraphy has continued to grow, increasing in complexity as new observations result in new discoveries in geoscience. Archaeology has also continued to grow, but in directions away from those followed by stratigraphy. Archaeology has maintained its ties with stratigraphy only by clinging to the stratigraphy of old. It has not kept pace with new stratigraphic developments. This loss of intimacy between the two disciplines has resulted in "site stratigraphies" within archaeological site reports that are locked in the format of eighteenth and nineteenth century stratigraphy. They do not reflect the pluralistic approach of modern stratigraphy and suffer as a consequence.

Stratification observed and analyzed in archaeological sites is the building block of culture history and site formation. As more and more sites are excavated, and even greater numbers of artifacts are collected, culture histories and site formation reconstructions become problematic. Difficulties arise because no standardization exists within archaeology concerning the method of stratigraphy, and stratigraphy is the methodology for examining both culture history and site formation processes. The solution to the crisis is to bring archaeology back together with stratigraphy, instituting an explicit stratigraphic nomenclature, a standard excavation procedure, and a pluralistic perspective in stratigraphic analysis. The necessary nomenclature and procedures have been proposed within modern stratigraphy. Archaeologists need

Deciphering a Shell Midden

only acquaint themselves with twentieth century stratigraphy, utilize the concept of multiple stratigraphic arrangements, and add a nomenclature peculiar to archaeology.

Stratigraphy is more than just describing the layering observed in the profile of a trench. Although strata can be described in terms of the physical attributes observed in the field, the same strata can also be divided using multiple stratigraphic arrangements. Strata can be grouped using the first or last appearance of an artifact, the grouping of an assemblage of certain kinds of artifacts, or the abundance of artifacts. Strata can be subdivided according to their age, their magnetic properties, their degree of weathering, or their fauna or flora. The division of strata into each of these groups results in sets of boundaries that for any arrangement do not necessarily coincide. The lack of coincidence in placement of the various boundaries is often the observation that allows site formation processes to be explained.

The archaeological procedures that most archaeologists have been following demands that we divide strata and group their contents with reference to only *one* arrangement of physical attributes, which somehow is thought to be the *real* stratigraphy. If the stratigraphic procedures proposed in modern stratigraphy are followed, one stratigraphic sequence cannot represent the site as a whole. The site has to be seen from a wider perspective and examined using many criteria. The approach is more useful for modern archaeology, because in the present century more than just human antiquity is being investigated.

In this chapter modern stratigraphy and its relationship to archaeology is discussed. The manner in which archaeological strata should be described, divided, and interpreted using multiple criteria is explained. The multiple criteria are divided into three areas: physical lithology, artifact content, and time. Initially, archaeological strata are divided into lithostratigraphic units (units defined by changes in their physical attributes). Second, the same archaeological strata are divided into ethnostratigraphic units (units divided by changes in their artifactual content). Finally, they are divided according to their age into chronostratigraphic units and geochronometric units. A set of standardized rules for archaeological stratigraphy is discussed, along with an example of their use on the Northwest Coast. This chapter is followed by a description of the British Camp shell midden (Stein *et al.*, Ch. 6, this volume). The systematic classification of archaeological stratigraphy at this Northwest Coast shell midden will demonstrate how to bring archaeology and stratigraphy back into an intimate linkage and allow archaeologists to communicate effectively with each other about issues of chronology and site formation.

II. Stratigraphy

The concept of stratigraphy was born when scientists observed that rocks of great antiquity had chronological significance, and that they contained sediments whose physical characteristics were similar to characteristics observed in modern environments (Ager 1981). Stratigraphy is now a branch of geosciences concerned with the description, origin, and the spatial and temporal relationships of layered rocks. Relationships of layered rocks are determined using the principles of superposition, cross-cutting relationships, inclusions, and unconformities (Matthews 1984). These principles work most clearly when applied on a local scale, however, for geographically separated sequences of rocks, fossils are used to determine relative ages of strata, environments of deposition, and evidence of evolution. Correlations of geographically separated layered rocks are also made using chronologic techniques such as dating with magnetic polarity and isotopic inclusions.

Stratigraphy accomplishes these directives by following a standardized nomenclature and methodology. Although the standardization is not internationally agreed upon, most nomenclatures resemble each other. Many countries have published their own guides of stratigraphic nomenclature. Two of the most valuable to North American archaeologists are the *International Guide to Stratigraphic Nomenclature* (Hedberg 1976) and the *American Code of Stratigraphic Nomenclature* (NACOSN 1983). These guides and codes divide sequences of rocks based on lithology (lithostratigraphic units), on fossil content (biostratigraphic units), and on the time periods in which rocks were deposited (chronostratigraphic, geochronologic, and geochronometric units) (Table 1). Fundamental to stratigraphy is the view that any sequence of strata can be arranged into multiple sequences, and that stratigraphic units are based on direct observations of the tangible features and characteristics of the strata. These observations are independent of interpretations regarding the significance of those observations (Krumbien and Sloss 1963:28).

Although archaeologists have considered stratigraphy for some time (Drucker 1972; Kenyon 1952; Pyddoke 1961), only recently have they become concerned with closing the gap between modern stratigraphy and archaeology by proposing a systematization of archaeological strata (Gasche and Tunca 1983; Stein 1987, 1990). This concern has resulted in the recent proposal of new archaeological stratigraphic nomenclature, including a new lithostratigraphic unit (layer, Table 1), the ethnostratigraphic unit (ethnozone or zone), and the ethno-chronostratigraphic unit (ethnochronozone). These terms will replace or add to the informally defined terms presently used in archaeology.

Most archaeological textbooks discuss stratigraphy in some fashion, and usually include definitions of a few terms (e.g., natural level, arbitrary levels, Law of Superposition, reversed stratigraphy, and context). These terms and their definitions have become embedded in archaeology in an informal way and emphasize the separation of archaeology and stratigraphy over the last century. For example, archaeologists created the term "reversed stratigraphy," using it to refer to the age of manufacture of objects in strata rather than to the stratigraphic position of a stratum (Stein 1987:350–351). The notion of a reversed stratigraphy is disallowed by the Law of Superposition. Yet within archaeology the term was useful because stratigraphy had become focused on artifacts instead of the enclosing deposits.

Another example of the separation of modern stratigraphy and traditional archaeological stratigraphy are the terms "primary and secondary context." These terms refer to an interpretation of the time of manufacture, use, and discard of artifacts, not to deposition of the objects in strata and stratification. The last episode of deposition is the only one to which stratigraphy refers. The location of use and place of manufacture of artifacts, to which primary and secondary context refers, is not related

Table 1 Stratigraphical Terminology

LITHOSTRATIGRAPHIC UNIT Subdivision of the rocks in the earth's crust, distinguished and delimited on the basis of lithologic characteristics: *formation, member, bed, layer* (archaeology)

Formation Fundamental stratigraphic unit, homogeneity of physical characteristics recognized on a scale convenient for mapping.

BIOSTRATIGRAPHIC UNIT Body of rocks characterized by content of fossils, which must be contemporaneous with the deposition of the strata: *biozone, sub-biozones* (ranked units)

Biozone[a] Named for assemblages of fossils based on either: (1) first or last appearance of fossils (*interval zones*); (2) the appearance of an association of two or more types of fossils (*assemblage zones*); or (3) the quantitatively distinctive maxima, minima, or relative abundance of one or more types of fossils (*abundance zones*).

TIME UNITS

Chronostratigraphic Unit Rocks formed during the same span of time. It may be based upon the time span of a biostratigraphic unit, a lithostratigraphic unit, an ethnostratigraphic unit, or any other feature of the rock record that has a time range: *eonothem, erathem, system, series, stage* (ranked units)

Chronozone A nonhierarchical, but commonly small, formal chronostratigraphic unit, with boundaries independent of those of ranked units. It may be based on lithostratigraphic, biostratigraphic, or ethnostratigraphic units.

Ethno-chronostratigraphic Unit Strata formed during the same span of time and based on artifacts in strata: *ethnochronozone*[a] (basic unranked unit)

Geochronologic Units Time span of an established chronostratigraphic unit: *eon, era, period, epoch, age* (ranked units)

Geochronometric Units Direct division of geologic time, expressed in years. Divisions of convenient magnitude.

ETHNOSTRATIGRAPHIC UNITS Deposits characterized by the presence of classes of artifacts they contain.

ethnozone[a] Named for assemblage of artifacts, based on either: (1) first or last appearance of artifacts (*interval ethnozone*); (2) the appearance of an association of two or more types of artifacts (*assemblage ethnozones*); or (3) the quantitatively distinctive maxima, minima, or relative abundance of one or more types of artifacts (*abundance ethnozone*).

[a]Modifiers (bio- or ethno-) used in formal names of the units need not be repeated in general discussions where meaning is evident from content.

to stratigraphy, but rather to inferences flowing from artifact analysis. These terms, described as archaeological stratigraphic terms, germinated informally within archaeology, without consideration of modern stratigraphy. The recent concern for the adoption of a more modern, pluralistic archaeological stratigraphy flows from the lackluster traditions of stratigraphy in archaeology.

Introductory archaeology texts also include terms introduced by McKern (1939) and Willey and Phillips (1958) that in a manner have acted as substitutes for stratigraphic nomenclature. The terms focus, component, phase, stage, period, horizon, and tradition (Table 2) were introduced in the middle of this century and used within the Americas in various ways. The definitions of these terms include some confusing language. For example, a focus is a "class of culture exhibiting characteristic peculiarities in the finest analysis of cultural detail" (McKern 1939:308). The meanings of "class of culture" and "finest analysis of cultural detail" are difficult to operationalize. A phase is "an archaeological unit possessing traits sufficiently characteristic to distinguish it from all other units similarly conceived, whether of the same or other cultures or civilizations, spatially limited to the order of magnitude of a locality or region and chronologically limited to a relatively brief interval of time" (Willey and Phillips 1958:22). The definition of phase includes references to not only traits and spatial limitations, but to chronological intervals as well.

Although the traditional archaeological terms in Table 2 have lost their usefulness in archaeological stratigraphy, they can be compared roughly to the stratigraphic terms in Table 1. The focus, component, phase, and horizon are terms related to ethnostratigraphic units (stratigraphic divisions based on artifactual content); stages and phases are related to chronostratigraphic units (strata formed during a certain time span and based on ethnostratigraphic units). Stages divide strata into time units based on technological attributes of artifacts. Phases divide strata into time units based on stylistic and functional attributes that can be correlated across a region. Periods and traditions are related to geochronologic units (time span of an established chronostratigraphic unit), and are based on ethnostratigraphic units constructed using historically relevant attributes of artifacts.

Even though the terms of Willey and Phillips and others can be related in a general way to the formal stratigraphic units, they have no utility in archaeology because they each carry regional connotations developed and communicated through regional traditions of the discipline, and they lack the rigor of modern stratigraphic terminology. In these definitions, a single term is used to define stratification that is based on a mixture of observations and analyses of artifacts, geography, and time. No

Table 2 Traditional Archaeological Stratigraphical Terms

COMPONENT An association of all the artifacts from one occupation level at a site (Fagan 1988:575). A manifestation of any given focus at a specific site, a focus being that class of culture exhibiting characteristic peculiarities in the finest analysis of cultural detail (McKern 1939:308).

PHASE An archaeological unit possessing traits sufficiently characteristic to distinguish it from all other units similarly conceived, whether of the same or other cultures or civilizations, spatially limited to the order of magnitude of a locality or region and chronologically limited to a relatively brief interval of time (Willey and Phillips 1958:22). Similar components from more than one site (Fagan 1988:504).

STAGES A technological subdivision of prehistoric time that has little chronological meaning but denotes the level of technological achievement of societies within it (e.g., Stone Age) (Fagan 1988:583).

PERIODS A major unit of prehistoric time that contains several phases and pertains to a wide area (Fagan 1988:581).

HORIZON A primarily spatial continuity represented by cultural traits and assemblages whose nature and mode of occurrence permit the assumption of a broad and rapid spread (Willey and Phillips 1958:33). A widely distributed set of culture traits and artifact assemblages whose distribution and chronology allow one to assume that they spread rapidly (Fagan 1988:578).

TRADITION A period during which a lasting artifact type, assemblages of tools, architectural styles, economic practices, or art styles exist. Longer than one phase or even the duration of a horizon (Fagan 1988:507). Temporal continuity represented by persistent configurations in single technologies or other systems of related forms (Willey and Phillips 1958:37).

single stratigraphic unit could ever separate adequately all the various possibilities for correlation. The introduction of these terms was symptomatic of the need within archaeology to do stratigraphy, as well as of the growing distance between the development of archaeology and that of modern stratigraphy.

The old archaeological stratigraphic terms need to be replaced by new, explicitly defined terms and by a standardized stratigraphic procedure. In the next century culture histories will be refined and greater precision will be needed to address the intricacies of the fine subdivisions of prehistory. In addition to culture histories, stratigraphy will be needed to address concerns with site-formation processes. Unlike our predecessors, we must be explicit in our stratigraphic practices so as to encourage the creation of more accurate regional correlations and more precise reconstructions of site formation. In this way we can reunite stratigraphy and archaeology.

III. Archaeological Stratigraphic Units

An exclusively archaeological stratigraphic guide, constructed and approved by a majority of archaeologists, does not yet exist and is not needed. One such guide was proposed in 1983 (Gasche and Tunca 1983), and comments have been forthcoming (Farrand 1984a, 1984b; De Meyer 1984, 1987; Harris 1989; Stein 1987, 1990). However, rather than writing an entire stratigraphic guide exclusively for archaeology, a more practical solution is to bring archaeology and modern stratigraphy together, adding to already existing stratigraphic guides those concepts that are relevant to archaeology. The stratigraphic needs of archaeology can be accommodated by three additions to existing stratigraphic codes: (1) a new lithostratigraphic unit of smaller dimensions useful for archaeologists; (2) the ethnostratigraphic unit (ethnozone or zone) that divides sequences of sediment according to their artifactual content; and (3) the ethno-chronostratigraphic unit (ethnochronozone) that divides strata formed during the same span of time and defined by any ethnozone useful for regional correlation. Each of these new additions will be discussed under the headings of lithostratigraphic units, ethnostratigraphic units, and time units.

A. *Lithostratigraphic Units*

In stratigraphy the units based on physical attributes of the strata are called lithostratigraphic units. They are "subdivisions of the rocks in the earth's crust, distinguished and delimited on the basis of lithologic characteristics" (NACOSN 1983). Although the definition used the term rock, unconsolidated sediment can be substituted for rock in the definition. The lithostratigraphic unit is generally, but not invariably, layered and tabular and conforms to the Law of Superposition. The limits of a lithostratigraphic unit are determined on the basis of its boundaries, defined by positions of lithologic change. The fundamental lithostratigraphic unit is the *formation*. A formation is defined by homogeneity of physical characteristics (lithologies). It has well-defined boundaries and is recognized on a scale convenient for mapping (Hedberg 1976; NACOSN 1983). A formation can be subdivided into members and beds.

Members are contained in formations and usually are more local in lateral extent. A bed is smaller than a member, the smallest unit that can be defined (Campbell 1967).

1. Small Excavation Lithostratigraphic Units

In archaeology, stratigraphic units based on physical attributes of the unconsolidated sediment have been called a variety of things. One term is *natural level*, defined as an "excavation unit corresponding to levels defined by stratigraphy, as opposed to arbitrary levels" (Sharer and Ashmore 1987:596). The natural level resembles the lithostratigraphic unit, as does the definition of an archaeological deposit, that is, "a three-dimensional unit that is distinguished in the field on the basis of the observable changes in some physical properties" (Schiffer 1983, 1987; Stein 1987:339). Archaeologists also use terms such as cuts (Fedele 1984), excavation levels, and facies. All these terms refer to the field descriptions of the smallest unit into which the physical lithology of the site can be divided during excavation.

Archaeologists define these terms with reference to excavation, a reference not observed in definitions of geological lithostratigraphic units. Geologists do not excavate as a rule. Strata are exposed already in natural outcrops or cores. Archaeologists, on the other hand, must be able to divide into some grouping the strata they are digging, because the strata are not exposed in vertical cliffs conveniently located near the roadside. Archaeological strata are being destroyed as they are being exposed. Natural levels are units defined by stratification that are observed only after exposure through excavation. Whereas, geologists always describe strata in terms of sections, archaeologists include plan maps as well as profiles, procedures defined in terms of the excavation process.

Arbitrary levels are also defined during excavation to create small groupings of strata before the material is removed and destroyed. The use of arbitrary levels is directly related to the fear that the smallest "natural" unit of analysis will be too large to detect changes in artifacts, plants, bones, or other analyzed material not observed in the excavation process. Excavation procedures destroy the "outcrop" from which archaeologists collect their data. Unlike a geologist who can return to an outcrop, an archaeologist must be sure to create units small enough to detect all changes, including those derived from artifactual or chronological data. In many cases arbitrary subdivisions of larger layers defined by (natural) lithology are needed. Thus, arbitrary levels are common in archaeological excavation and are a function of the destructive nature of the excavation technique.

No formal term has been proposed for these small lithostratigraphic units defined during excavation. They are the analytical unit on which all later groupings are based. They are usually defined by small changes in lithology, but also on arbitrary thicknesses, for example, 10 cm. I have proposed the term facies (Stein and Rapp 1985) for these small excavation units. The term facies, however, suffers from overuse in geology in a manner similar to the terms focus and phase in archaeology. Deposit (Stein 1987:339) has also been proposed, as well as others (Jacobsen and Farrand 1987). Whatever term is selected for these small lithostratigraphic units, they must be defined explicitly in the description of excavation procedures.

2. Large Lithostratigraphic Units Constructed after Excavation

In addition to small lithostratigraphic units, which are defined during excavation and based on observable changes in lithology, larger lithostratigraphic units that are based on lithologic changes are also crucial for archaeological stratigraphy. Every archaeologist lumps the small excavation layers into larger layers that are mappable across the entire site. The large layers are defined after excavation, usually from profiles, and are more frequently called by archaeologists layers or zones. These larger lithological units are the ones described in "site stratigraphy" sections of publications and are often the grouping devices for creating artifact assemblages.

These larger layers are the only unit formally proposed in the archaeological guides (and called layer), and are mappable across the entire site. A layer is "a three-dimensional body characterized by the general presence of a . . . (dominant) . . . lithologic type" (Gasche and Tunca 1983:329). Although no scalar requirements are associated with the original definition, Stein (1990:515) discusses the implications of defining layer on the scale appropriate for mapping across a site.

The formally proposed stratigraphic unit of layer has crucial differences from the terms natural level and deposit. Layers are a synthetic unit, and may be larger than the units used during excavation. They are usually described in the field on the basis of physical properties, created by grouping the smaller units used in excavation. They may not be the smallest observable differences in lithologies, but rather the smallest observable differences that are convenient for mapping between excavation units and across the entire site. All formal layers must be formally described, using precisely defined terms. All descriptions are accompanied by either references to literature where the precise definition of the word is given, or explicit definitions of the terms included. Terms that are qualitative and imprecise, (e.g., shelly, dark, rocky, mottled) are inappropriate unless accompanied by the criteria used to define the term.

For example, in pedology the term "mottled" has a precise definition (irregularly marked with spots of different colors, usually indicative of poor aeration and lack of good drainage). Archaeologists describe strata as mottled, but they do not always mean what the pedologists' definition of the word implies. Another example is "silty," which refers to the percentage of grains between .0625 mm and .0039 mm in a sediment (Folk 1980:22–28). This sedimentological definition differs from most archaeologists' use of the term, where the exact size of the particle is not known. Last, "shelly" as an adjective describing a shell midden has no quantitative definition. Describing a deposit as shelly at a shell midden in New England may be different from what is described as shelly in a shell midden in California. If shelly is to be part of the description of a layer, it should be accompanied by a quantitative definition of how much shell is indicated by the term shelly.

Both small and large lithostratigraphic units must be described on the basis of lithological homogeneity. The unit has particles within its boundaries that have a consistent homogeneous mixture or sequence of sizes, shapes, and composition, whatever the combination of light/dark, big/small, round/angular, or whole/broken. Homogeneity does not suggest that within the unit the lithology must be perfectly mixed. Rather, the lithology within the unit must contrast sharply with the lithology of adjacent units. When the lithology changes, a new unit is defined. The description includes a reference to the magnitude of change required to define a new stratum. That magnitude will presumably be smaller for small lithostratigraphic units than that required for defining a change in large lithostratigraphic units. In both cases, however, the criteria and magnitude must be explicitly stated. The descriptions of both small and large lithostratigraphic units are the basis for further interpretations and correlations. Interpretations are subject to falsification or reinterpretation, but basic descriptions are unchangeable. They are used to evaluate interpretations.

3. Northwest Coast Lithostratigraphic Units

In most reports of Northwest Coast shell midden excavations, stratification is described and interpreted in sections labeled "stratigraphy of site" (Blukis Onat 1985; Campbell 1981; Capes 1977; Carlson 1979; Chapman 1982; Ham 1982; Luebbers 1978; Matson 1976; McMillan and St. Claire 1982; Mitchell 1971, 1988; Murray 1982). The section usually begins with descriptions of what is defined here as layers, using a mix of qualitative and quantitative terms. The layers are given labels of either alphabetic letters or sequential numbers. In most cases, layers are divided and defined after excavation, and are based on observations of excavation profiles. The layers are selected because they have strongly contrasting characteristics, are observed over a large part of the site, and are useful for discussing interpretations of site formation.

For example, Mitchell (1971:83–88) divides the Montague Harbour shell midden site (Galiano Island, British Columbia) into three layers. He calls them zones: Zone A containing considerable quantities of whole shell, Zone B containing relatively little shell (that which can be seen is fragmentary and decomposing), and Zone C containing yellow-brown gravel and sand that is devoid of artifacts. The Shoemaker Bay site (McMillan and St. Claire 1982) and the O'Connor site (Chapman 1982) are also divided into two or three similarly described stratigraphic layers.

Although excavation occurs in small-scale strata, Mitchell (1971:89) has lumped these small-scale strata together into large-scale zones for purposes of discussing the site as a whole. He states that "numerous lenses and layers of shell—some composed of fragmented, some of whole valves—and many thin strata of humus soil, ash, and charcoal, were observed within Zone A." The site obviously was excavated in units smaller than zones. Although Mitchell refers to Sub-Zone A1, A2, and A3, he describes only the larger lithostratigraphic unit; the one convenient for mapping across the site.

Although Northwest Coast archaeologists recognize that defining the stratigraphic "zones" at a scale equal to the smallest lithostratigraphic unit would not be appropriate for correlating across a site or across a region, they need to recognize that they still must include a discussion of the manner in which the smallest units were defined. The small lithostratigraphic unit is the basic unit of analyses, the most crucial for lumping and splitting various arrangements of strata. Small-scale layers may not be best for correlating strata from trench to trench, but they are the cornerstone of all later stratigraphic arrangements. In fact, as archaeologists become more and more concerned with site-formation processes, the importance of the small-scale lithostratigraphic units will increase. Today archaeologists may discuss site formation only in terms of the larger lithostratigraphic layer. In the future, however, site-formation processes may be interpreted at the scale of the smaller lithostratigraphic unit, making the description of these units crucial to the research objectives.

B. *Ethnostratigraphic Units*

1. Definitions

Sequences of layered strata can also be divided according to their artifactual content, resulting in ethnostratigraphic units. A formal definition of an ethnostratigraphic unit is an archaeological deposit characterized by the presence of classes of artifacts it contains (Table 1). The formal unit defined by Gasche and Tunca (1983), is a zone. However, to bring modern stratigraphy and archaeology together, rather than creating our own brand of archaeostratigraphy, we should follow the example of the North American Commission on Stratigraphic Nomenclature (NACOSN), and propose the term ethnozone. NACOSN suggests biozone as the term for biostratigraphic units (subdivisions of rocks based on fossil content), rather than the term of the earlier International Code (Hedberg 1976), which was the zone. The paleontological literature is full of references to zones; people are accustomed to using it. Introducing the term biozone clarifies the referent of the zone. In ethnostratigraphy, the proposal by Gasche and Tunca of the term zone is confusing, as the term has already been used in biostratigraphy. Rather, ethnozone should be the formal name for the fundamental ethnostratigraphic unit. Once the ethnozone is established, and no further possibility of confusion with biostratigraphic or other kinds of units is possible, then the shorter form zone can be used.

The ethnozone, the basic ranking unit of ethnostratigraphy, is identified by either: (1) the first and/or last appearance of an artifact type (interval ethnozone); (2) the appearance of an association of two or more types of artifacts (assemblage ethnozone); or (3) the quantitatively distinctive maxima, minima, or relative abundance of one or more types of artifacts (abundance ethnozone). This definition is based on the definition of the biostratigraphic units: the interval biozone, assemblage biozone, and abundance biozone (NACOSN 1983:862–863). The term artifact is here defined as *anything* that displays one or more attributes as a consequence of human activity (Dunnell 1971:117; Spaulding 1960:438). Artifacts come in any scale and have attributes that are physical or locational properties (Dunnell and Stein 1989:32). This definition includes subsistence remains and all material used in tool production.

There are many attributes on which classifications of artifactual material can be based. The attributes depend on the research questions being asked. Dividing a site into ethnostratigraphic units is done for many reasons, one of which is regional correlation. Regional correlations can in turn be based on many different types of artifacts. In some regions the shape of ceramic vessels is the historically diagnostic artifact attribute; in others, burial types, architecture, biface bases, or bone tools are definitive types. In some cases, the appearance, disappearance, or cooccurrence of these types are the relevant observation. All these attributes are significant for regional correlation. For example, technological attributes were one of the first sets of attributes used to divide strata into groups. The groups were called stages (Table 2) and were defined as "a technological subdivision of prehistoric time that has little chronological meaning but denotes the level of technological achievement of societies within it" (Fagan 1988:583). The stages (e.g., Stone Age, Iron Age) are comparisons of "technological achievements of societies," which are kinds of technological attributes of artifacts from various sites that are compared from region to region. In this case, ethnostratigraphic units were defined using "technological" attributes of artifacts. Ethnostratigraphic units can also be based on debitage frequency, on subsistence remains, burial practices, architectural styles, or any other classification of artifactual material. They are constructed for purposes of regional and intrasite correlation.

Archaeologists have been correlating artifacts across regions for centuries. The ethnostratigraphic approach proposed here is new in that the ethnozones are defined explicitly, and separated clearly from each other lithostratigraphic unit, and from time units. They are not mixtures of form, space, and time all in a single unit, as is the phase (Willey and Phillips 1958). The ethnostratigraphic unit is only the division based on artifacts. Many different divisions can be constructed, each based on different attributes (interval, assemblage, or abundance ethnozones) of the same artifacts in the same strata. However, each new ethnozone division is described clearly and used for a specific purpose.

2. Northwest Coast Ethnostratigraphic Units

Dividing a site according to its artifact content has been a primary focus of Northwest Coast archaeologists for decades. The procedure followed has not always been systematic, has never been explicitly stated, and has even been suggested to be inappropriate (Abbott 1972; Mitchell 1971; Murray 1982). However, the format for reporting artifact groupings most frequently followed in site reports is (1) to define lithostratigraphic units, and (2) to group the artifacts (including debitage) from those units into artifact assemblages of similar scale. This procedure is followed in areas other than the Northwest Coast as well.

In Northwest Coast sites, the units defined by groups of artifacts are frequently called components. The artifact groups are defined by the appearance, disappearance, or abundance of one or more artifacts. For example, at Montague Harbour, Mitchell reports "the components were separated by plotting the distribution of *key artifact classes* (emphasis mine) on the stratigraphic profiles, in this manner determining which of the numerous breaks in physical stratigraphy were significant" (1971:99). Key artifact classes are matched with physical stratigraphy and discussed in relation to those lithostratigraphic units, or subdivisions of those units. Mitchell used small (lithostratigraphic) units to group material during excavation, and gave 3-point provenience to diagnostic artifacts. In this way he was able to relate the artifacts to the lithostratigraphic units and subdivide the large lithostratigraphic units into artifact subdivisions. Thus, the lithostratigraphic unit boundaries (layers) were used to arrange the artifacts into kinds of ethnostratigraphic units (ethnozones).

This methodology requires that the dimensions of the large-scale ethnostratigraphic units, into which the artifacts are grouped, are related directly to the boundaries of the large lithostratigraphic units defined in excavation unit profiles. Most artifactual materials (other than diagnostic tools) are brought out of the field grouped by small excavation units. The excavator assumes that the objects are deposited together and that little or no mixing has occurred across the small excavation unit boundaries. At some point during analysis the artifacts are grouped into larger units. The actual dimensions of this larger unit flows from the dimensions of the large-scale lithostratigraphic unit. These large stratigraphic units, traced across the entire site in the profiles of all excavated units, become the analytical unit for the artifacts. Thus lithostratigraphic boundaries define the boundaries and groupings of artifacts, that is, ethnostratigraphic units.

Defining ethnostratigraphic units by grouping artifacts according to the lithostratigraphic boundaries can result in the creation of inappropriate ethnostratigraphic units. One cannot assume that all physical properties observed in a layer during excavation were the result of depositional events. The physical properties may be a result of postdepositional alterations. At British Camp, where the groundwater produced what excavators believed was a new lithostratigraphic unit, grouping artifacts into ethnostratigraphic units on the basis of the boundaries of this lithostratigraphic units would create cultural units that are inappropriate and that will confuse regional correlation. Groundwater has inundated many sites in the Northwest (Mitchell 1971:88; Whittaker and Stein, Ch. 2, this volume), and in other parts of North America (Stright 1990). The inundation has created lithostratigraphic units that are not related to cultural or natural deposition. Mitchell noted this at Montague Harbour in two lithostratigraphic units (one with shell and one without) that had similar artifactual material. He lumped these two lithostratigraphic units into one artifact unit. However, the potential for inappropriate groupings of artifacts that are grouped, analyzed, and compared using boundaries defined by lithology is great. If such a mistake occurs, the ethnostratigraphic divisions may not be relevant for correlating prehistoric cultural changes.

C. *Time Units*

1. Definitions

Within stratigraphy there are three kinds of time units (Table 1): (1) those that refer to the rocks forming in a period of time (chronostratigraphic units); (2) those that refer to the time it took those rocks to form (geochronologic units); and (3) those that are a direct division of time, usually in years (geochronometric units). Each of these time units has a separate nomenclature.

Chronostratigraphic units are bodies of rocks that serve as material referents for all rocks formed during the same span of time (NACOSN 1983:868), for example, the Cambrian System of rocks initially described in Wales. The chronostratigraphic unit may be based upon the accumulation of a lithostratigraphic, biostratigraphic, magnetostratigraphic, or ethnostratigraphic unit, or any other feature of the rock record that has a time range (Stein 1987:345). But it must be defined by the rock (or unconsolidated sediment), not the time.

A chronostratigraphic unit, particularly useful for archaeology, is the chronozone. A chronozone is a nonhierarchical chronostratigraphic unit of short duration, whose boundaries may be independent of those of ranked units (eonothem, erathem, system, series, and stage) (NACOSN 1983:869). A chronozone is an isochronous unit, and it may be based on a biostratigraphic, lithostratigraphic, magnetopolarity, or ethnostratigraphic unit. Modifiers (litho-, bio-, polarity, and ethno-) used in formal names of the unit need not be repeated in general discussions where the meaning is evident from the context. The base and the top of a chronozone corresponds to an observed physical or artifactual boundary. They extend to other areas by means of recognizing the defining feature of the chronozone.

Geochronologic units are divisions of time that correspond to the time span of an established chronostratigraphic unit (e.g., the Cambrian Period of time). They are independent of the material referents. They are the time in which the rock referent accumulated. These time units were the only means of creating geological time scales before isotopic and other numerical ("absolute") dating methods were discovered.

Last, the geochronometric unit (NACOSN 1983:872) is time expressed in years (e.g., the Archean Eon begins 3.8 billion years ago, and ends 2.5 billion years ago). This geochronometric unit has been useful in subdividing the Precambrian where litho-, bio-, and chronostratigraphy cannot be easily applied.

In stratigraphy there is a growing use of absolute dates for regional correlation (Odin 1982). Since the discovery of isotopic, fission track, and other dating methods, which are independent of chronostratigraphic, lithostratigraphic, and biostratigraphic units, stratigraphers have had an independent means to check correlations. These absolute dates, because of their independence from rock references, are increasingly useful in regional correlation.

2. Rates of Deposition and Time

Stratigraphic time units do not refer to the rate or absolute duration (years) of the deposition, only to the total period of time represented in the rocks or unconsolidated strata. Determining a rate of deposition for lithostratigraphic units is difficult, because assumptions about gradual versus catastrophic deposition must be made (Ager 1981; Berggren and Van Couvering 1984; Cross 1990; Einsele and Seilacher 1982). A century ago geologists believed that the basins of the world filled slowly and steadily. They are now recognizing that some basins have filled rapidly (catastrophically) during one event. Such models affect calculations of rates of deposition. The fact that a stratum is physically very thick has no relationship to the amount of time it took that stratum to be deposited. It could have been deposited over a few seconds, a few days, or a few millennia.

A stratigraphic unit in archaeology also has no direct relationship to time, and calculating rates of deposition in archaeological deposits, especially rates based on dating artifacts, is as difficult as is calculating rates (and dates) of lithostratigraphic units (Ferring 1986; Butler and Stein 1988; Dunnell and Readhead 1988) or pollen accumulation in lake sediments (Bradley 1985). The problem is accentuated in shell middens (Bowdler 1976; Claassen 1991; Erlandson and Rockwell 1987; Sanger 1985; Waselkov 1987) where site volumes are large and artifact densities low. In the Northwest Coast the object dated is often a particle that itself has an age of death very different from its age of deposition. The age of wood of a large Western Red Cedar can vary by hundreds of years if the wood came from the inside versus the outside of a tree. Wood could be burned in one location, then moved several times as new midden layers are deposited. The rate of deposition of any lithostratigraphic or ethnostratigraphic unit in archaeology is difficult to determine (Gladfelter 1985).

3. Definition of Archaeological Time Units

Archaeologists have not yet proposed a formal unit for the time represented by the accumulation of archaeological deposits. Gasche and Tunca (1983) proposed a unit called the "phase" as a chronostratigraphic unit, defined as a grouping of adjacent strata of anthropic origins with a separate grouping of adjacent strata for those of natural origins. This term has been rejected (Stein 1990:518) for two reasons. The ethnocentric aspects of their definition as expressed in the cultural/natural dichotomy makes the term problematic. Whether a sediment has anthropic or natural origins (sources) has little to do with stratigraphy. If the layer contains artifacts, then it can be divided on the basis of those artifacts into an ethnostratigraphic unit. Then it can be arranged in time as a chronostratigraphic unit on the basis of those artifacts. However, strata that do not contain artifacts (i.e., natural) can also be divided into layers, and arranged in time as chronostratigraphic units. The cultural versus natural origins of the strata in archaeological sites are not criteria useful for defining chronostratigraphic units. The source of the strata is an interpretation of the strata's depositional history (Stein 1987). The cultural/natural dichotomy unfortunately emphasizes the separation of modern stratigraphy and archaeology in the last century. Time units in archaeological sites need not be defined differently for cultural or natural units. The difference alluded to by Gasche and Tunca is merely that the time unit is based on ethnostratigraphic units rather than lithostratigraphic units.

Another reason not to accept Gasche and Tunca's suggestion of the term "phase" as a chronostratigraphic unit is the long history and connotations associated with the term "phase" in American archaeology. Willey and Phillips (1958) proposed the term phase for "archaeological units possessing traits sufficiently characteristic to distinguish them from all other units similarly conceived, whether of the same or other culture or civilization, spatially limited to the order of magnitude of a locality or region and chronologically limited to a relatively brief interval of time" (1958:22). Willey and Phillips' definition is not useful for archaeological stratigraphy because a phase is based on artifact assemblages, time, and space. It is a combination of many stratigraphic criteria. To use it as the formally proposed term for a chronostratigraphic unit would just compound the already confusing language. No one would be sure to which definition of phase one referred.

Perhaps even more problematic for using phase as a stratigraphic unit is the common practice in archaeology of confusing interpretations with the definition of phases. Definitions should contain descriptions of empirical phenomena. Interpretations flow from these descriptions and are subject to falsification. Interpretations include an archaeologist's inference of prehistory based on the descriptions of assemblages of artifacts, architecture, or burial practices, all of which are usually found at a few closely spaced sites. The definition of the phase in archaeology almost always contains both description and interpretation. For example, it includes interpretations of the probable technology, subsistence, settlement, and ideology of the people represented by the phase. The phase has not been a description of empirical phenomena, but rather an interpretation of the meaning of those empirical phenomena. The phase is not the time represented by an ethnostratigraphic unit or a lithostratigraphic unit. It is closer to an interpretation or reconstruction of prehistoric life during the time represented by certain artifacts. Used in this way, phase is not a time stratigraphic or an ethnostratigraphic unit. The multiple definitions and connotations of "phase," as defined by either Gasche and Tunca or Willey and Phillips, make the term unsuitable as a chronostratigraphic unit. If these problems are not enough, the term phase has also been suggested as one of the hierarchial terms for the diachronic unit as defined by the NACOSN (1983:870)

The next obvious term to use in archaeological chronostratigraphy is the lowest ranking unit already proposed in chronostratigraphy. In both the *International Stratigraphic Guide* (Hedberg 1976) and the *North American Code of Stratigraphic Nomenclature* (NACOSN 1983), the stage is the lowest ranking unit. Stage, like phase, is a term laden with connotations and history in archaeology. As mentioned previously, a stage has been used in archaeology as a subdivision of time based on technological achievements (Table 2). Therefore, it would suffer from many of the same drawbacks as does the term phase, and is therefore inappropriate as the formal ranking chronostratigraphic unit in archaeology.

I propose that archaeologists adopt a chronostratigraphic unit, following the example of geophysicists who study magnetic-polarity properties. They proposed a polarity-chronostratigraphic unit, based on the magnetostratigraphic units (NACOSN 1983:861, 869), to define subdivisions of time based on magnetic-polarity properties. Using this example, subdivisions of time based on artifactual material (i.e., ethnostratigraphic units) would be named an ethno-chronostratigraphic unit, with the *ethnochronozone* as the basic unit. The ethnochronozone is based on artifact attributes that are rigorously defined, dated, and found useful for correlation across regions. Any artifact attribute or combination of attributes that is acceptable for defining the original ethnostratigraphic unit (the ethnozone) is acceptable for defining the ethnochronozone.

Ethnochronozone is a chronozone (Table 1) in that it is not a ranked unit. In stratigraphy the hierarchy of chronostratigraphic units begins at a scale appropriate for mapping geological phenomena. Archaeological phenomena are usually more local in scale than geological phenomena. By using a nonhierarchical chronostratigraphic unit in archaeology, the "scale appropriate for mapping" in geology does not conflict with that of archaeology.

The ethnochronozone conforms to the suggestion made earlier (Stein 1990) that archaeological stratigraphers should not write an entirely new code. The stratigraphic codes were written for all fields dealing with stratigraphy. Archaeology in the eighteenth century was closely related to other earth sciences and stratigraphy. To standardize archaeological stratigraphy a lower ranking lithostratigraphic unit, an ethnostratigraphic unit, and a new ethno-chronostratigraphic unit need to be defined. In this way, archaeology could be part of stratigraphy again and be incorporated into the North American Stratigraphic Code or any other stratigraphic code.

In summary, the archaeological literature is now full of references to terms implying time that are used in a manner inappropriate for stratigraphy. The chronological terms are defined using descriptions of cultures, lithologies, time, technology, subsistence, and geographic areas. They have become useless by overuse and misuse. Ethno-chronostratigraphy provides a new beginning for archaeology and stratigraphy, and a means by which the culture history we have amassed can be unraveled and site-formation processes can be studied.

4. Northwest Coast Time Units

In the Northwest Coast, archaeologists have constructed two time units: (1) the continuous interval represented by radiocarbon years; and (2) the phase.

a. Radiocarbon Ages A geochronometric unit in archaeology is defined usually on the basis of the radiocarbon dates of samples from the strata. In site reports, the absolute temporal placement of the strata and of the artifacts is described in a section entitled "Chronology." First, the ^{14}C dates are reported from one site that contains certain artifacts, and then compared to the ^{14}C dates reported for the same kinds of artifacts at other locations. For example, the discussion of the radiocarbon ages of the Montague Harbour site (Mitchell 1971:221) included the kind of organic material dated, the mean year reported by the lab, the standard error of the count, the lab number, the context of each sample dated, and Mitchell's general impression of what the date means. He uses these ages to place each artifact assemblage and layer into radiocarbon years.

b. Culture History of the Northwest Coast Subdivisions of the prehistory for the San Juan Islands and Gulf Islands have been proposed by a number of archaeologists (Borden 1951, 1968, 1970; Carlson 1960, 1970, 1983; Fladmark 1983; King 1950; Mitchell 1971, 1990; Thompson 1978). The sequences proposed by Carlson (1983) and Mitchell (1971, 1990) are most often referred to or relied upon, even though they differ in certain ways.

Mitchell divides the culture history into "culture types." Youngest to oldest these are: Gulf of Georgia, Marpole, Locarno Beach, St. Mungo (or Charles), and Old Cordilleran. Carlson divides the culture history into "phases." Youngest to oldest these are: San Juan Phase, Late Marpole Phase, Marpole Phase, Locarno Beach Phase, and Mayne Phase. The radiocarbon ages of these divisions indicate that the British Camp shell midden should fall somewhere between Mitchell's *Gulf of Geor-*

gia culture type (which includes Carlson's *San Juan Phase*) and Mitchell's *Marpole culture type* (which includes Carlson's *Late Marpole Phase* to *Marpole Phase*).

The Marpole Phase was named by Borden (1950, 1954, 1968, 1970), and later described in detail by Carlson (1960, 1970, 1983), Mitchell (1971, 1990), Matson (*et al.* 1980), and Burley (1980, 1989). Mitchell (1990:344–346) summarizes the Marpole Phase as assemblages with a wide variety of bone and stone tools, flaked stone points in a number of forms, including leaf-shaped, stemmed, and unstemmed points, and small triangular forms, microblades and microcores, large ground-slate points, thin ground-slate fish knives, sandstone whetstones and abraders, celts, leaf-shaped and smaller triangular ground-slate points, disk beads, labrets and earspools, hand mauls, perforated stones, shale or clam shell sculpture, bone needles, awls, nontoggling unilaterally barbed harpoon points of antler and bone, antler wedges, antler sculpture, and use of native copper for ornaments. Also included are midden burials, human skull deformation and occasional trepanation, large post molds, and house outline.

Burley (1980:30) evaluated artifact lists of the Marpole phase from a variety of publications, concluding that all these artifacts are defined in terms of their frequencies as compared to preceding or following phases. He points out that defining a phase in terms of changing frequencies of artifact types is problematic because of sample size effects. These assemblages are defined on the basis of frequencies (especially the frequencies of artifacts types which are rare) and are based on changes from preceding assemblages (which in turn are also frequencies). In addition, the Marpole Phase is defined not only by the presence of certain artifact classes, but also by the absence of artifact classes that were present in the preceding phase. The problems of frequencies and the continuity of other less rare artifact types (e.g., utilized flakes) has led Mitchell (1971), Abbott (1972), Burley (1980), and Murray (1982) to suggest that all these phases may be continuations of each other.

The San Juan Phase is described by Mitchell (1971:47; 1990:346–348) and Carlson (1960, 1970, 1983). The artifact types distinctive of the phase include: small triangular chipped points (found rarely), thin triangular ground-slate points (many with thinned bases), thin ground-slate knives, large celts, flat-topped hand mauls, irregular abrasive stones, unilaterally barbed bone points, numerous single- and double-pointed bone objects, split or sectioned bone awls, composite toggling harpoon valves, antler wedges, decorated bone blanket or hairpins, and triangular ground sea-mussel points (for the Western Gulf). Also included are loosely flexed midden burials and aboveground preparation of the dead, skull deformation, large post molds, and house outlines. Many of these features continue from the Marpole Phase.

As was the case for artifacts of the Marpole Phase, problems exist for the San Juan Phase. In fact, the problem of frequencies of artifact types is even greater in this phase, because one of the definitions of the phase refers to the decrease in frequencies of all chipped stone artifacts, including utilized flakes and debitage. In Chapter 6, this volume, the frequency problem is discussed, with a suggestion for its resolution.

In summary, regional correlations have been successfully constructed in archaeology, and for the Northwest Coast. As long as investigated sites are widely distributed both spatially and temporally, and thus appear very different from one another, the descriptions of strata and misinterpretations of the purposes and rules of stratigraphy do not affect the construction of culture histories. As more archaeologists excavate more closely spaced sites, and move into new field areas without the common training of one influential archaeologist who has been excavating for decades in that region, the correlations between sites and of phases become extremely difficult to make. The rules for excavating and grouping assemblages, often passed on through a mentor system through one influential archaeologist, are not explicitly described in a published format. When new people begin working in a new area, they have difficulty finding the criteria on which correlations were made. One way to combat this trend is to follow the stratigraphic procedure discussed in Chapter 6, this volume. Another is to encourage all archaeologists to follow a stratigraphic guide and impose a strict stratigraphic nomenclature. In this way all archaeologists will know exactly the basis for the correlations and regional chronologies.

IV. Conclusions

Archaeologists could improve their ability to make regional correlations and reconstruct site-formation processes by closing the gap that has developed between modern stratigraphy and archaeology. Closing this gap requires that archaeologists understand the need for explicit descriptions of stratigraphic subdivisions and the need to use multiple arrangements of the same strata. Sites are excavated in small lithostratigraphic units, which are the smallest units of analysis and which must be described adequately in site reports. The strata are then rearranged in two different ways: once using their physical characteristics (larger lithostratigraphic units; layers) and once using the artifacts (ethnostratigraphic units; ethnozones). These arrangements are two of many possible subdivisions of the strata in the site, and provide the basis for interpreting the manner in which sites were formed and the manner in which artifacts change within the site. The artifacts can also be divided into units so as to relate to other sites in the region (ethno-chronostratigraphic units; ethnochronozones). Finally, radiocarbon dates are used to determine the time of the ethnochronozone.

This standardized approach is rarely used in archaeology today. Presently in archaeology only one stratigraphic sequence is described in site reports, in sections entitled site stratigraphy. The strata observed in the profiles of excavation units are arranged into layers on the basis of their physical categories, then the layers are described and the material (bone, stone tools, debitage, ceramics, plant remains) in those layers are discussed in terms of their technological changes, their subsistence implications, and their cultural relevance. The boundaries of these groupings of artifacts are usually made to coincide with the boundaries of layers defined on the basis of their physical properties. Sometimes the layers are subdivided, other times layers are combined. However, in most site reports all stratigraphic discussions refer back to the physical layers observed in the profiles.

Archaeologists encounter problems in regional correlations because they have not separated the various stratigraphic classifications in their studies. They have forced the changes noted in any material, either discovered during field observations or analysis, into the boundaries of a single kind of stratigraphic unit, those based on physical observation made in the field. Obviously the physical attributes of strata are not used in regional correlations. They do not necessarily correlate with the physical attributes of strata from other sites. The artifacts in the strata and their datable material are the observations that are expected to correlate between sites. These are the data used in regional correlations between more than one site. However, regional correlations of artifacts and datable materials are effective only if the multiple sequences of site strata are described clearly, and the stratigraphic classifications are independent.

The stratigraphic approach proposed here encourages archaeologists to examine the archaeological sites from many perspectives. This encouragement follows a similar plea by Thomas (1989:6) who advocates a cubist perspective in connection with the study of the Spanish Borderlands. Thomas suggested that archaeologists and historians who study the consequences of Columbus's voyages should follow the example of twentieth-century artists in Paris who wanted to enlarge the spectator's vision to include multiple, simultaneous views of the subject. As Thomas encourages scholars to broaden their perspective of the Spanish Borderlands, I encourage scholars to broaden their perspective of stratigraphy. Archaeological stratigraphy requires that sites are seen in multiple perspectives, described and analyzed using explicit standardized nomenclature. This archaeological stratigraphy allows many research questions to be addressed clearly and culture histories to be unraveled. Adopting the pluralistic perspective of modern stratigraphy will allow archaeology to reestablish its ties to stratigraphy.

Acknowledgments

The ideas concerning stratigraphy that are expressed here have been developed through conversations with Ann Ramenofsky, William Farrand and W. Hilton Johnson. This material has been influenced significantly by discussions with Kimberly Kornbacher, Angela Linse, and Sarah Sherwood. The chapter was improved with the helpful comments of Steve Cole, Kimberly Kornbacher, and especially William Farrand, Ann Ramenofsky, and Donald Mitchell.

References

Abbott, D. N.

1972 The utility of the concept of phase in the archaeology of the southern Northwest Coast. *Syesis* **5**, 267–278.

Ager, D. V.

1981 *The nature of the stratigraphical record, 2nd ed.* New York: John Wiley and Sons.

Berggren, W. A., and J. A. Van Couvering

1984 *Catastrophes and earth history: the new uniformitarianism.* Princeton, New Jersey: Princeton University Press.

Blukis Onat, A. R.

1985 The multifunctional use of shellfish remains: from garbage to community engineering. *Northwest Anthropological Research Notes* **19**, 201–207.

Borden, C. E.

1950 Notes on the prehistory of the southern North-West Coast. *British Columbia Historical Quarterly* **XIV**, 241–246.

1951 Fraser Delta archaeological findings. *American Antiquity* **16**, 263.

1954 Some aspects of prehistoric coastal - interior relations in the Pacific Northwest. *Anthropology in British Columbia* **4**, 26–32.

1968 Prehistory of the lower mainland. In *Lower Fraser Valley: Evolution of a cultural landscape*, edited by A. H. Siemens. B.C. Geographical Series 9, University of British Columbia, Vancouver. Pp.9–26.

1970 Cultural history of the Fraser-Delta region: An outline. *B. C. Studies* No. 6–7, 95–112.

Bowdler, S.

1976 Hook, line, and dilly bag: An interpretation of an Australian coastal shell midden. *Mankind* **10**, 248–258.

Bradley, R. S.

1985 *Quaternary paleoclimatology: Methods of paleoclimatic reconstruction.* Boston: Allen and Unwin.

Burley, D. V.

1980 Marpole: Anthropological reconstruction of a prehistoric Northwest Coast culture type. Simon Fraser University, Department of Archaeology, Publication No. 8, Burnaby, British Columbia.

1989 Senewelets: Culture history of the Nanaimo Coast Salish and the False Narrows midden. Royal British Columbia Museum, Memoir No. 2, Victoria, British Columbia.

Butler, V. L., and J. K. Stein

1988 Comment on "Changing Late Holocene flooding frequencies on the Columbia River, Washington" by J. C. Chatters and K. A. Hoover. *Quaternary Research* **29**, 186–187.

Campbell, C. V.

1967 Lamina, laminaset, bed and bedset. *Sedimentology* **8**, 7–26.

Campbell, S. K.

1981 The Duwamish No. 1 Site: A lower Puget Sound shell midden. Office of Public Archaeology, Institute for Environmental Studies, University of Washington, Research Reports, No. 1, Seattle.

Capes, K. H.

1977 Archaeological investigations of the Millard Creek site, Vancouver Island, British Columbia. *Syesis* **10**, 57–84.

Carlson, C.

1979 The early component at Bear Cove. *Canadian Journal of Archaeology* **3**, 177–194.

Carlson, R. L.

1960 Chronology and culture change in the San Juan Islands, Washington. *American Antiquity* **25**, 562–586.

1970 Excavations at Helen Point on Mayne Island. *B.C. Studies* No. 6–7, 113–125.

1983 Prehistory of the Northwest Coast. In *Indian Art traditions of the Northwest Coast*, edited by R. L. Carlson. Archaeology Press, Simon Fraser University, Burnaby, British Columbia. Pp.13–32.

Chapman, M. W.

1982 Archaeological investigations at the O'Connor site, Port Hardy. In *Papers on central coast archaeology*, edited by P. M. Hobler. Simon Fraser University, Department of Archaeology, Publication No. 10, Burnaby, British Columbia. Pp.65–132.

Claassen, C.

1991 Normative thinking and shell-bearing sites. In *Archaeological method and theory, vol. 3*, edited by M. B. Schiffer. Tucson, Arizona: University of Arizona Press. Pp.249–298.

Cross, T. A.

1990 *Quantitative dynamic stratigraphy*. Engelwood Cliffs, N.J.: Prentice Hall.

De Meyer, L.

1984 *Stratigraphica archaeologica*, vol. 1. (editor) The A.C.T. Workshop, The University of Gent, Gent.

1987 *Stratigraphica archaeologica*, vol. 2. (editor) The A.C.T. Workshop, the University of Gent, Gent.

Drucker, P.

1972 *Stratigraphy in archaeology: An introduction.* Modules in Anthropology 30. Reading, Massachusetts: Addison-Wesley.

Dunnell, R. C.
1971 *Systematics in prehistory*. New York: Free Press.

Dunnell, R. C., and M. L. Readhead.
1988 The relation of dating and chronology: Comments on Chatters and Hoover (1986) and Butler and Stein (1988). *Quaternary Research* **30**, 232–233.

Dunnell, R. C., and J. K. Stein
1989 Theoretical issues in the interpretation of microartifacts. *Geoarchaeology: An International Journal* **4**, 31–42.

Einsele, G., and A. Seilacher
1982 *Cyclic and event stratification*. Berlin: Springer-Verlag.

Erlandson, J. M., and T. K. Rockwell
1987 Radiocarbon reversals and stratigraphic discontinuities: the effects of natural formation processes on coastal California archaeological sites. In *Natural formation processes and the archaeological record*, edited by D. T. Nash and M. D. Petraglia. London: BAR International Series 352. Pp.51–73.

Fagan, B. M.
1988 In the beginning: An introduction to archaeology, 6th ed. Glenview, Illinois: Scott, Foresman.

Farrand, W. R.
1984a Stratigraphic classification: living within the law. *Quarterly Review of Archaeology* **5**, 1.
1984b More on stratigraphic practice. *Quarterly Review of Archaeology* **5**, 3.

Fedele, F. G.
1984 Towards an analytical stratigraphy: stratigraphic reasoning and excavation. *Stratigraphica Archaeologica* **1**, 7–15.

Ferring, C. R.
1986 Rates of fluvial sedimentation: implications for archaeological variability. *Geoarchaeology: An International Journal* **1**, 259–274.

Fladmark, K. R.
1983 Comparison of sea-levels and prehistoric cultural developments on the east and west coasts of Canada. In *The evolution of maritime culture on the northeast and northwest coasts of America*, edited by R. J. Nash. Simon Fraser University, Department of Archaeology, Publication No. 11, Burnaby, British Columbia. Pp.65–76.

Folk, R. L.
1980 *Petrology of sedimentary rocks*. Austin, Texas: Hemphill Publishing Company.

Gasche, H., and O. Tunca
1983 Guide to archaeostratigraphic classification and terminology: definitions and principles. *Journal of Field Archaeology* **10**, 325–335.

Gladfelter, B. G.
1985 On the interpretation of archaeological sites in alluvial settings. In *Archaeological sediments in context*, edited by J. K. Stein and W. R. Farrand. Center for the Study of Early Man, University of Maine, Orono. Pp.41–52.

Ham, L. D.
1982 *Seasonality, shell midden layers, and coast Salish subsistence activities at the Crescent Beach site, DgRr 1*. Unpublished Ph.D. dissertation, Department of Anthropology, University of British Columbia, Vancouver.

Harris, E. C.
1989 *Principles of archaeological stratigraphy, 2nd ed.* New York: Academic Press.

Hedberg, H. D.
1976 *International stratigraphic guide: a guide to stratigraphic classifications, terminology and procedure*. New York: Wiley.

Jacobsen, T. W., and W. R. Farrand
1987 *Franchthi Cave and Paralia: maps, plans, and sections*. Bloomington, Indiana: Indiana University Press.

Kenyon, K. M.
1952 *Beginning in archaeology*. London: Phoenix House.

King, A. R.
1950 *Cattle Point: A stratified site in the southern Northwest Coast region.* Society for American Archaeology, Memoir 7.
Krumbein, W. C., and L. L. Sloss
1963 *Stratigraphy and sedimentation.* San Francisco: Freeman.
Luebbers, R.
1978 Excavations: Stratigraphy and artifacts. In *Studies in Bella Bella prehistory*, edited by J. J. Hester and S. M. Nelson. Simon Fraser University, Department of Archaeology, Publication No. 5, Burnaby, British Columbia. Pp.11–66.
McKern, W. C.
1939 The midwestern taxonomic method as an aid to archaeological culture study. *American Antiquity* **4**, 301–313.
McMillan, A. D., and D. St. Claire
1982 *Alberni prehistory: Archaeological and ethnographic investigations on western Vancouver Island.* Penticton, British Columbia. Theytus Books.
Matson, R. G.
1976 The Glenrose Cannery site. Archaeological Survey of Canada, Mercury Series No. 52. National Museum of Man, Ottawa.
Matson, R. G., D. Ludowicz, and W. Boyd
1980 Excavations at Beach Grove (DgRs 1) in 1980. Unpublished manuscript, Laboratory of Archaeology, University of British Columbia, Vancouver.
Matthews, R. K.
1984 *Dynamic stratigraphy: an introduction to sedimentation and stratigraphy, 2nd ed.* Englewood Cliffs, New Jersey: Prentice-Hall.
Mitchell, D. H.
1971 Archaeology of the Gulf of Georgia area, a natural region and its culture type. *Syesis* **4**, suppl. 1, 1–228.
1988 The J. Puddleduck site: a northern Strait of Georgia Locarno Beach component and its predecessor. Contributions to Human History, No. 1, Royal British Columbia Museum, Victoria.
1990 Prehistory of the coasts of southern British Columbia and northern Washington. In *Handbook of North American Indians*, Northwest Coast: volume 7, edited by W. Suttles. Washington, D.C.: Smithsonian Institution. Pp.340–358.
Murray, R.
1982 Analysis of artifacts from four Duke Point area sites near Nanaimo, B.C.: An example of cultural continuity in the southern Gulf of Georgia. Archaeological Survey of Canada, Mercury Series No. 113. National Museum of Man, Ottawa.
NACOSN (North American Commission on Stratigraphic Nomenclature)
1983 North American stratigraphic code. *American Association of Petroleum Geologists Bulletin* **67**, 841–875.
Odin, G. S.
1982 *Numerical dating in stratigraphy.* New York: John Wiley & Sons.
Pyddoke, E.
1961 *Stratification for the archaeologist.* London: Phoenix House.
Sanger, D.
1985 Seashore archaeology in New England. *Quarterly Review of Archaeology* **6**, 3–4.
Schiffer, M. B.
1983 Toward the identification of formation processes. *American Antiquity* **48**, 675–706.
1987 *Formation processes of the archaeological record.* Albuquerque: University of New Mexico Press.
Sharer, R. J., and W. Ashmore
1987 *Archaeology: Discovering our past.* Palo Alto, California: Mayfield.
Spaulding, A. C.
1960 The dimensions of archaeology. In *Essays in the science of culture in honor of Leslie A. White*, edited by G. E. Dole and R. L. Carneiro. New York: Crowell. Pp.437–456.

Stein, J. K.
1987 Deposits for archaeologist. In *Advances in archaeological method and theory, vol. 11*, edited by M. B. Schiffer. Orlando, Florida: Academic Press. Pp.337–393.
1990 Archaeological stratigraphy. In *Archaeological geology of North America*, edited by N. Lasca and J. Donahue. Geological Society of America, Centennial Special Volume 4, Boulder, Colorado. Pp.513–523.

Stein, J. K., and G. Rapp, Jr.
1985 Archaeological sediments: a largely untapped reservoir of information. In *Contributions to Aegean archaeology*, edited by N. C. Wilkie and W. D. E. Coulson. Center for Ancient Studies, University of Minnesota, Publications in Ancient Studies 1, Minneapolis. Pp.143–159.

Stright, M. J.
1990 Archaeological sites on the North American continental shelf. In *Archaeological geology of North America*, edited by N. Lasca and J. Donahue. Geological Society of America, Centennial Special Volume 4, Boulder, Colorado. Pp.439–465.

Thomas, D. H.
1989 Columbian consequences: The Spanish borderlands in cubist perspective. In *Columbian consequences: Archaeological and historical perspectives on the Spanish borderlands west, volume 1*, edited by D. H. Thomas. Washington, D.C.: Smithsonian Institution Press. Pp.1–14.

Thompson, G.
1978 Prehistoric settlement changes in the southern Northwest Coast: A functional approach. University of Washington, Department of Anthropology, Reports in Archaeology No. 5, Seattle.

Waselkov, G. A.
1987 Shellfish gathering and shell midden archaeology. In *Advances in archaeological method and theory, vol. 10,* edited by M. B. Schiffer. Orlando, Florida: Academic Press. Pp.93–210.

Willey, G. R., and P. Phillips
1958 Method and theory in American archaeology. Chicago: University of Chicago Press.

6

British Camp Shell Midden Stratigraphy

Julie K. Stein
Kimberly D. Kornbacher
Jason L. Tyler

I. Introduction

The stratigraphy of the British Camp shell midden involves the description of the layers in the site, the arrangement of those layers and their content into multiple sequences, the explanation of the origin of those sequences, and the correlation of those sequences with other sites in the region. Northwest Coast shell middens are large, stratigraphically complex sites, usually located very close to the shoreline. Descriptions, explanations, and correlations of strata are difficult, especially given the influence of fluctuating sea levels and progradation of shorelines. The methodology used at British Camp has allowed the complex stratigraphy of the shell midden to be unraveled.

Within the twentieth century, excavations have shifted from large-scale expeditions aimed at building basic cultural chronologies to small-scale operations aimed at refining chronologies and addressing new sets of questions concerning technology, subsistence, and site formation. In the first half of this century, as culture historians were building the basic chronologies of the New World, large-scale excavations were necessary. Large, stratigraphically complex shell middens in the Northwest were excavated using arbitrary levels, a procedure which mixed artifacts from distinctly different depositional contexts. The mixing was justified because archaeologists were following a culture historical paradigm. To obtain sufficiently large samples of artifacts, large volumes of sites needed to be excavated. The use of arbitrary levels and the mixed context of artifacts were inevitable consequences of the culture historical paradigm. In the second half of this century archaeologists have employed natural levels in excavation more frequently than arbitrary levels. It was worth the difficulty encountered during excavation and the additional time required to excavate in natural levels to achieve the control gained over artifact provenience. The new research questions required greater control over the context of stone tools, animal bones, and plant remains.

This recent shift in excavation strategy, however, has brought new problems for archaeologists. Field procedures, descriptive attributes, and reporting styles are not standardized. The criteria used to define a "natural level" are not agreed upon by the discipline. Every excavator creates a new set of criteria and is trusted to do it "correctly." "Correctly" flows from a logical normative system defined in the field. If excavators are asked to describe the manner in which the natural levels are defined, they turn to the profile and define contrasting lithologies. One stratum looks different from the other. The standard method of reporting field procedures in site reports does not require that the criteria used to define natural levels be published in the text. Rather, one just includes the statement "the site was excavated in natural levels." Presumably, everyone knows what that means. The shift to natural levels in excavation has occurred without a parallel shift in standardization of describing and reporting.

Perhaps the reason for the lack of standardization and exclusion of natural levels' descriptions is that "traditional" site reports and excavation strategies flowed from the culture history paradigm and did not include them. Using natural levels was a new strategy, but the procedure we use to report excavation methods flows from the methodology developed in the early half of the twentieth century, from culture history. Arbitrary levels do not demand explicit description of the manner in which they are selected, and do not need to be described extensively in site reports. However, natural stratification is more complex. If it is used as a grouping device for artifacts, then the descriptions of the natural levels must be included in the site report.

In this chapter, stratigraphy from the British Camp shell midden is discussed, including the stratigraphic divisions selected for excavation, site correlation, artifact comparisons, and regional correlation. The procedure followed is stratigraphy, as described in the Code of Stratigraphic Nomenclature (NACOSN 1983; Stein, Ch. 5, this volume). Initially, the excavation divisions are defined (the smallest lithostratigraphic units, Table 1, Ch. 5, this volume), followed by two schemes for grouping these small units into larger lithostratigraphic units. One scheme is based on field descriptions and was created for purposes of correlating strata between adjacent excavation units. The second scheme is based on laboratory analysis (measurement of percent carbonate in fine-grained sediment fraction) and was created to detect post-depositional weathering phenomena, especially those associated with groundwater inundation. Second, divisions of site strata created by noting changes in artifactual content are described (ethnostratigraphic units). Abundance ethnozones (Table 1, Ch. 5, this volume) are created by noting a change in the number of objects of chipped stone in each small lithostratigraphic unit. Last, geochronometric and ethnochronostratigraphic units are discussed for purposes of regional correlation. By using this pluralistic (cubist) perspective (see Ch. 5, this volume) the results are sufficiently explicit for other Northwest Coast archaeologists to determine exactly the criteria on which the British Camp site excavation was based, to evaluate the site's stratigraphy, and to make any appropriate correlations with other sites.

The results reported in this chapter indicate that:

1. Four layers including a lower dark layer (Layer I, II, III, and IV) are observed across the whole excavation.
2. Lower percentages of carbonate are found in the matrix of the deeper portions of the site (Layer L) relative to the shallow portions (Layer H).
3. The abundance of chipped stone is greater in the lower portion of the midden (Ethnozone I) than in the upper portion (Ethnozone II).
4. The boundary of the dark layer (Layer I) does not always coincide with the boundary between Ethnozone I and Ethnozone II.
5. The lithostratigraphic boundaries defined by field observation (Layer I, II, III, or IV) do not coincide (except in two locations) with the boundaries defined by laboratory-generated measures of weathering (Layer H and L).
6. The radiocarbon age of the boundary between the two ethnozones is between 670 and 1062 B.P.
7. The age of the boundary of the lithostratigraphic unit is meaningless because the boundary was created after datable material was deposited.
8. The ethno-chronostratigraphic units for the British Camp site are defined by the abundance ethnozones. Their correlations await examination of the abundance of objects of chipped stone at other sites in the region.
9. The explanation for these results is that the site has been altered postdepositionally by groundwater, creating physical stratification and chemical signatures that are not related to the stratification defined by the abundance of objects made of chipped stone.

In the rest of this chapter, each of these stratigraphic divisions and conclusions will be described in detail.

II. The Small Lithostratigraphic Units at British Camp: The Facies

Attributes used to define the smallest units used in excavation at the British Camp shell midden were selected in 1983. The term facies was chosen for these small lithostratigraphic units to emphasize that they are the minimal unit of observation at the site, and are described using standardized descriptions (Stein and Rapp 1985). Although these facies are very small lithostratigraphic units (much smaller in scale than geological facies) (Reading 1978; Weller 1960) and do not always conform to the concept implied in Walther's Law of Facies (Stein 1987, 1990), the term emphasizes visible lithological changes that occur vertically and horizontally.

Facies are subdivisions of the shell midden that are distinguished and delimited on the basis of lithologic characteristics. Facies represent an event over time that resulted from transport agents bringing material from similar sources and depositing them in the shell midden. A facies represents an "individual dumping event" when prehistoric people brought material together from some combination of activities and deposited them in one location. The depositional event could have occurred over hours, days, years, or centuries. The amount of time over which the facies accumulated is not part of the definition, only the fact that the same lithology was continuously or sporadically deposited without an intervening change. One person dumping one basket of shellfish in the same place every year for 25 years could produce a facies, or 25 people dumping 25 baskets in one hour could produce a facies. The duration of the dumping event is not the criterion used to define facies, rather, the kinds of materials brought together and the manner in which they are dropped (which produced the lithology) is the criterion observed to define facies. Archaeologists have only the archaeological record in front of them. Distinguishing and delimiting facies is one way to subdivide that record.

A. *Description of Facies Taxonomy*

To increase the control over horizontal provenance in the excavation block, the 6 × 8-m excavation area (referred to as Operation A) was divided into twelve 2 × 2-m excavation units. Each excavation unit is named by the grid coordinate at its NW corner (Figure 1). Facies are excavated and lettered separately in each 2 × 2-m excavation unit. Using this provenience format, facies that are larger than 2 × 2 m and extend into other excavation units are given different names. Their "equivalence" is recorded and all facies that belong together are analyzed together.

Facies were described in the field by reference to the following attributes: size of particles, composition of particles, abundance of each compositional type of particle, condition of particle, and color. These attributes, however, were not described

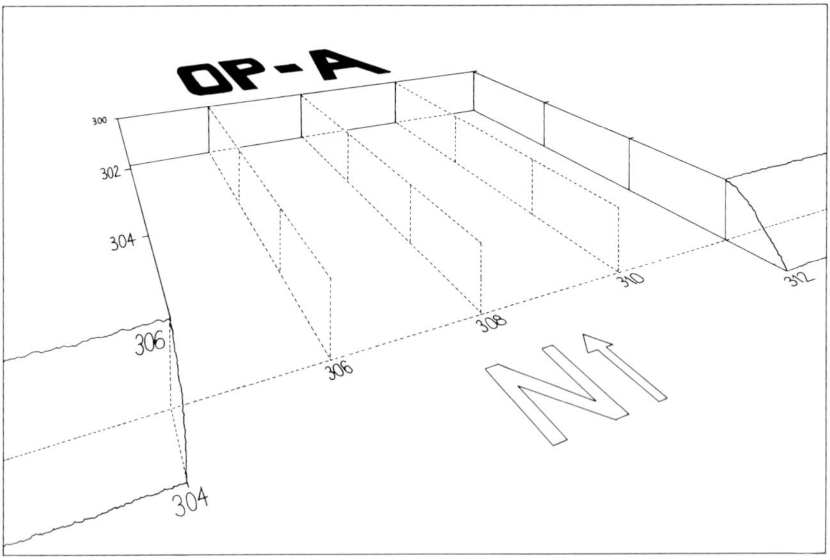

Figure 1 Three-dimensional representation of Operation A, showing its orientation to wave-cut bank and the grid coordinates. Vertical lines on walls of Operation A and dashed vertical lines in center mark the outlines of the twelve 2 × 2-m units excavated. Each of these units is named by its NW coordinate, with East/West coordinate given first (e.g., E306/S300). Drafted by Timothy D. Hunt.

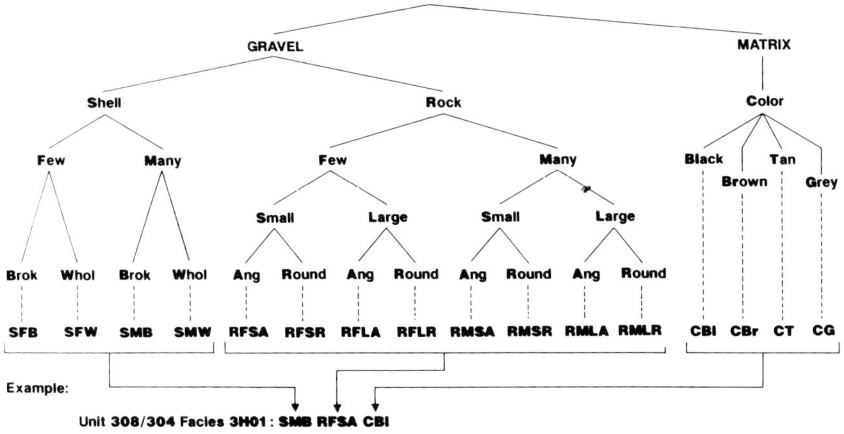

Figure 2 Classification of facies based on lithology. Brok is abbreviation for Broken; Whol for Whole; Ang for Angular. Note that for each facies, the gravel is described on the basis of shell and rock, and the sand, silt, and clay (matrix) is described on basis of color only. The example of Unit 308/304 facies 3H01 is described as containing many broken shells (SMB), few small angular rocks (RFSA) and black fine-grained sediment (CBl). Each facies excavated is described using these lithological attributes. Drafted by Timothy D. Hunt.

for every particle in every facies. One set of attributes was used for gravel-sized particles (objects caught on the 1/8-inch or 3-mm screen), another set of attributes for sand-, silt-, and clay-sized particles.

Gravel-sized particles were first examined in the field for their compositional types, dividing them into shell and rock (Figure 2). Although bone and culturally modified stone (chipped stone or ground stone) are two other compositional types of gravel-sized particles noted in the midden, the vast majority of the large particles is shell and rock. Therefore, the abundance of the rare stone and bone is not used in the field description of a facies. For each facies the percentage of shell is estimated (using visual comparators) as less than or greater than 50% (labeled Few or Many). Within each of these divisions the abundance of broken (defined as a shell that has any biologically produced portion of it removed) or whole shells is estimated. The 50% estimate was used as a cutoff because it is roughly the mean value of shell observed. In this way the shell for each facies is described (e.g., SFB = shell few broken).

Rock (labeled Few or Many) is described by estimating (using visual comparators) its abundance as less than or greater than 25%. The 25% estimate was used as a cutoff because it is roughly the mean value of rock observed. The size of the gravel-sized rock is further described as Small (less than 30 mm) or Large (greater than 30 mm). The rock is also described as angular (defined by the presence of at least one angular edge) or rounded (defined as all edges being rounded). The rock for each facies is thus described (e.g., RFLA = rock few large angular).

The particles in each facies that are smaller than gravel cannot be identified easily in the field. Their size makes identification of compositional types difficult without magnification (Dunnell and Stein 1989). This small-sized fraction is called matrix and is described using only one attribute, that of color. One of four possible colors is noted for each facies: Black, Brown, Tan, or Grey. Although shades and hues change with moisture content and lighting conditions, all facies are described in the field (slightly moist condition) within these categories.

The attributes noted in Figure 2, along with the abbreviations for those attributes, allow each facies to be described in the field in a standardized way as it is being excavated. The descriptions are first made of the facies as they appear in plan view, and modified as the facies is removed from the floor. There are no baulks from which to observe the facies in profile, thus descriptions must be made before and after facies are removed. For example, Unit 308/304 facies 1K01 is described as SMB/RFLA/CBl. The abbreviation means that it contains shells that are Many (more than 50%), and Broken; rocks that are Few (less than 25%), Large (greater than 30 mm), and Angular; and that the matrix is the Color Black.

A new facies is designated when one or more of the attributes used in these descriptions changes significantly. "Significant" changes are determined by the field supervisors and, although admittedly difficult to systematize, are defined when one or more attributes on the description form shifted from one category to another. For example, when the percentage of shell changes from less than 50% to greater than 50%, or when the color of the matrix changes from brown to tan the excavator designates the stratum a new facies and fills out new forms. Each facies is given an alphabetic designation (e.g., 1A, 1B, 1C . . . . 2A, 2B . . . . 3A, 3B, etc.). In those cases where a facies is very thick, it is divided into 10-cm levels (e.g., 1A01 refers to facies A level 01, and 1A02 to facies A level 02).

The field descriptions of the facies are quantified through a screening operation. The facies are excavated using trowels and dust pans. All archaeological material is collected in 8-liter buckets lined with plastic bags. The volume of each facies is then calculated by counting the number of 8-liter buckets removed from each facies. These bags are weighed, then poured into a set of four nested screens [25 mm (1 inch), 13 mm (1/2 inch), 6 mm (1/4 inch), and 3 mm (1/8 inch); Figure 3]. The material retained on the two large screens is washed and sorted in the field. The material retained on the 1/4- and 1/8-inch screens is sorted in the lab. Bone, charcoal, and chipped stone are sorted from the smallest two screens (Figure 4), and the shell, rock (both rounded and angular), bone, charcoal, and lithics are sorted from the largest two screens. The weights of all these materials are recorded separately. All bone and the stone artifacts are retained for analysis. This system allows the weight and the percentage of every major compositional type of particle to be quantified for each facies. In the discussion that follows, the reference to "number of buckets" or "liters" is the quantification unit for facies derived from the data generated during screening.

B. *The Harris Matrix*

In the mid-1970s, Harris (1975, 1977, 1979) described a method of recording archaeological stratigraphy called the Harris Matrix, which is very useful for recording strat-

Figure 3 Sediment is washed through four nested screens. From top to bottom is 25 mm (1 inch), 13 mm (1/2 inch), 6 mm (1/4 inch), and 3 mm (1/8 inch). Material from the two largest screens is sorted and weighed in the field, and that from the two smallest screens is sorted and weighed in the laboratory. Water was collected in a holding pond and recycled into a historic cistern. Photo by Ken Alford.

ification in large stratigraphically complex sites such as shell middens. The Harris Matrix method reduces each excavation layer to a rectangular box. The boxes are arranged according to the superpositional order of the excavation layers, into a concise graphic display. The procedure by which this is accomplished has been outlined in detail by Harris (1989). He developed the method while working on a complex "urban site" in Winchester, England. It is particularly useful for recording complex stratigraphy (e.g., caves, rockshelters, and shell middens). Some discussion of the

Figure 4 Material caught on the 6-mm and 3-mm screens is sorted in the laboratory. Chipped stone flakes, other artifacts (e.g., microblades and beads), bone, and charcoal are removed and weighed. An average of 90 to 95% of the fish bone recovered in excavation is found in the 3-mm screen. Photo by Ken Alford.

method and its theoretical underpinnings may be found in Ambrosiani (1977), Barker (1977), Lynch (1977), Paice (1991), Rance (1977), Schofield (1977), and Stein (1987, 1990).

Correlating strata observed in profiles with strata excavated from the unit (or the Harris Matrices of these strata) is extremely difficult. The small-scaled shell midden strata used in the excavation of shell-bearing sites are often not continuous across a substantial portion of an excavated area, even across a 1 × 1-m unit, and may not appear in any profile. On the other hand, a small layer may appear in the profile of the excavation unit, but it does not extend into the excavation area. In other words, a layer distinguished in the excavation may have no recognizable counterpart in a Harris Matrix of a profile, or vice versa. Thus, arranging small lithostratigraphic units from a profile into a temporal framework for excavation units may be very difficult, even if a Harris Matrix is used as a recording device.

1. Use of the Harris Matrix on the Northwest Coast

The difficulty of correlating small excavation units with layers observed in profiles has been encountered by other Northwest Coast archaeologists. The application of the Harris Matrix to Northwest Coast shell middens began with Stucki (1981, 1983). She used the technique to sort out the stratigraphic sequence of hundreds of separate deposits at the Hoko River rockshelter (45CA21) on the Washington coast. Stucki used a derivation of Harris diagrams to order the deposits visible on the vertical walls of an excavated trench. All deposits on a profile were drawn and numbered and their stratigraphic relationships represented schematically on a Harris Matrix diagram.

Stucki encountered difficulties in using the Harris Matrices to discuss general site stratigraphy and to correlate stratigraphy with analysis of other materials (e.g., fauna and artifacts) (Wigen and Stucki 1988). The stratigraphic units represented on the Harris diagrams (derived from the profiles) were not correlated with the individual deposits actually removed in excavation. However, Stucki grouped the small stratigraphic units observed in the profile into large lithostratigraphic units called depositional periods (DP). These depositional periods are the same as large lithostratigraphic units described in Chapter 5, this volume. They are large-scale units observed across the entire excavated area. Stucki discusses the frequency of various "layer types" found within each DP, and Wigen correlates the faunal material found at the site with the DPs. Although the Harris Diagrams are used to record the stratigraphic order of the small-scale layers, the large-scale stratigraphic units (DP) are the analytical units.

Ham (1982) developed a variation of the Harris Matrix to illustrate functional differences among shell midden layers at the Crescent Beach site (DrRr1) in British Columbia. Harris matrix diagrams of actual layers removed during excavation were not correlated with any strata visible in the profiles. Although the Harris matrix allowed Ham to record the stratification, at the Crescent Beach site the small-scale stratigraphic units used in excavation and recorded in the Harris diagrams were lumped into larger scale units for analytic purposes.

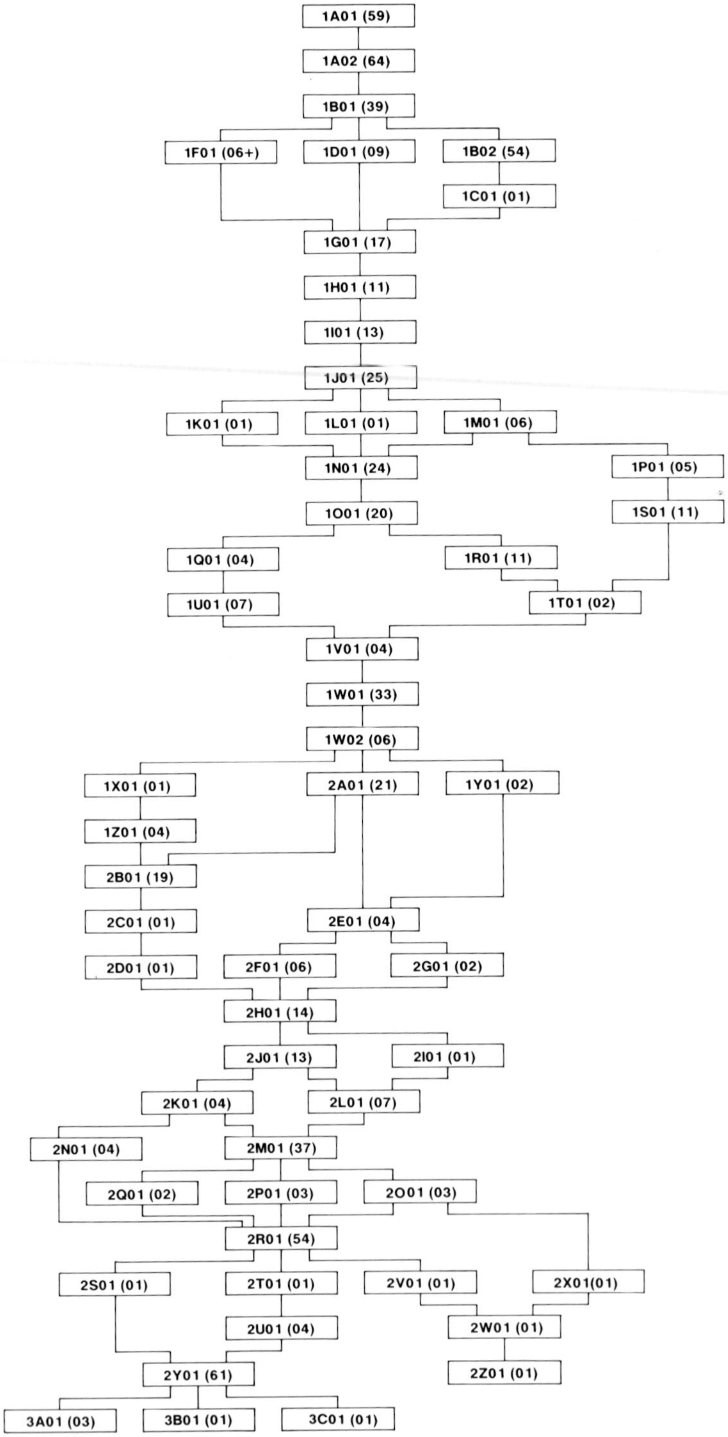
1A01 (59)
1A02 (64)
1B01 (39)
1F01 (06+)
1D01 (09)
1B02 (54)
1C01 (01)
1G01 (17)
1H01 (11)
1I01 (13)
1J01 (25)
1K01 (01)
1L01 (01)
1M01 (06)
1N01 (24)
1P01 (05)
1O01 (20)
1S01 (11)
1Q01 (04)
1R01 (11)
1U01 (07)
1T01 (02)
1V01 (04)
1W01 (33)
1W02 (06)
1X01 (01)
2A01 (21)
1Y01 (02)
1Z01 (04)
2B01 (19)
2C01 (01)
2E01 (04)
2D01 (01)
2F01 (06)
2G01 (02)
2H01 (14)
2J01 (13)
2I01 (01)
2K01 (04)
2L01 (07)
2N01 (04)
2M01 (37)
2Q01 (02)
2P01 (03)
2O01 (03)
2R01 (54)
2S01 (01)
2T01 (01)
2V01 (01)
2X01(01)
2U01 (04)
2W01 (01)
2Y01 (61)
2Z01 (01)
3A01 (03)
3B01 (01)
3C01 (01)

2. Construction of the Harris Matrix for Each Excavation Unit

At British Camp, K. Kornbacher and J. Tyler drew the Harris diagrams to record the stratigraphic relationships of the facies and to define large lithostratigraphic units. The process required three phases: (1) construction of Harris Matrices to record superpositional relationships of all facies in each 2 × 2-m excavation unit; (2) comparison of Harris Matrices to construct larger lithostratigraphic units; and (3) correlation of large lithostratigraphic units with the profiles (Kornbacher 1989a, 1989b).

For the first three years (1984–1986) of the San Juan Archaeological Project the stratigraphic relationships of the deposits were not recorded using the Harris Matrix symbols. Throughout the excavation detailed plan maps of the maximum extent of facies were drawn before the excavation of the next facies. Thus, the superpositional relationship of each facies, based on the degree of overlap indicated on the maps, could always be determined. The maps, however, were not as convenient as are Harris Matrices. Therefore, the facies from 1984–1986 were all arranged in Harris Matrices using the information on the maps and in field notes. After 1986, Harris Matrices were used to record the superpositional relationship of facies in the field.

These relationships are diagramed in Harris Matrices for each unit. The letter designation of each facies is represented in the matrix by a rectangular box with the letter inside (Figure 5). If a facies was excavated in two arbitrary levels (because it exceeded 10 cm in thickness), it is represented on the matrix as two boxes. In addition to the facies name, each box on the diagram also includes the volume of the facies expressed in numbers of 8-liter buckets. The superpositional relationship of two facies is graphically depicted by drawing two boxes, one on top of the other, connected by a straight line. If the stratigraphic relationship of two facies cannot be determined, the facies are drawn as separate boxes on the diagram with no lines connecting them.

3. Correlation of the Facies

Once the determinable superpositional relationships of the facies are recorded, the relationships between facies in different excavation units are defined. As mentioned before, the strategy of excavating one large area, composed of twelve contiguous 2 × 2-m excavation units (Figure 6) uninterrupted by baulks, results in two or more portions of the same facies being removed separately and given different letter designations. These equivalent facies are correlated between units through the compilation of a variety of data, including observations made by excavators regarding the continuity of facies, field descriptions of facies, profile drawings, and elevations. Portions of contiguous facies are identified on Harris Matrices using symbols and colors, and in this way the Harris Matrices from different excavation units can be linked.

Figure 5 Harris Matrix of unit 308/302. Each rectangle represents a facies, the name of which is in the box along with the volume (given in number of 8-liter buckets). For example, facies 1A01 filled 59 8-liter buckets (a volume of 472 liters). The Harris Matrix is a graphic representation of the superpositional relationship of all the facies. Drawn by K. Kornbacher and drafted by Jo Linse.

Figure 6 Operation A at end of season in 1985. Note that nine (2 × 2-m) units are being excavated continuously without baulks (separated by string). The three units farthest away from the photographer were excavated with baulks until the end of 1985, at which time the baulks were removed. The reconstructed historic building in the background is the Block House. The beach is on the right, north is to the left. Photo by Ken Alford.

Once the Harris Matrices are correlated from area to area, the small lithostratigraphic units (facies) are lumped into larger lithostratigraphic units. The abbreviations of the field description of the facies are plotted on the Harris Matrices (Figure 7). The abbreviations referring to shells are noted above the upper left corner of each box, the ones for rock are inside the box on the left, and color is inside the box on the right. In Figure 7, with only a portion of the excavation unit displayed, the facies have brown- and grey-colored fine-grained sediment, with some tan. The tan facies are usually small, as indicated by the many adjacent facies. Most facies have few broken shells and few small angular rocks. The abundance of one facies type over others is the criterion used to create the large lithostratigraphic units.

The Harris Matrix is the most logical graphic technique used to display data for this research program. The facies are given sequential alphabetic labels, yet they are not superimposed on each other. Because facies are often adjacent to one another, and vary in size from one bucket (8 liters) to many buckets, listing data in tables or plots according to alphabetical order is misleading. All data are therefore displayed using the Harris Matrix format.

C. *Profiles*

At the end of each field season, excavators drew profiles of the exposed walls of the excavation. Baulks were not retained between units, so profiles exist only for the outer edges of the Operation A excavation. Every visible rock, shell, bone, or piece

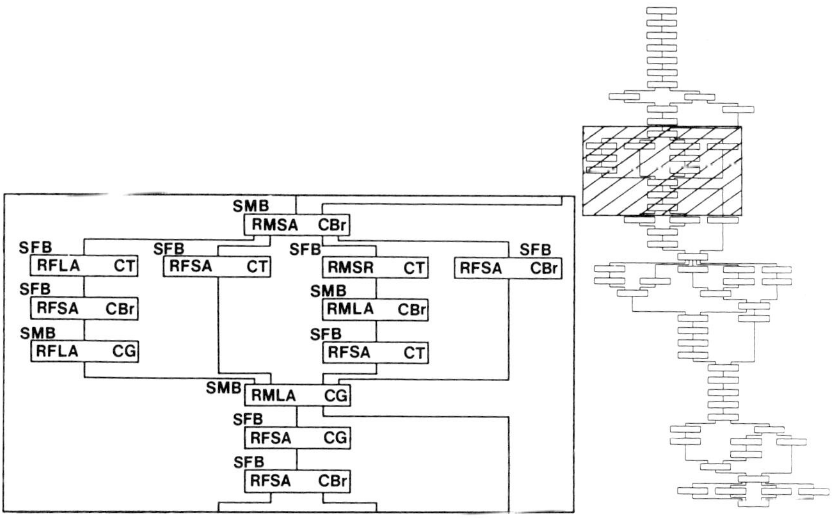

Figure 7 Harris Matrix of a portion of unit 308/304, with classification of facies (based on lithology) provided for some of the facies. The key for the abbreviations are given in Figure 2. Note that in this portion of unit 308/304, most of the facies have brown- and tan-colored fine-grained sediment, with gravel fractions containing few small rocks and few broken shells. The tan facies are small and discontinuous and usually are depicted in Harris diagrams as many layers lying adjacent to each other (note additional series of tan facies in two other lower areas of the figure). Drafted by Timothy D. Hunt.

of charcoal 1 cm or larger was drawn (Figure 8). Excavators did not describe the layering they saw nor draw any lines on the profiles. They simply recorded outlines of the large (greater than 1 cm) objects. Any layering observed in the profiles is not matched with the facies described during excavation because most facies are not represented in the profile (they do not extend across the entire excavation area to the profile).

The profiles are used, however, in constructing the large-scale lithostratigraphic units described below (Layer I, II, III, and IV). Large variations in such things as shell density, angular rock density, and matrix color are noted on the profiles. These descriptions were not used in Harris Matrix construction, but were helpful in determining the location of boundaries for the large stratigraphic units that were correlated across the entire excavated area.

III. The Large Lithostratigraphic Units at British Camp: The Layers

The facies defined during excavation are grouped into larger lithostratigraphic units using physical attributes. Field descriptions, profiles, and elevations were used to group the excavation facies into larger lithostratigraphic units useful for correlating across the excavated portion of the site. Laboratory analysis of carbonate was used to create a second lithostratigraphic scheme, conducted to find evidence of weathering. The layers (Gasche and Tunca 1983; Ch. 5, this volume, Table 1) used for site correlation will be described first, followed by the layers created to investigate weathering.

A. *Layers Used for Correlation of Excavation Units*

The shell midden at British Camp is divided into four layers, each of which is observed across most of the excavated area (Kornbacher 1989a). For example, in Figure 9 the fourfold large-scale lithostratigraphic units are drawn on a profile of one excavation unit 308/300. Layer I is the lowest division with Layer IV the one closest to the surface.

Layer IV (the uppermost Layer) is a heterogeneous, disturbed layer that is the "plow zone." Layer IV has a high frequency of facies with grey matrix (CG), many broken shells (SMB), and few large angular rocks (RFLA). Nearly all the historic artifacts (glass, ceramics, and metals) recovered in the excavations are from Layer IV. In several excavation areas a concentration of hard-packed burned facies was

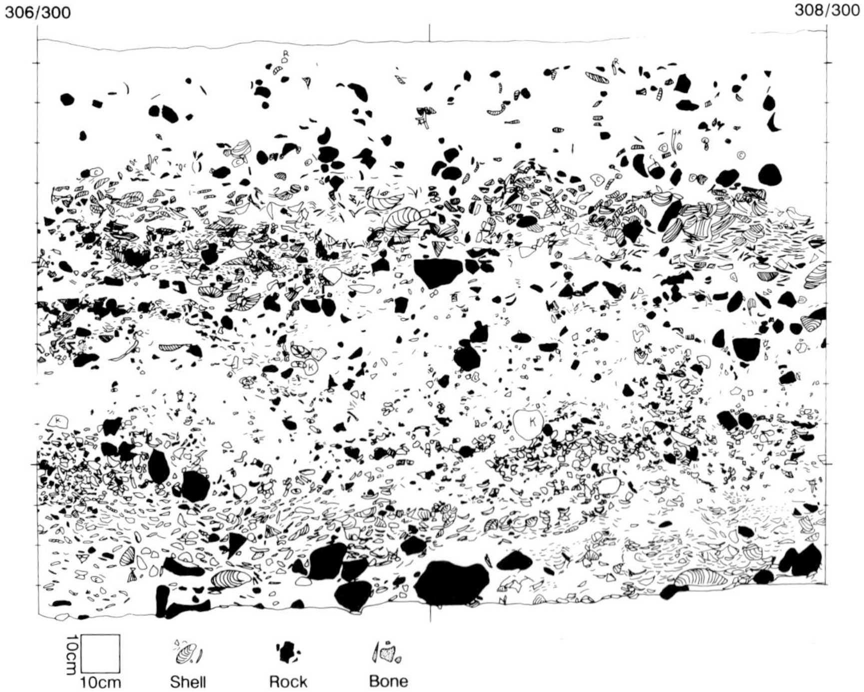

Figure 8 Profile of unit 306/300, north wall (two meters from grid coordinate 306/300 to 308/300). Each object in the profile larger than 1 cm is drawn at the end of each summer's excavation. Shells are drawn in their correct orientations, as are rocks and bones. Because no differences in color are noted on the profiles, no tan facies are recorded in these drawings. Drafted by Timothy D. Hunt.

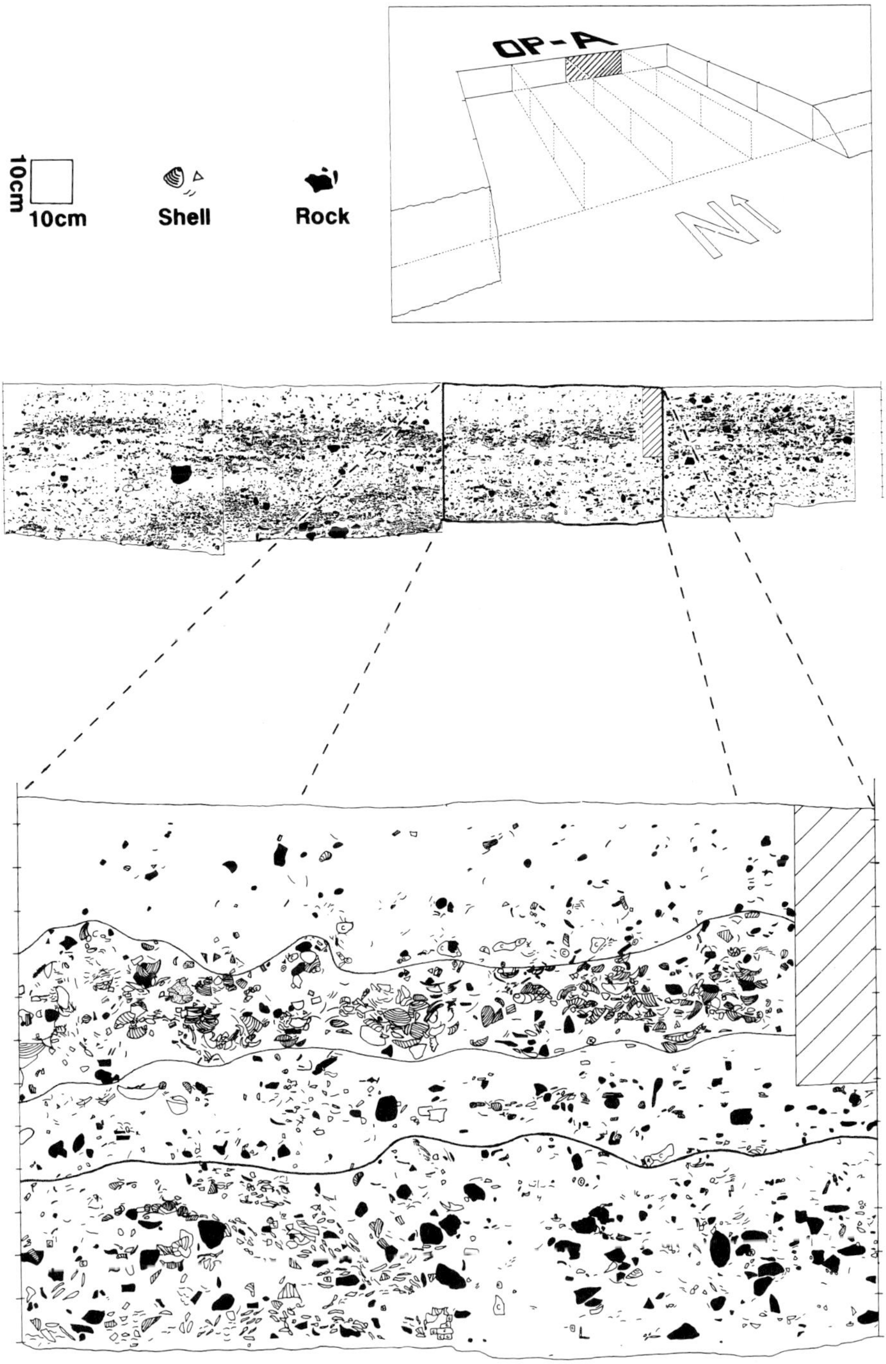

Figure 9 The complete stratigraphic section of Operations A's north profile (45SJ24) is displayed, with a close-up of unit 308/300. The location of the profile within Operation A is noted in the inset. On the close-up the boundaries of the fourfold lithostratigraphic layers are sketched. The lowest lithostratigraphic layer is Layer I, the layer nearest the surface is Layer IV. On the close-up the area of cross hatching is the location of the baulk removed in 1985. Drafted by Timothy D. Hunt.

noted within Layer IV. In addition, a small charcoal facies was removed from the eastern part of the excavation. The fine-grained sediment in Layer IV is dry.

Layer III is large, distinct, and contains abundant shell, most of which is whole. Layer III has a high frequency of facies with grey matrix (CG), many whole shells (SMW), and few small angular rocks (RFSA). It contains shells of a variety of taxa (Ford, Ch. 13, this volume). In excavation and in profile, the upper and lower boundaries of this layer are abrupt (Buol *et al.* 1989). The fine-grained matrix is dry.

Layer II lies directly under the shelly Layer III and is defined on the basis of large amounts of charcoal and angular rock. Layer II has a high frequency of facies with brown matrix (CBr), few broken shells (SFB), and many large angular rocks (RMLA). Although facies included in Layer II were excavated from most areas in Operation A, this layer extends to the profiles in only two excavation units. The upper boundary of Layer II is abrupt, the lower boundary diffuse. The fine-grained matrix is dry.

Layer I is a large, ubiquitous layer in the lowest levels of the excavation. Layer I has a high frequency of facies with black and brown matrix (CBr and CBl), few broken shells (SFB), and few small angular rocks (RFSA). Near the upper diffuse boundary, Layer I is usually described as Brown, and in the lower portions as Black (often described as "greasy" when smeared with fingers). The percentage of shell is variable within the layer and seems to decrease from the upper to the lower portions of this layer. The fine-grained matrix of Layer I is moist. Although the excavations did not reach the lowest extreme of Layer I, augering indicates that glacial drift lies underneath Layer I.

Throughout the excavation of Layers III, II, and I, the excavators encountered facies that are described as Tan (Figure 10). Tan facies are strikingly similar to each other, in both appearance and content. They are usually very small (8–16 liters) and are described in the field as ashy, with burned bone and shell. They are particularly lacking in gravel-sized (larger than 3 mm) objects (especially rock and shell). These small facies are grouped within the lithostratigraphic units of Layer III, II, and I because they are distributed across the three layers and are small and discontinuous. In this manner they are similar to lenses, lamina, or beds within rock formations (Campbell 1967; Stein 1987, 1990).

B. *Layers Used for Detection of Postdepositional Weathering*

The shell midden at British Camp can also be divided using a second scheme, one created to detect postdepositional weathering. If the lowest layer observed in the field (Layer I) was produced by postdepositional inundation, then the presence of water should have affected the chemistry of Layer I in a manner different from the other layers. The water should have caused dissolving of the carbonate and even leaching of carbonates from the deposit (Stein, in press). To investigate the nature and impact of this weathering process the amount of carbonate in the fine-grained matrix was measured (see Ch. 7, this volume).

Strata that have been weathered are usually assigned pedostratigraphic units (NACOSN 1983). A pedostratigraphic unit is a body of rock that consists of one or

Figure 10 The west wall of unit 304/304 with many tan facies noted. Although many of these facies with tan-colored fine-grained sediment had large-sized shell and rock in them, most contained less than 20% gravel. Many of them also contained burned bone and shell as well as small pieces of charcoal. Photo by Ken Alford.

more pedologic horizons developed in one or more lithostratigraphic, allostratigraphic, or lithodemic units and that is overlain by one or more formally defined lithostratigraphic or allostratigraphic units. It is a buried, traceable, three-dimensional body of rock that consists of one or more differentiated pedologic horizons (NACOSN 1983:864). The weathered zone at the base of the midden was not assigned as a pedostratigraphic unit, because the process is not exactly pedological. Weathering is the process being investigated, yet not the weathering usually associated with soil genesis. Weathering, as part of soil genesis, usually occurs at the surface and is associated with the accumulation of organic matter in the presence of water. The process investigated at British Camp shell midden is not classic soil genesis, and does not fit the description of the pedostratigraphic unit. Therefore, we have not called the stratigraphic subdivision, which is based on the weathering of the lower portion of the shell midden, a geosol (the fundamental unit of the pedostratigraphic unit).

Postdepositional weathering (Stein, Ch. 7, this volume) is proposed as the process by which the British Camp shell midden obtained the dark-colored layer at its base with its light-colored layer above. Weathering alters the texture and composition of sediment and requires water derived from either the surface as rain, or the subsurface as groundwater to produce the alteration. Material susceptible to dissolution in water can be measured to detect weathering. The material chosen for susceptibility to dissolution at British Camp is carbonate. Its percentage in the fine-grained sediment (less than 3 mm) was measured using loss-on-ignition (Stein, Ch. 7, this volume) for samples from every facies. The results are used to group facies into two layers, one with low percentages of carbonate and another with high percentages. This lithostratigraphic division that resulted is based on laboratory analysis, instead of field descriptions.

The criterion used to place a facies into this kind of layer was the carbonate percentage obtained in analysis and the superpositional order of the facies. The layers created to detect weathering are presumably the result of a process that is either controlled by rainwater at the surface or controlled by the level of groundwater in the subsurface. Both these sources of water would weather facies in stratigraphical order from top down, or from bottom up. Thus, superpositional arrangements of the facies, plus the amount of carbonate lost are both considered.

The boundary of the layer defined by carbonate percentages was defined at the place on the Harris Matrix where the majority of facies had less than the mean value of carbonate. Carbonate values range between 70% and 3%, with the mean value of 35%. The value of 35% carbonate was selected as the cutoff for high versus low carbonate value. The location of this boundary is then compared to the boundaries of layers created from field descriptions (Layer I, II, III, and IV), or any other stratigraphic unit.

The layer with the high percentages of carbonate is called Layer H (for high), characterized by the grouping of facies with greater than 35% carbonate. The name Layer H is used to avoid confusion with Layer I, II, III, and IV that are based on field descriptions. The layer with the low percentages of carbonate is called Layer L (for low), characterized by the grouping of facies with less than 35% carbonate.

In Unit 310/300 (Figure 11) the carbonate percentages change from values between 40 and 70%, to values ranging from 34 to 25% in the middle of the Harris diagram. The boundary occurs about 70 cm below the present ground surface. In Unit 310/304 (Figure 12) the carbonate percentages change from values between 41 and 70%, to values ranging from 19 to 33% near 110 cm below the ground surface. More carbonate data are provided in Chapter 7, this volume.

IV. Ethnostratigraphic Units: The Ethnozone

The facies identified by lithologic descriptions during excavations at British Camp also contain artifactual material, which can be analyzed and used to stratigraphically subdivide the site. Although artifacts are recovered during excavation and after screening, they are not described in the field and are not usually used to define facies. Artifacts are analyzed later and described using attributes that are related to the research and to regional correlation. At British Camp, relative abundance of chipped stone tools and debitage was selected as the attribute on which one ethnostratigraphic division was based. This attribute results in an abundance ethnozone (Ch. 5, Table 1, this volume). Relative abundance of chipped stone was selected as the defining characteristic of the ethnozone because a decline in chipped stone abundance

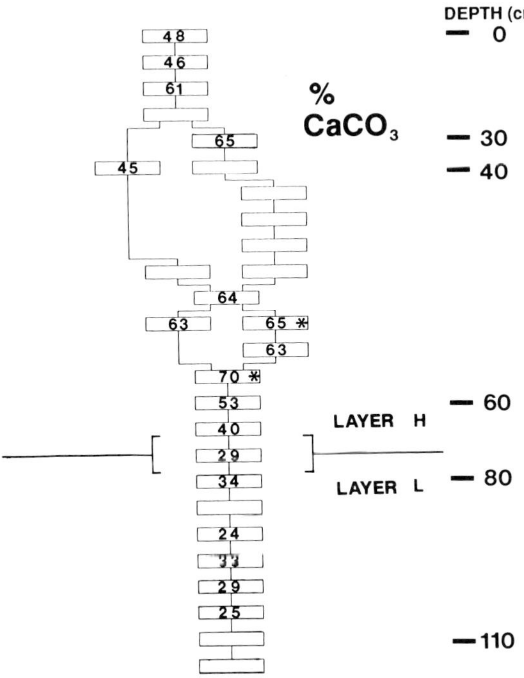

Figure 11 Carbonate percentages of facies of unit 310/300, displayed in their correct superpositional arrangement on a Harris Matrix. Asterisks (*) indicate facies with tan-colored fine-grained sediment. These facies often have higher carbonate percentages. Facies left blank were not analyzed using loss-on-ignition. Location of the boundary of Layer H and Layer L is indicated.

has been observed at other Northwest Coast sites (Carlson 1960:581) in deposits about 1500 years old (400 A.D.) (Mitchell 1990:340). These references make chipped stone abundance a useful attribute for regional correlation. The majority of facies at British Camp contain objects of chipped stone (bifaces, utilized flakes, and debitage). Thus the abundance of these objects can be compared throughout the British Camp shell midden and at other sites as well.

Interval ethnozones and assemblage ethnozones could have been used if the first or last appearance of one or more traditionally described tools (e.g., biface or harpoon; see list in Ch. 5, this volume) were used to define the ethnozone. These artifacts, however, are rare in most Northwest Coast sites and eliminate the possibility of intrasite comparisons (on the scale of facies or layers), impeding regional intersite comparisons. Rather than basing our regional correlations on rare items, the purposes of ethnostratigraphy are better served if all objects of chipped stone are used to define ethnozones. Using the first or last appearance of rare artifacts types, or the association of two or more rare types of artifacts will never work because the results will always suffer from sample size problems (see discussion in Burley 1980). Rather, we should be developing chronologies based on the common (ubiquitous) material. Thus, the abundance ethnozone was selected for ethnostratigraphy at British Camp.

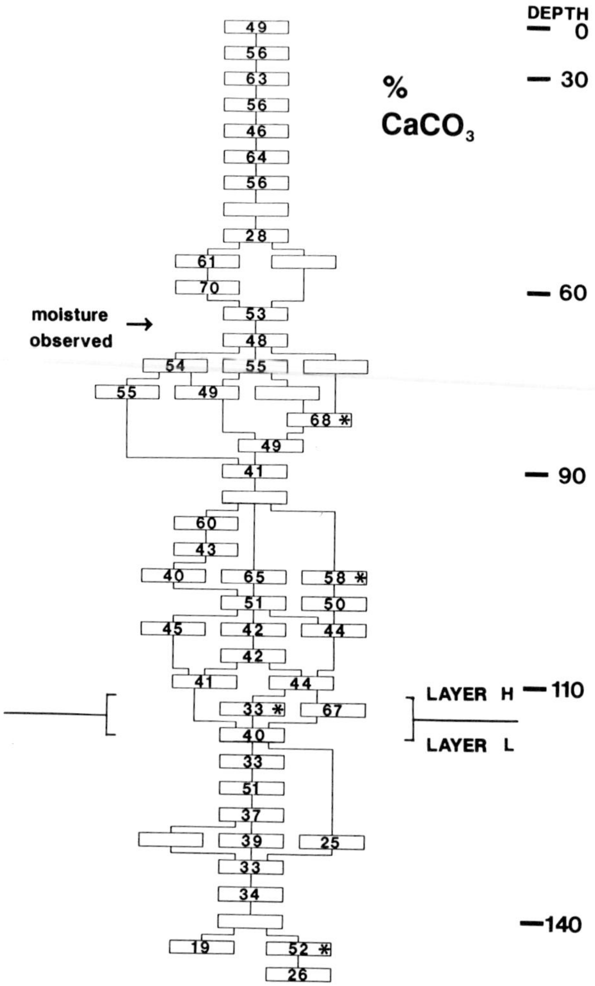

Figure 12 Carbonate percentages of facies of unit 310/304, displayed in their correct superpositional arrangement on a Harris Matrix (see Figure 11 for further explanation). Note the location where moisture was observed during excavation (the boundary of Layer I and II based on field observations). Yet the boundary of Layer H and Layer L is located much lower. The two boundaries were expected to coincide. However, in this unit the lack of coincidence is caused by uneven distributions of fine-grained shell, which affects the carbonate percentages in the central portion of the unit.

A. The Chipped Stone Data

The relative abundance of chipped stone tools and debitage (lithics) at British Camp is given in Table 1. The number of objects of chipped stone, which are greater than 13 mm (one-half inch), were counted for each facies in each excavation unit and standardized to an arbitrary volume of 100 liters. For example, a facies that fills 10

Table 1 Lithic Counts for Unit 308/300

Facies	*In situ* (count)	25-mm screen (count)	13-mm screen (count)	Total lithics (count)	Facies volume (liters)	Count per 100 liters
A			1	1	344	0.3
A2	1	4		5	640	0.8
C	1			1	224	0.4
C2	1			1	136	0.7
C3		1	4	5	120	4.2
E		1	4	5	80	6.3
H*		1	1	2	56	3.6
J	1	5	12	18	328	5.5
K	1	1		2	24	8.3
M	1			1	24	4.2
O	7	14	77	98	320	30.6
R*		1	6	7	64	10.9
S*			2	2	24	8.3
T	8	3	45	56	408	13.7
W	2	3	46	51	304	16.8
X	1	1	11	13	40	32.5
Y		1		1	24	4.2
Z	5	7	57	69	240	28.8
2A	1	1	21	23	88	26.1
2E	1	2	4	7	32	21.9
2F	3	2	9	14	64	21.9
2G			1	1	8	12.5
2I	1	1	7	9	16	56.3
2J	5	6	60	71	296	24.0
2K			10	10	24	41.7
2M	1		1	2	8	25.0
2P	2	2	34	38	56	67.9
2R		1		1	48	2.1
2S	4			4	200	2.0
2W	1			1	32	3.1
2X	2			2	112	1.8

Note: Asterisks indicate facies with tan-colored fine-grained matrix; blank spaces indicate facies with no lithic material.

buckets (each bucket is 8 liters), and has 24 objects of chipped stone (which are greater than 13 mm in size) is recorded as containing 24 lithics/80 liters, and standardized to 30 lithics/100 liters. Counts were used instead of weights because most lithics are roughly the same size range (they are sorted into equal sizes in nested screens), and counting was the most expedient way to express their frequency.

The number of lithics found in each facies of five excavation units (2 × 2-m areas) are plotted on Harris diagrams (Figures 13–17) to give an idea of the range of variability found at British Camp. The change in lithic frequency is noted in each figure, and defines the boundary between Ethnozone I (the lowest stratigraphically) and Ethnozone II (near the surface). The boundary is defined at the location where

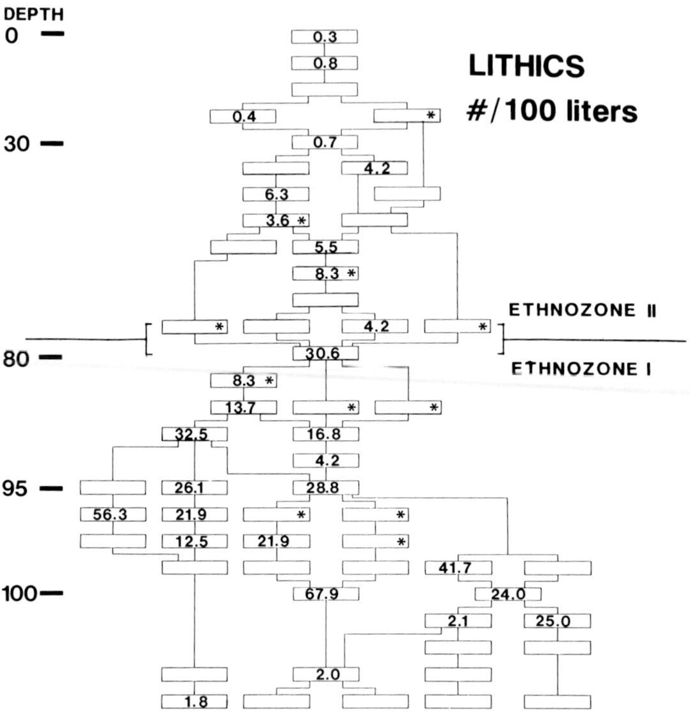

Figure 13 The number of lithics greater than 13 mm (per 100 liters) is noted for each facies of unit 308/300. The count is placed on the Harris Matrix in the rectangle representing the facies from which the count is obtained, and thus displayed in their correct superpositional arrangement. Blank facies contain no lithic material greater than 13 mm. Asterisks indicate facies with tan-colored fine-grained sediment. These facies often contain no artifactual material. Location of the boundary of the ethnostratigraphic units is indicated, along with the location of Ethnozone I and Ethnozone II.

the lithic counts change from below the range of 8–15 lithics/100 liters to above that value. This range was chosen after examining the counts of lithics and noting that most values were either less than 8 or greater than 15. This range was a natural break observed in the data and selected accordingly.

In Unit 308/300 (Figure 13) the number of lithics increases from a range of 0.3–6.3, to a range of 22–68, near 80 cm below the surface. In Unit 308/302 (Figure 14) the change also occurs at a depth near 80 cm below the surface, although there are three facies with high numbers of lithics (35.7, 11.4, and 21.6) in the top of the diagram. The facies with 11.4 lithics/100 liters is a tan facies, as indicated by the asterisk. This facies is unusual in that few other tan facies have high numbers of lithics. The other facies with counts of 35.7 and 21.6 also have anomalous values, but were included in Ethnozone II nonetheless. In Units 308/304 (Figure 15) and 310/304 (Figure 16) the change from low to high numbers of lithics occurs closer to a depth of 100 cm below the surface. In Unit 310/300 (Figure 17) the change occurs at a depth near 60 cm below the surface, but the numbers of lithics in Ethnozone I never reach the high numbers seen in the other units.

B. *Discussion of the Chipped Stone Data*

The changes in the abundance of objects of chipped stone can be explained in technological and depositional terms. The technological significance of the changes in lithic frequencies is discussed elsewhere (Kornbacher 1989a, Ch. 8, this volume; Madsen, Ch. 9, this volume). However, the location of the boundary between Ethnozone I and II in the various areas of the site can also be evaluated in terms of depositional events. The boundary is located at different depths in different areas. In Units 308/304 and 310/304 (Figures 15 and 16, respectively), the boundary is nearly 110 cm below the surface. This depth of the boundary is probably related to the younger age of facies in these units. The facies in the southeastern portion of Operation A (Units 310/304 and 308/304) were deposited later than facies from similar depths in other units. As discussed by Whittaker and Stein (Ch. 2, this volume) the

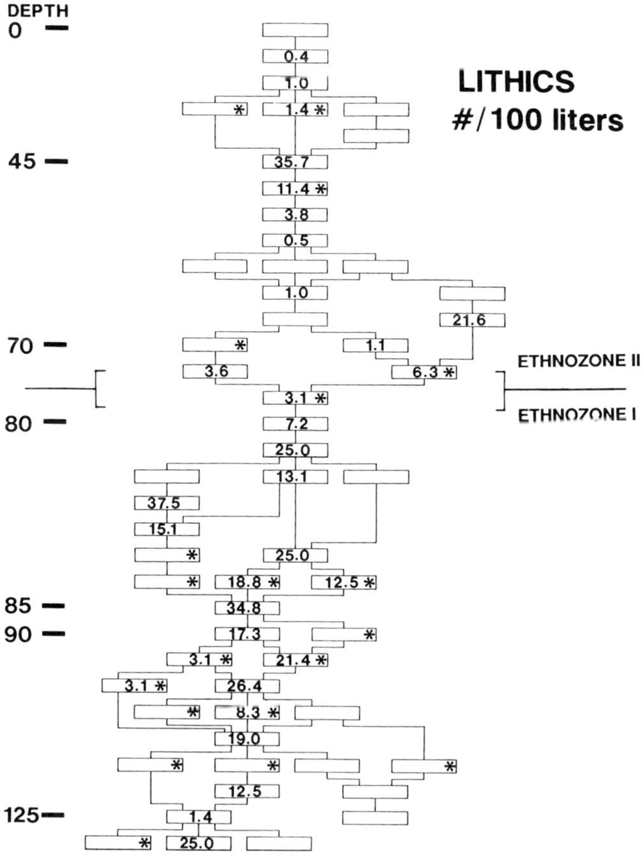

Figure 14 The number of lithics greater than 13 mm (per 100 liters) is noted for each facies of unit 308/302 (see Figure 13 for further explanation). Location of the boundary of the ethnostratigraphic units is indicated, along with the location of Ethnozone I and Ethnozone II.

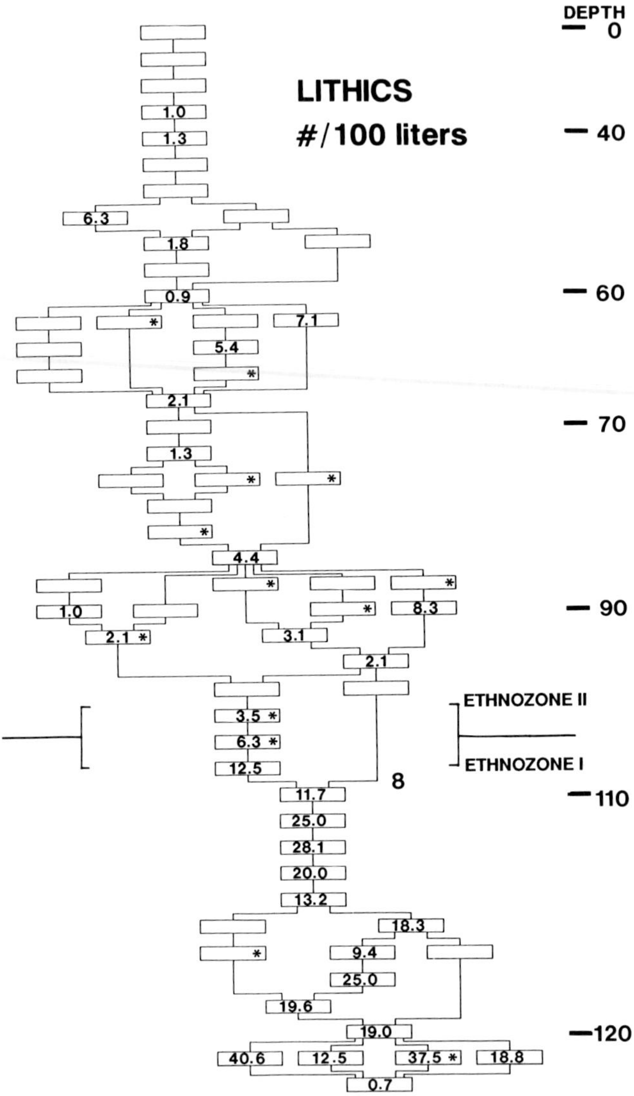

Figure 15 The number of lithics greater than 13 mm (per 100 liters) is noted for each facies of unit 308/304 (see Figure 13 for further explanation). Location of the boundary of the ethnostratigraphic units is indicated, along with the location of Ethnozone I and Ethnozone II.

radiocarbon ages of samples from facies in the southeastern unit are younger than the radiocarbon ages of samples at the same depth below surface from other units. Evidently, a low spot on the surface was later filled (during a prehistoric period) with additional shell midden that contained few lithics. These depositional events result in facies with few lithics found in deeper positions in these two units.

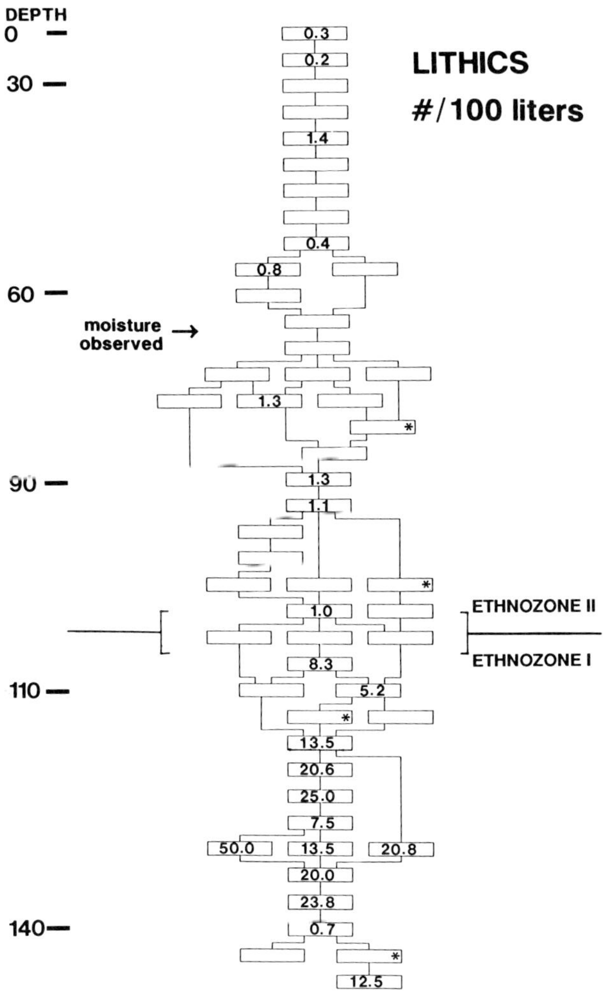

Figure 16 The number of lithics greater than 13 mm (per 100 liters) is noted for each facies of unit 310/304 (see Figure 13 for further explanation). Location of the boundary of the ethnostratigraphic units is indicated, along with the location of Ethnozone I and Ethnozone II. Note the location where moisture was observed during excavation (the boundary of Layer I and II based on field observations).

In Unit 310/300 fewer lithics were found in all facies. The low counts of lithics in the deeper portions (from 60 to 120 cm below the surface) of Unit 310/300 (Figure 17) may be the result of a collection bias. Unit 310/300 was excavated entirely in 1984 and 1985. It was the first unit excavated and was dug to expose the stratigraphy in this part of Operation A. All other units were excavated only 20 to 40 cm below the surface in 1984 and 1985, while Unit 310/300 was excavated to a depth of 120 cm below the surface. Even though there are fewer lithics in the lower facies of Unit 310/300 than in other units, there is still an observable increase in lithic counts

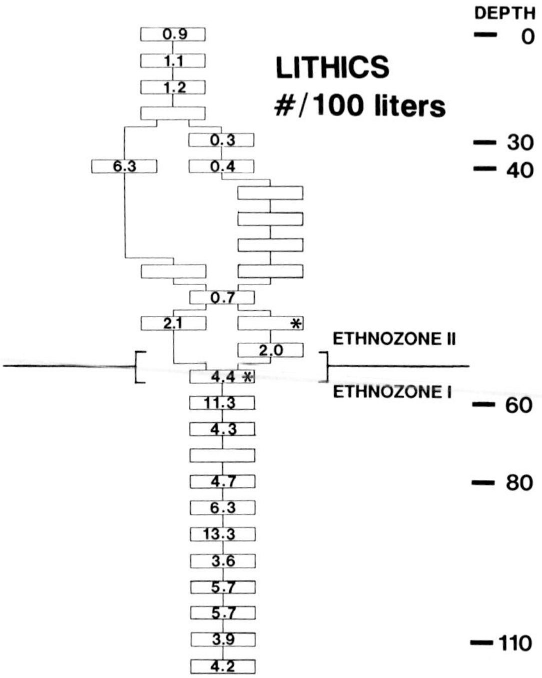

Figure 17 The number of lithics greater than 13 mm (per 100 liters) is noted for each facies of unit 310/300 (see Figure 13 for further explanation). Location of the boundary of the ethnostratigraphic units is indicated, along with the location of Ethnozone I and Ethnozone II. Although the counts of lithics increase in Ethnozone I from the counts of lithics obtained in facies in Ethnozone II (at the boundary), the number of lithics in Ethnozone I are lower than the counts of lithics in Ethnozone I from all other units.

with depth. This change is used to define the ethnostratigraphic boundary between Ethnozone I and II.

The change in the abundance of lithics at various depths of every excavation unit at British Camp may be explained in terms of activity areas over the whole site. In other words, lithic manufacture may have occurred only in certain areas of the site at different times in the past (Kornbacher, Ch. 8, this volume; Stein 1984). The deposition of lithics may have shifted as the location of the shoreline shifted. Prehistoric activities may have been tied to locations near to or far from the shoreline, or tied to locations of structures that in turn were tied to the shoreline. We know that the location of the shoreline shifted in the past (Ch. 2, this volume), prograding seaward over time. Most activities in Northwest Coast villages occur within structures or between the structure and the shoreline (Suttles 1951, 1990), and structures are located near the shore. If the shoreline progrades seaward, then the location of the structure should also.

Our excavation at British Camp investigated only one small part of the site. The vertical arrangements of facies in Operation A may be a sample of a transgressive/regressive sequence. The lower facies may contain remains of activities deposited while the shoreline is close to Operation A, the upper facies may contain remains of activities deposited after the shoreline prograded seaward. This interpretation suggests that the facies in Operation A are a vertical sample (through time) of material deposited from activities in front of structures, within structures, and behind structures. The shift in abundance of lithics seen in the area of Operation A may be tied to shifts in the location of activity areas across the site, relative to the shift in the location of the shoreline.

V. Discussion of Lithostratigraphic and Ethnostratigraphic Units

The boundary between the two ethnostratigraphic units (Ethnozone I and Ethnozone II), and the boundaries between the two kinds of lithostratigraphic units (Layer H and Layer L, and Layer I, II, III, and IV) do not correspond across the area excavated in Operation A. The lack of coincidence of some boundaries provides information for reconstructing site-formation processes.

For example, in Unit 310/300 the ethnostratigraphic boundary (between Ethnozone I and II) is located near a depth of 60 cm below the surface (Figure 17). The lithostratigraphic boundary (between Layer H and L) based on carbonate percentages is located closer to 80 cm below the surface (Figure 11). The lithostratigraphic boundary based on observed lithology (Layer I and II) is also 80 cm below the surface. The boundaries of the ethnostratigraphic and the two lithostratigraphic units are separated by thick facies (totalling a volume of 64 buckets, or 512 liters). In Unit 310/304 the ethnostratigraphic boundary is located at a depth of nearly 110 cm below the surface (Figure 16). The lithostratigraphic boundary based on carbonate percentages is located closer to 120 cm below the surface (Figure 12), and the lithostratigraphic boundary based on field observations is closer to 80 cm below the surface. The elevations of the surface of each of these units is nearly identical, making the difference in the depth below surface significant.

Given a pluralistic stratigraphic perspective (cubist perspective, Ch. 5, this volume), the position of the various stratigraphic unit boundaries and their relationships to each other are not expected to coincide. The differences in the placement of those various boundaries provides explanation of cultural and natural phenomena that would otherwise have been missed. That variation is lost if only one stratigraphic division is used for the whole site. For example, as discussed in Chapter 7, this volume, the carbonate percentages in Unit 310/304 are influenced by the presence of abundant sand-size shell in the fine-grained sediment in the facies between 60 and 110 cm below the surface. Yet the lithostratigraphic boundary based on color of the matrix observed during excavation is at a depth of 70 to 80 cm below the surface (noted in Figure 12). The increase in shell is a consequence of the depositional sequence, and the dip of the facies to the southeast (Figure 18) (see discussion of dipping layers in Whittaker and Stein, Ch. 2, this volume). These dipping strata have physical characteristics more closely related to facies in adjacent units that are higher than 40 cm below the surface. These shell-rich, deeply buried facies are found only in the southeastern excavation units. The large amount of shell elevates the carbonate

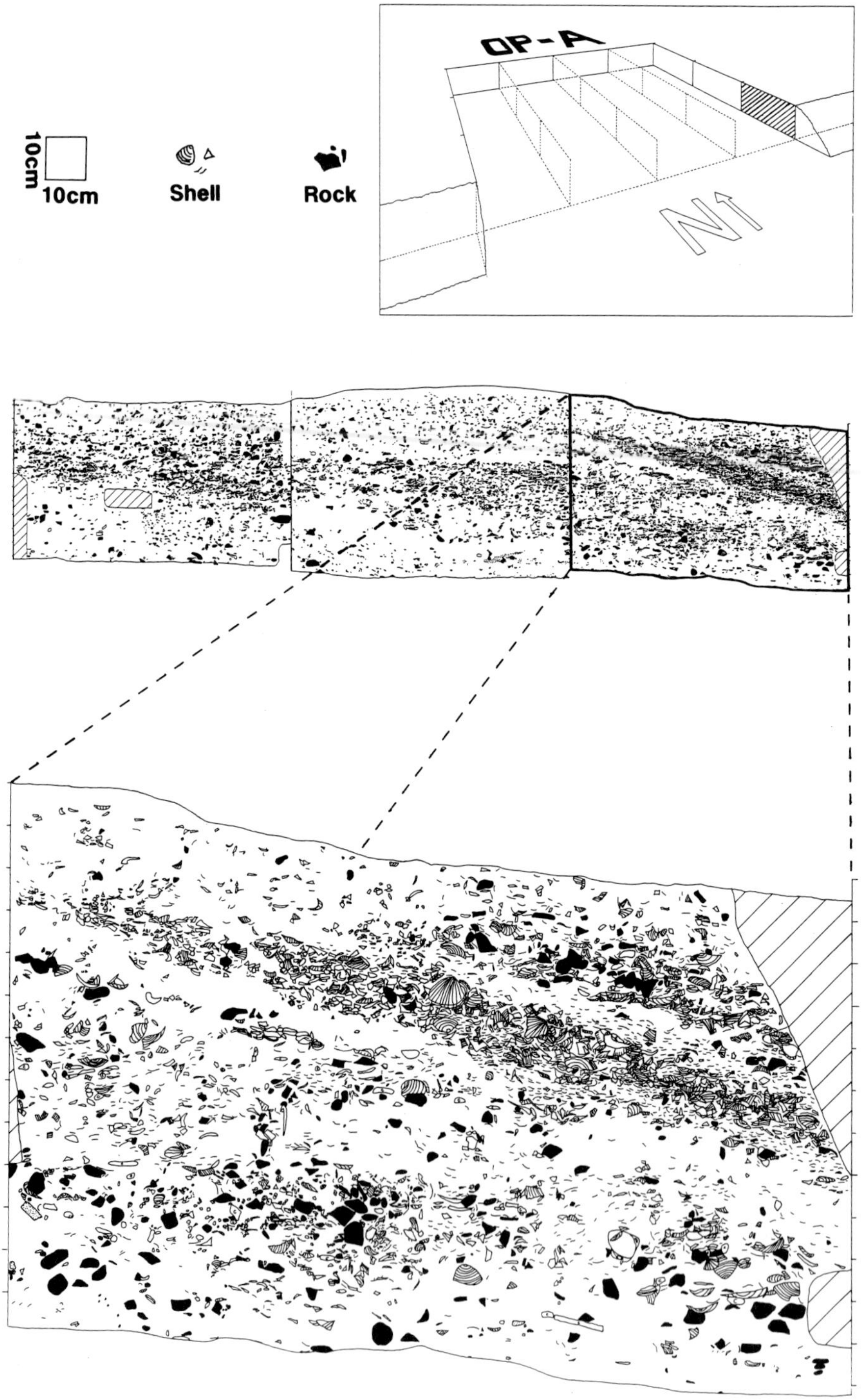

Figure 18 The complete stratigraphic section of Operations A's east profile (45SJ24) is displayed, with a close-up of unit 310/304. The location of the profile within Operation A is noted in the inset. On the close-up the cross-cutting relationship of recent strata is evident. The facies in the recent strata have high percentages of shell in the fine-grained matrix. Drafted by Timothy D. Hunt.

percentages and complicates the lithostratigraphic division proposed for analyzing weathering. These facies with the large amount of shell have been weathered, however (i.e., they have lost carbonate through leaching). They started with more carbonate than other facies, and thus have more to remove.

In most excavation units at British Camp, the location of the boundaries between lithostratigraphic *layers* and ethnostratigraphic *ethnozones* does not coincide. The difference in the location of those boundaries may be only a matter of several centimeters in one area of the excavation, or a matter of .5 m in another. However, that vertical difference in the stratigraphic boundaries has the potential to be compounded over larger horizontal distances in the site. Archaeologists do not usually subdivide their sites into different stratigraphic units (e.g., layers or ethnozones). Rather they use only one kind of stratigraphic unit to group all the data extracted from every excavation area over the entire site. All artifacts, bones, shell, and radiocarbon ages, from all areas of the site, are discussed in terms of lithostratigraphic units observed in profiles and during excavation. All data are expressed in tables and graphs according to the layers used to correlate physical descriptions of strata across the site. Separating various kinds of lithostratigraphic units from ethnostratigraphic units, mapping the placement of their boundaries separately using explicit criteria, and comparing the placement of those boundaries horizontally across the excavation area of the site, is essential to the genetic interpretation of the stratigraphic sequence.

VI. Chronostratigraphic and Geochronometric Units

Two kinds of chronologic units in stratigraphy are (1) geochronometric units, which are the direct division of geologic time usually expressed in years, and (2) chronostratigraphic units, which are strata formed during a given span of time, and based on either lithostratigraphic or ethnostratigraphic units. The British Camp midden is discussed in terms of each unit.

A. *Geochronometric Units of the British Camp Shell Midden*

Charcoal samples from the British Camp strata have been dated using radiocarbon analysis (Table 2). The dates reported by the labs are calibrated using the computer program available from the Quaternary Isotope Lab at the University of Washington (Stuiver and Becker 1986; Stuiver and Pearson 1986; Stuiver and Reimer 1986). All the dates are on wood charcoal.

The dates of samples from Unit 310/300 range from approximately 600 calibrated B.P. near the surface to about 1500 calibrated B.P. near the base. All dates conform to the expected superpositional sequence. The sample from facies 1N (970 calibrated B.P.) is located just above the ethnostratigraphic boundary between Ethnozone I and II (Figure 17, facies 1N has 2.0 lithics/100 liters), and provides an estimate of the age of that boundary in this excavation unit.

Samples taken from Unit 310/302 yield dates that are much older than expected given their depth below surface. An explanation for the anomalous dates is related to the outline of a large pit visible in the lower portion of the profile exposed in the

Table 2 Radiocarbon Ages of Charcoal Samples from British Camp

Unit	Facies	Layer[a]	Ethnozone[b]	Lab.	Age B.P.[c]	SD[c]	Calibrated age B.P.[d]	Calibrated range B.P.[d]	Wood ID
294/270	1F			WSU-3154	680	135	667	720–540	NA[e]
310/300	1B01	H	II	WSU-3517	535	80	542	670–470	NA
310/300	1B02	H	II	WSU-3518	670	70	665	720–540	NA
310/300	1C	H	II	WSU-3152	885	65	786	940–680	NA
310/300	1N	L	II/I	WSU-3151	1070	80	970	1062–929	NA
310/300	1R03	L	I	WSU-3519	1585	70	1512	1549–1398	NA
310/302	1D01	H	II	WSU-3520	1150	90	1061	1280–920	NA
310/302	1D02	H	II	WSU-3521	1250	70	1178	1277–1070	NA
310/302	1E	H	II	WSU-3522	1690	60	1579	1730–1500	NA
310/304	1B	H	II	WSU-3514	160	60	271/12	290–64	NA
310/304	fea2	H	II	WSU-3153	355	50	461	494–315	NA
310/304	1D	H	II	WSU-3515	370	70	470	508–314	NA
310/304	1D03	H	II	WSU-3516	450	50	510	528–489	NA
310/306	44 cm	H	NA	WSU-2917	modern				NA
310/306	104 cm	L	NA	WSU-2918	580	70	622/555	661–532	NA
310/306	135 cm	L	NA	WSU-2919	630	55	650/576	668–553	NA
306/300	1F	H	II	QL-4153	430	40	509	520–482	*Pinus*
306/300	1L	H	II	QL-4154	810	80	727	792–672	*Thuja*
306/300	1W	H	II	QL-4155	1000	40	942	971–796	*Pinus*
306/300	2E	L	I	QL-4156	830	70	735	930–670	*Pseudotsuga*
306/302	1M	H	II	QL-4157	900	40	814	929–727	*Pseudotsuga*

[a]Lithostratigraphic unit (Layer) based on carbonate percentage (Layer H = high %, Layer L = low %).
[b]Ethnostratigraphic unit (Ethnozone) based on number of chipped stone material (Ethnozone II = low number, Ethnozone I = high number).
[c]Age B.P. is the count reported by the laboratory, along with the standard deviation (SD).
[d]The calibrated age B.P. is the place along the calibration curve where the laboratory-reported age crosses, and the calibrated range of that age. Calibrations according to Stuiver and Reimer 1986.
[e]NA, Data not available.

Figure 19 Photograph of the east wall of unit 310/302 (defined by the outer vertical strings). In the lower portion of the profile note the orientation of the tan facies, outlining the base of an intrusive pit. The pit was excavated from a depth on the profile about 20 to 30 cm below the surface. All objects in the pit-fill (charcoal) have an age of death (or manufacture) that is older than the age of their deposition. The upper portion of this unit does not seem to be affected by the intrusion. Also note that the color of the fine-grained matrix is darker in the lower portion of the profile. Photo by Ken Alford.

eastern wall of Unit 310/302 (Figure 19). The pit seems to originate from a prehistoric surface somewhere in the center of the profile and is expressed most clearly by the angle of dip in the tan facies at the base of the profile. The pit, which extended westward over 75% of Unit 310/302, was not identified as a pit until 1987, and contains charcoal of greater antiquity than the charcoal of strata surrounding the pit. The age of death for the tree (the charcoal) does not coincide with the age of deposition. Therefore, the dates from Unit 310/302 were not used to build the chronology of the site.

Ages obtained from other excavation areas support the suggestion that the shell midden was deposited over the last 1500 years. The ages of samples from Unit 310/304 are from facies near the surface and range from 271/12 calibrated B.P., to 510 calibrated B.P. The samples from 310/306 are not from an excavation unit, but rather from a profile exposed when the wave-cut bank was cleared at the southern edge of Operation A. Samples collected at 310/306 are labeled with reference to their depth below surface. Note that the sample 135 cm below the surface has an age range of 668–553 calibrated B.P. As discussed in the chapter by Whittaker and Stein (Ch. 2, this volume), this age supports the suggestion that facies in the southeastern portion of Operation A were deposited more recently than facies from similar depths in other units.

In excavation Unit 306/300, the radiocarbon dates of charcoal are consistent with the ages from Unit 310/300. The sample from facies 2E in Unit 306/300 (735 cal. B.P.) is located just above the ethnostratigraphic boundary defined by changes in the frequency of objects of chipped stone. This age is younger than the age of the boundary in Unit 310/300. The ages from Unit 310/300 (970 cal. B.P.) and from Unit 306/300 (735 cal. B.P.) suggest that the boundary between Ethnozone I and Ethnozone II is between 670 and 1062 calibrated B.P. at this site.

In unit 306/300 three of the radiometric dates obtained from charcoal fall within an overlapping range. The dates are 727 calibrated B.P., 942 calibrated B.P., and 735 calibrated B.P., and the calibrated ranges are 792–672 B.P., 971–796 B.P., and 930–670 B.P. The charcoal from all of these facies have ages ranging from 971 to 670 B.P. One possible explanation for thc overlapping ages of samples from unit 306/300 is the problem associated with dating samples from very long-lived trees. On the Northwest Coast, archaeological sites contain wood charcoal from trees with long lifespans. The date from facies 2E in Unit 306/300 is derived from a charcoal sample, identified by M. Nelson and K. Kornbacher as Douglas Fir (*Pseudotsuga menziesii*), a tree that can live to be 1000 years old. The sample from facies 1L in Unit 306/300 is Western Red Cedar (*Thuja plicata*), also capable of living over 1000 years (Fowells 1965). Thus, a charcoal sample derived from the inner portion of the tree could yield a radiocarbon date much older than a sample from the outer, younger portion of the same tree.

Other possible explanations for the overlapping dates in Unit 306/300 are: faster or variable rates of sedimentation; rearrangement of midden material by prehistoric and historic inhabitants (Blukis Onat 1985; Claassen 1991a, 1991b); and charcoal moving either across the surface before burial (Erlandson and Rockwell 1987), or downward through the profile after deposition (Sanger 1981) . The sediments at British Camp are porous and permeable, and the surfaces at any given time in prehistory were probably undulating. Also, the statistical nature of radiocarbon dating may be responsible. Any combination of the above conditions are possible explanations for the overlap in radiocarbon ages of samples from Unit 306/300.

B. *Chronostratigraphic Units of the British Camp Shell Midden*

An ethno-chronostratigraphic unit represents the ethnostratigraphic units accumulated during a certain time period, and can be useful for correlating sites within a region. Ethno-chronostratigraphic units at British Camp are defined on the basis of the ethnostratigraphic unit, which in turn is based on the abundance of chipped stone materials in facies. In the Northwest, however, no other excavation has defined ethnostratigraphic units in this manner. Rather, sites are divided into stratigraphic layers, and then the frequency and presence/absence of stone and bone tools in those layers or sublayers are described (Borden 1950; Burley 1980, 1989; Carlson 1960, 1970, 1983; Matson 1976; Mitchell 1971, 1990, Murray 1982).

Most of the same kinds of stone and bone tools described for other sites are found at British Camp. In Tables 3 and 4, the frequencies and presence/absence of tools are presented for both ethnozones. Leaf-shaped, stemmed, unstemmed, and small tri-

Table 3 Frequency of Artifact Types by Ethnozone

	Ethnozone	
	I	II
	N	*N*
CHIPPED STONE		
Small biface	61	6
Large biface	9	1
Microblade	23	0
Preform	1	1
Retouched flake	63	6
Utilized flake	129	13
GROUND STONE		
Knife or fragment	2	5
Unshaped abrader	41	8
Shaped abrader	0	1
Adze/celt	2	2
Beads	9	0
PECKED STONE		
Net weight	0	1
Hammerstone	4	0
Wedge	1	0
Anvil	1	0
Pendant	0	1
BONE/ANTLER		
Wedge or fragment	10	6
Awl or fragment	5	12
Single-pointed bone object	11	23
Bone bipoint	1	6
Faceted bone point	2	7
Bilaterally barbed object	1	1
Unilaterally barbed object	0	7
Harpoon valve	5	3
Needle	0	1
Bone bowl/container	0	1
Bone bead	11	20
Bone tube	1	1
Pendant	0	1
Miniature bone club	0	1

Note: Names of tool types are from Mitchell (1971) and described in Kornbacher (1989a; Ch. 8, this volume).

angular forms of chipped stone bifaces, along with utilized and retouched flakes and abundant debitage are found in the facies of Ethnozone I. Microblades and microcores, slate knives, unshaped abraders, adze/celts blades, hammerstones, shell and stone disc beads, bone awls (Figure 20), wedges (Figure 21), and harpoon valves are also found. No unilaterally barbed fixed antler points are found in Ethnozone I. Small triangular chipped points, ground slate knives and points, shaped and unshaped

Table 4 Presence/Absence of Artifact Types by Ethnozone[a]

	Ethnozone I	Ethnozone II
	N	*N*
CHIPPED STONE		
Small biface	X	X
Large biface	X	X
Microblade	X	
Preform	X	X
Retouched flake	X	X
Utilized flake	X	X
GROUND STONE		
Knife or fragment	X	X
Unshaped abrader	X	X
Shaped abrader		X
Adze/celt	X	X
Beads	X	
PECKED STONE		
Net weight		X
Hammerstone	X	
Wedge	X	
Anvil	X	
Pendant		X
BONE/ANTLER		
Wedge or fragment	X	X
Awl or fragment	X	X
Single-pointed bone object	X	X
Bone bipoint	X	X
Faceted bone point	X	X
Bilaterally barbed object	X	X
Unilaterally barbed object		X
Harpoon valve	X	X
Needle		X
Bone bowl/container		X
Bone bead	X	X
Bone tube	X	X
Pendant		X
Miniature bone club		X

[a]"X" indicates presence.

abrasive stones, net weights, unilaterally barbed objects (Figure 22), bone bipoints, awls, wedges, nonutilitarian decorated bone and antler objects, harpoon valves, and abundant small single-pointed bone objects and bone bipoints (or gorges) (Figure 23) are found in the facies of Ethnozone II.

These data suggest that for purposes of regional correlation the stone and bone tools found in the facies of Ethnozone I and II at British Camp are similar to those

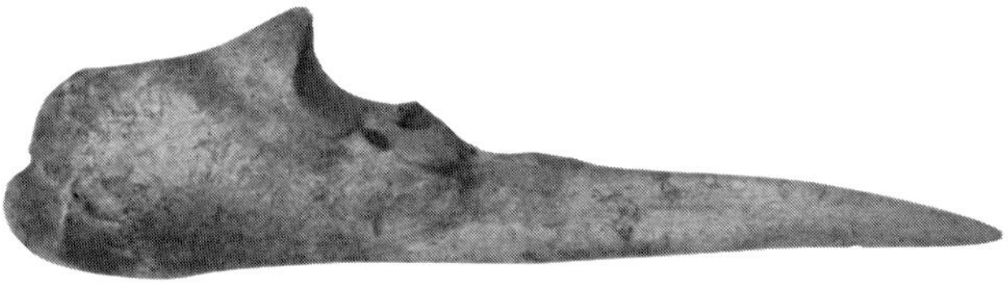

Figure 20 An ulna awl found at British Camp shell midden.

found in other Northwest sites. The difference in stratigraphic techniques used at British Camp from those used at other excavations prevents any more detailed comparison using the artifacts listed in Tables 3 and 4. However, comparisons can be made easily in the future by counting chipped stone material in the collections from other excavations and comparing the results to those from British Camp.

The decline in numbers of chipped stone material as was measured in Ethnozone II is a characteristic reported for some other sites of roughly the same age and is suggested here as a good attribute on which to make correlations. At the British Camp site one ethnochronozone (the oldest) is here named the Marpole Ethnochronozone and represents Ethnozone II, containing a high frequency of objects of chipped stone (Marpole Abundance Ethnochronozone). The younger ethnochronozone is here

Figure 21 Bone wedges found at British Camp shell midden.

Figure 22 Unilaterally barbed bone objects found at British Camp shell midden.

named the San Juan Ethnochronozone and represents Ethnozone I, containing a low frequency of objects of chipped stone (San Juan Abundance Ethnochronozone).

Neither of these ethnochronozones are here proposed formally, because only some of the requirements needed for such a formal proposal have been provided. If ethnostratigraphy was patterned after NACOSN, then the formal proposal of an ethnochronozone should include: (1) statement of intention to designate such a unit; (2) selection of name; (3) statement of kind and rank of unit; (4) statement of general concept of unit including historical background, synonymy, previous treatment, and reasons for proposed statement; (5) description of characterizing physical features; (6) designation and description of boundary type section and other units on which ethnochronozone is based; (7) correlation of age relation; and (8) publication in recognized scientific medium. Instead, the ethnochronozones are proposed informally as an example of the usefulness of ethnochronozones, and as a suggestion to be reviewed by other archaeologists in the region.

VII. Summary

Lithostratigraphic, ethnostratigraphic, geochronometric, and chronostratigraphic divisions are analytic units created to serve different investigative needs. The processes of delineating stratigraphic units as outlined here departs radically from stratigraphic

Figure 23 Miscellaneous bone objects found at British Camp shell midden.

analyses presented in other Northwest Coast site reports. The crucial difference lies in the clear explanation of the stratigraphic process and the use of a specific nomenclature. The analysis of the British Camp stratigraphy demonstrates that archaeologists can create stratigraphic units using slight revisions of the present stratigraphic code as long as the criteria used to distinguish them are defined explicitly.

The stratigraphy at British Camp indicates that a change in the abundance of chipped stone artifacts occurs within the site. The boundary between the ethnozones where that change takes place does not always coincide with the boundary between the lithofacies drawn on the basis of the change in physical attributes of the strata observed in the field. Rather, some of the physical observations were created by post-depositional weathering as a result of groundwater infiltration of the shell midden. The weathering by the inundation produced a lower layer with dark-colored, fine-grained sediment and with deficiencies in carbonate caused by leaching. The relative position of the boundaries of the ethnostratigraphic and various lithostratigraphic boundaries in each excavation unit shifts across the site. The shift is related to the behavior of the groundwater, the amount of shell in the fine-grained matrix, and the depositional history of the site (which influenced where the chipped stone material was deposited on an undulating surface). To interpret this site-formational history (which is expanded in Ch. 7, this volume), and to create units appropriate for regional artifact correlations (expanded in Ch. 8, this volume), these stratigraphic units had to be separated and interpreted independently.

The excavation and analytical strategy used at British Camp follows the procedure and nomenclature of stratigraphy and allows the strata to be examined from diverse perspectives using multiple arrangements. The procedures separate clearly the definition of strata based on lithology, chemistry, artifact content, and age. As demonstrated, the boundaries between different kinds of stratigraphic units do not always coincide. At British Camp the effects of weathering caused by groundwater cross-cut the facies, shifting the relative positions of lithostratigraphic and ethnostratigraphic boundaries in different areas of the site. The explicit procedures used at British Camp allow archaeologists (both ones involved in this project and others who later may want to use the data from this project) to separate the stratification caused by cultural events from that caused by depositional and postdepositional processes. The procedures provide a clear methodology that results in better interpretations of site formation and clearer regional correlations.

Acknowledgments

Many people helped with this chapter. Kimberly Kornbacher and Jason Tyler summarized excavator's descriptions and maps to determine the superpositional relationships of the facies, and to define the fourfold lithostratigraphic scheme. Pam Ford, Meg Nelson, Pat McCutcheon, Fran Whittaker, Kris Wilhelmsen, and Kim Kornbacher were teaching assistants supervising the excavation of facies and field descriptions. Kris Wilhelmsen, Jana McAnally, John Linse, and Tim Hunt drafted profiles. Kim Kornbacher, Jason Tyler, Mary Parr, Jo Linse, and Tim Hunt drafted the Harris Matrix in their various versions. Steve Cole counted the lithics and entered the counts into the computer. Angela Linse, Mark Madsen, Lisa Nagaoka, Kris Wilhelmsen, and Steve Cole processed and entered the screening record data. A special thanks goes to Timothy D. Hunt whose artistic eye made this chapter easier to write. This manuscript was greatly improved by the comments of William Farrand, Donald Mitchell, and especially Ann Ramenofsky.

References

Ambrosiani, B.

1977 Comments on units of archaeological stratification. *Norwegian Archaeological Review* **10**, 95–97.

Barker, P.

1977 *Techniques of archaeological excavation.* London: Batsford.

Blukis Onat, A. R.

1985 The multifunctional use of shellfish remains: from garbage to community engineering. *Northwest Anthropological Research Notes* **19**, 201–207.

Borden, C. E.

1950 Notes on the prehistory of the southern North-West Coast. *British Columbia Historical Quarterly* **XIV**, 241–246.

Buol, S. W., F. D. Hole, and R. J. McCracken.

1989 *Soil genesis and classification, 3rd ed.* Ames, Iowa: Iowa State University Press.

Burley, D. V.

1980 Marpole: Anthropological reconstruction of a prehistoric Northwest Coast culture type. Simon Fraser University, Department of Archaeology, Publication No. 8, Burnaby, British Columbia.

1989 Senewelets: Culture history of the Nanaimo Coast Salish and the False Narrows midden. Royal British Columbia Museum, Memoir No. 2, Victoria.

Campbell, C. V.

1967 Lamina, laminaset, bed and bedset. *Sedimentology* **8**, 7–26.

Carlson, R. L.

1960 Chronology and culture change in the San Juan Islands, Washington. *American Antiquity* **25**, 562–586.

1970 Excavations at Helen Point on Mayne Island. *B.C. Studies* No. 6–7, 113–125.

1983 Prehistory of the Northwest Coast. In *Indian art traditions of the Northwest Coast*, edited by R. L. Carlson. Archaeology Press, Simon Fraser University, Burnaby, British Columbia. Pp.13–32.

Claassen, C.

1991a Normative thinking and shell-bearing sites. In *Archaeological method and theory, vol. 3*, edited by M. B. Schiffer. Tucson, Arizona: University of Arizona Press. Pp.249–298.

1991b Gender, shellfishing, and the Shell Mound Archaic. In *Engendering archaeology: Women and prehistory*, edited by M. W. Conkey and J. M. Gero. Cambridge, Massachusetts: Basil Blackwell. Pp.276–300.

Dunnell, R. C., and J. K. Stein

1989 Theoretical issues in the interpretation of microartifacts. *Geoarchaeology* **4**, 31–42.

Erlandson, J. M., and T. K. Rockwell

1987 Radiocarbon reversals and stratigraphic discontinuities: the effects of natural formation processes on coastal California archaeological sites. In *Natural formation processes and the archaeological record*, edited by D. T. Nash and M. D. Petraglia. B.A.R. International Series No. 352, Oxford, England. Pp.51–73.

Fowells, H. A.

1965 Sylvics of forest trees of the United States. U.S. Department of Agriculture, Handbook No. 271, Washington D.C.

Gasche, H., and O. Tunca

1983 Guide to archaeostratigraphic classification and terminology: definitions and principles. *Journal of Field Archaeology* **10**, 325–335.

Ham, L. D.

1982 Seasonality, shell midden layers, and coast Salish subsistence activities at the Crescent Beach site, DgRr 1. Unpublished Ph.D. dissertation, Department of Anthropology, University of British Columbia, Vancouver.

Harris, E. C.

1975 The stratigraphic sequence: A question of time. *World Archaeology* **7**, 109–121.

1977 Units of archaeological stratification. *Norwegian Archaeological Review* **10**, 84–94.

1979 *Principles of archaeological stratigraphy*. London: Academic Press.

1989 *Principles of archaeological stratigraphy, 2nd ed.* London: Academic Press.

Kornbacher, K. D.

1989a Shell midden lithic technology: An investigation of change at British Camp (45SJ24), San Juan Island. Unpublished Master's thesis, Department of Anthropology, University of British Columbia, Vancouver.

1989b A methodological stride in shell midden archaeology. Paper presented at the 54th Annual Meetings of the Society for American Archaeology, Atlanta, Georgia.

Lynch, F.

1977 Comments on units of archaeological stratification. *Norwegian Archaeological Review* **10**, 97–98.

Matson, R. G.

1976 The Glenrose Cannery site. Archaeological Survey of Canada, Mercury Series No. 52, National Museum of Man, Ottawa.

Mitchell, D. H.

1971 Archaeology of the Gulf of Georgia area, a natural region and its culture type. *Syesis* **4**, suppl. 1, 1–228.

1990 Prehistory of the coasts of southern British Columbia and northern Washington. In *Handbook of North American Indians, Northwest Coast: volume 7*, edited by W. Suttles. Washington, D.C.: Smithsonian Institution. Pp.340–358.

Murray, R.

1982 Analysis of artifacts from four Duke Point area sites near Nanaimo, B.C.: An example of cultural continuity in the southern Gulf of Georgia. Archaeological Survey of Canada, Mercury Series No. 113, National Museum of Man, Ottawa.

NACOSN (North American Commission on Stratigraphic Nomenclature)

1983 North American stratigraphic code. *American Association of Petroleum Geologists Bulletin* **67**, 841–875.

Paice, P.

1991 Extensions to the Harris Matrix system to illustrate stratigraphic discussion of an archaeological site. *Journal of Field Archaeology* **18**, 17–28.

Rance, A. B.

1977 Comments of units of archaeological stratification. *Norwegian Archaeological Review* **10**, 99–100.

Reading, H. G.
1978 Facies. In *Sedimentary Environments and facies*, edited by H. G. Reading. New York: Elsevier. Pp.4–14.

Sanger, D.
1981 Unscrambling messages in the midden. *Archaeology of Eastern North America* **9**, 37–42.

Schofield, J.
1977 Comments on units of archaeological stratification. *Norwegian Archaeological Review* **10**, 101–102.

Stein, J. K.
1984 Interpreting the stratigraphy of Northwest shell middens. *Tebiwa* **21**, 26–34.
1987 Deposits for archaeologist. In *Advances in archaeological method and theory, vol. 11*, edited by M. B. Schiffer. Orlando, Florida: Academic Press. Pp.337–393.
1990 Archaeological stratigraphy. In *Archaeological geology of North America*, edited by N. Lasca and J. Donahue. Geological Society of America, Centennial Special Volume 4, Boulder, Colorado. Pp.513–523.
1992 Formation processes of coastal sites: View from a Northwest Coast shell midden. In *Proceedings of the Circum-Pacific Prehistory Conference: IIIC. Maritime Societies in Western North America*, edited by A. R. Blukis Onat. Washington State University Press, in press.

Stein, J. K., and G. Rapp, Jr.
1985 Archaeological sediments: a largely untapped reservoir of information. In *Contributions to Aegean archaeology*, edited by N. C. Wilkie and W. D. E. Coulson. Center for Ancient Studies, University of Minnesota, Publications in Ancient Studies 1, Minneapolis. Pp.143–159.

Stucki, B.
1981 Culture and Context: Shell midden research on the Northwest coast. Unpublished manuscript, Department of Anthropology, Washington State University, Pullman.
1983 Geoarchaeology at the Hoko River rockshelter: The anatomy of a shell midden. Paper presented at the 48th Annual Meetings of the Society for American Archaeology, Pittsburgh, Pennsylvania.

Stuiver, M., and B. Becker
1986 High-precision decadal calibration of the radiocarbon time scale, A.D. 1950–2500 B.C. *Radiocarbon* **28**, 863–910.

Stuiver, M., and G. W. Pearson
1986 High-precision calibration of radiocarbon time scale, A.D. 1950–500 B.C. *Radiocarbon* **28**, 805–838.

Stuiver, M., and P. J. Reimer
1986 A computer program for radiocarbon age calibration. *Radiocarbon* **28**, 1022–1030.

Suttles, W.
1951 The economic life of the coast Salish of Haro and Rosario Straits. Ph.D. dissertation, Department of Anthropology, University of Washington. Published in 1974 by Garland Press, New York.
1990 Central coast Salish. In *Handbook of North American Indians, volume 7, Northwest Coast*, edited by W. Suttles. Washington, D.C.: Smithsonian Institution. Pp.453–475.

Weller, J. M.
1960 Stratigraphic principles and practice. New York: Harper and Bros.

Wigen, R. J., and B. R. Stucki
1988 Taphonomy and stratigraphy in the interpretation of economic patterns at Hoko River rockshelter. In *Research in economic anthropology, supplement 3, prehistoric economies of the Pacific Northwest Coast*, edited by B. Isaac. Greenwich, Connecticut: JAI Press. Pp.87–146.

7

Sediment Analysis of the British Camp Shell Midden

Julie K. Stein

I. Introduction

The most striking stratigraphic feature at the British Camp shell midden is a large-scale division of light-colored matrix in the upper portion of the midden, and dark-colored matrix in the lower portion. This stratigraphy has been observed at other Northwest Coast middens, but never analyzed in detail. The difference in strata can be explained in two ways. Either different cultural activities occurred during the deposition of the lower and upper strata, or postdepositional events altered a homogeneous lithology resulting in the two-toned stratigraphy. If cultural activities are responsible for the stratification, then the sources, transport agents, or depositional environments of all the sediment in the site shifted over time. If postdepositional phenomena produced the stratification, then depositional events (human activities) did not shift and one original stratum appears as two lithologically distinct strata.

These two explanations can be tested by examining the artifacts and sediments in the shell midden. If the stratification is caused by depositional (cultural) events, then the artifacts, fauna, and flora within the two strata should change. If the stratification is caused by postdepositional events, then artifacts should remain the same and evidence of weathering should be found in one of the two strata observed.

Postdepositional alterations and evidence of weathering are the focus of this chapter, with artifactual, faunal, and floral evidence discussed in respective chapters of this book. Many kinds of postdepositional alterations can occur at archaeological sites, those producing horizonation (weathering) and those producing homogenization (e.g., faunalturbation, floralturbation, argilliturbation, etc.) (Hole 1961; Wood and Johnson 1978). Although some tree roots and burrows of the European hare (introduced in the historic period) are found in the excavation, extensive evidence of homogenization is not observed in the British Camp midden. The boundaries of

Deciphering a Shell Midden

facies are distinct, no earthworms are found because of saline conditions (Stein 1983), and the orientation of shells are parallel with bedding planes (Gorski, in press). Horizonation or weathering seems to be the most likely postdepositional alteration responsible for the stratification at British Camp.

Weathering alters the texture and composition of sediment (Birkeland 1984; Buol *et al.* 1989; Holliday 1990), and requires water derived from either the surface as rain, or the subsurface as groundwater to produce the alterations. The composition of the sediment in the midden that is most susceptible to weathering is carbonate (the main constituent of shell). If the midden has been weathered by surface water percolating through the deposit, then carbonate percentage should be low at the surface and higher in the deeper portion of the midden. If the midden has been weathered by groundwater inundation, then carbonate percentages should be low in the deeper facies.

Although rain falls on the surface of the British Camp site throughout the year, within the midden the surface rain seems not to contribute to hydration of the midden. Surface water is either held within the surface root mass, or permeated rapidly through the porous midden. The upper portion of the midden is dry. Only the sediments in the lower portion of the midden are moist, pointing to the groundwater as the source of water, rather than the rain. If groundwater saturation of the lower portion of the midden produced the two-layered stratification, then the sediments of the lower portion should be weathered.

In this chapter the sediments of the British Camp midden are examined for evidence of weathering. The results indicate that weathering has occurred from groundwater saturation, because carbonate percentages are low at the base of the midden and high near the surface. The groundwater flows through the midden along the hydraulic gradient, dissolves the carbonate, and actually leaches it from the midden. The groundwater is responsible for the two-layered stratification observed at the site, without which the British Camp shell midden would have a homogeneous lithology. Artifactual changes are also noticed in the shell midden (Stein *et al.*, Ch. 6, this volume; Kornbacher, Ch. 8, this volume), however, the artifactual change does not always coincide with the location of the stratigraphic change. These results suggest that this same two-toned stratification observed at other shell middens may also be the product of postdepositional alteration, and not culturally significant.

II. Sediment Analysis in Shell-Bearing Sites

Clastic sediments are any particulate matter that was transported by some process from one location to another. These clastic sediments include large- and small-sized objects, of any compositional type. Shell-bearing sites contain shell as one compositional type of both large and small clasts (Claassen 1991; Waselkov 1987; Widmer 1989). The shell at British Camp, as a sediment, was transported (Ford, Ch. 13, this volume) along with a variety of other types of sedimentary clasts. These clastic sediments have depositional histories that include their sources, transport agents, depositional environments, and postdepositional alterations (Stein 1985, 1987; Stein and

Rapp 1985). The attributes of the clastic sediment allow interpretations to be made about the depositional history of the shell-bearing site.

Clastic sediments can be any size from boulders to clay particles. At British Camp the analytical techniques used vary with the size of the particle. Large sediments (over 6 mm) are separated and weighed in the field, then analyzed by different specialists. Small sediments (under 6 mm) are dried and returned to the Archaeological Sediment Laboratory for separate analyses. In this chapter the analysis of the finest grained sediment (less than 1.5 mm) is reported.

In archaeology, the fine-grained sediment is called the matrix and is defined as the "physical medium that surrounds, holds, and supports archaeological material" (Sharer and Ashmore 1987). The distinction between matrix and artifacts in archaeology is based on scale; matrix is all the material that is too small to see without magnification. Artifacts, in this traditional distinction, have to be large enough to be held by the matrix. In sedimentology the term matrix is meaningless because clasts are described by their grain size using the terms gravel, sand, silt, and clay (Folk 1980), and not differentiated into artifact and nonartifact distinctions. Archaeologists know that matrix not only surrounds and supports artifacts, but also contains small artifacts within it (Dunnell and Stein 1989; Fladmark 1982; Rosen 1986; Stein and Teltser 1989). This new research on small-sized artifacts suggests that the traditional distinction (one based on practical and not theoretical considerations) between matrix and artifact is misleading. Therefore, the use of the sedimentological nomenclature (gravel, sand, silt, and clay) is preferable to the traditional archaeological terminology (matrix) and will be used in this research.

III. Process of Groundwater Saturation

The saturation of the lower portion of a shell midden by groundwater is not the most obvious process of weathering expected in most temperate regions of the world. A much more common process involves water falling as rain onto the surface, hydrating the upper portion of the solum. This process is part of the basic weathering cycle characterized by soil genesis (Buol *et al.* 1989) and results in soil horizonation. Shell middens, however, are often extremely porous and permeable, allowing surface water to travel rapidly through the strata. The organic matter and clay of the upper portion of the shell midden (within the zone of aeration) never hydrate. The porosity and permeability of the archaeological sediments negate any affects resulting from the common process of water falling as rain on the surface.

The less common process of inundating the shell-bearing site from below by groundwater is often the means most capable of producing extreme chemical weathering in a shell midden. Shell-bearing sites are often located near shorelines, proximate to the water table. In the portions of the shell midden that intersect the water table (intersect the zone of saturation) the organic matter and clay hydrate. The hydration causes the color of the organic matter to darken, and the consistency of the organically draped clays to feel greasy. The hydration provides the soil solution in which weathering occurs, and the vehicle through which dissolved ions are leached.

The actual process of alteration caused by the hydration involves chemical weathering. Organic acids released during the decomposition of organic matter (Stein, 1992) dissolve the carbonates of the shell in the midden. Although all carbonate grains are affected simultaneously, the grains with the greatest surface area (those smallest in size) disappear first. The dissolved carbonates are flushed from the shell midden as the water table flows along the hydraulic gradient and eventually discharges into the bay. Although the carbonate within the fine-grained clasts disappears first, eventually the larger grains also dissolve completely. The archaeological record changes from an alkaline shell-bearing site, to an acidic non-shell-bearing site.

The dynamics of the groundwater result in more than just two zones (one weathered, and one not weathered) being observed in the shell-bearing site. There is another (third) zone that is subtly expressed. At most locations where shell middens are found, water tables fluctuate with seasonal and climatic changes. The location of the boundary between the zone of aeration and zone of saturation shifts. The area of the shell midden that alternates between being aerated and saturated with groundwater will have characteristics different from either the shell midden above, which is always dry, or the shell midden below, which is always wet. This middle zone of the midden is also within the capillary fringe of the water table. The capillary fringe is a zone above the water table in which water is lifted by surface tension into openings. In a shell midden, this fringe is influenced by the sizes of the pores in the various small lenses located near the top of the water table. Fluctuating water table and the capillary fringe both produce a zone within the midden that is transitional between the aerated and saturated portions. These dynamics result in three zones in a shell-bearing site that has been inundated by groundwater.

The weathering process observed at British Camp occurs only if a combination of attributes are found in the shell midden (Stein in press):

1. The shell midden must intersect the groundwater table, and be porous enough to hold sufficient water and permeable enough to transmit the water through it. The groundwater must only affect the lower portion of the midden. The water table cannot rise to the top of the midden, inundating the entire midden. If this occurs, the carbonate will be leached and the entire shell midden will look the same; no stratification will result.
2. The shell-bearing site must contain organic matter in a sufficiently large quantity to dissolve all the carbonate in the site. If no organic matter is present, there will be only the naturally occurring acid in the groundwater (which is dependent on local geological factors). The water will flow around the shells and not affect the surface.
3. The groundwater cannot be supersaturated with calcium (as is most seawater). No carbonate from the shell will be dissolved if the groundwater cannot transport any more calcium. In this case the organic matter and clay will hydrate, the organic matter will decompose, and the clay minerals will weather. The shell surfaces will not be etched, however, because carbonate will not be dissolved in the solution.
4. In addition, silt and clay must be present in the sediment. Clay minerals bond with organic matter, absorb the soil solution, and provide essential cations for

various chemical reactions. The silt and clay hold the organic matter and water within the site, preventing removal of organic matter from the shell midden by the groundwater. In sites with large amounts of gravel and sand, and little clay and silt, very little organic matter is retained.

5. Lastly, the groundwater inundation must occur over a sufficiently long period of time to be observed and measured. In some shell-bearing sites the abundance of fine-grained shell is so great that to measure an appreciable change in the percentage of carbonate in the sand, silt, and clay fraction, centuries (if not millennia) of leaching has to occur. The inundation and leaching is a continuous process. Depending on the age of the site, the exterior of shells might display slight signs of etching, and the fine-grained sediment still high in carbonate, or both the carbonate in the fine-grained sediment and the large shells, will be completely leached.

IV. The Site and Methods

The British Camp sediments discussed here were collected from one location (Operation A) of the site excavated by the University of Washington during the summers from 1983 to 1989. As each unit is excavated, the deposits are divided into facies according to their lithology (Stein *et al.*, Ch. 6, this volume). From each facies one sample (called a Bio Sample) is taken for flotation and sediment analysis (see Nelson, Ch. 11, this volume; Ford, Ch. 13, this volume). This sample, although not reflecting the variety within a facies, is taken as representative of the entire facies. If facies are large, two or even three Bio Samples are taken.

The sample used in the sedimentological analysis contains only the small-sized clasts that pass through a 1.5-mm (1/16-inch) screen, roughly correlating with sands, silts, and clays (Folk 1980). These sediments are split into equal portions, half being returned to the Bio Sample with the larger sized grains for further processing in the float tank, the other half being retained for sediment analysis. The portion retained for sediment analysis is air dried thoroughly and stored in Zip Lock plastic bags.

The techniques used in this analysis are selected to measure texture and chemistry. Particle-size analysis follows the sieve and pipette technique (Folk 1980). Pretreatment of samples includes removal of soluble salts and organic matter using hypochlorite oxidation (Jackson 1969:37). Carbonate is not removed during pretreatment because a significant portion of the mineral clasts are limestone, derived from local bedrock and glacial drift (Brandon *et al.*. 1988). Samples from a section of glacial drift act as a control sample and are analyzed texturally before and after pretreatment.

Chemical analysis includes determining the percentage of organic matter and carbonate using the loss-on-ignition (LOI) method (Stein 1984a; Holliday and Stein 1989). This quick and inexpensive method is recommended for carbonate analysis in samples with high carbonate content and with low clay content (Dean 1974). The accuracy is less than that in those methods that measure evolved carbon dioxide gas, and the results do not differentiate between the various kinds of carbonates. LOI also does not differentiate the carbonate derived from shell from that derived from limestone. To make this differentiation, the sand fraction has to be examined using microscopy, and the grains counted following the procedure described in Stein and Teltser (1989). The organic matter percentages resulting from loss-on-ignition are also not as accurate as the results using a total-carbon analyzer or the Walkley–Black technique, which use oxidizing agents to measure organic carbon (Holliday and Stein 1989:349). However, the sediments from the British Camp site have high percentages of organic matter and carbonate warranting the use of the loss-on-ignition technique.

Other related analyses are pH, measured in a sediment:water ratio of 1:1 using a Beckman pH meter with digital display and autocalibration, and following the recommended procedure in Jackson (1969). For our samples, pH is measured repeatedly until three readings agree within a value of 0.1 (see Linse, Ch. 14, this volume). Total phosphorus is measured using perchloric acid digestion according to the procedure described in Sjoberg (1976) and Jackson (1969).

Graphic display of sediment data is difficult, because unlike standard soil samples that are usually taken in regular-spaced vertical sequences from a profile, these sediment samples are taken from facies. Facies are labeled in alphabetical order, yet many facies are not superimposed on top of each other. They are adjacent to each other. Graphing the data in a vertical sequence and arranging them in alphabetical order, gives the impression that the facies are stratigraphically superimposed. The Harris Matrix of unit 308/300 (Figure 1) demonstrates the complex organization of facies. Because the stratigraphic arrangement of the samples is not in a systematic vertical arrangement, data are displayed on Harris diagrams, providing the most stratigraphically accurate graphic technique.

V. The Data

A. *Grain Size*

In Appendix 1 the grain size statistics for samples of the shell midden from excavation units are given. Although individual facies within the midden have slightly different amounts of sand, silt, and clay, overall the differences are slight. The overall mean grain size for all the samples from the shell midden is 4.11 ϕ.

Also in Appendix 1 are the grain size results for the non-pretreated samples from the control profile. These samples were taken at 10-cm intervals starting from the surface to a depth of 2 m (Figure 2). The collecting site for the control profile is an exposure of glacial drift eroded by storm waves. The location is on Garrison Bay west of the parade grounds and near 45SJ25 (Carlson 1960), but has no shell midden on the surface. Individual samples within the glacial drift have similar amounts of sand, silt, and clay with fluctuations as predicted by soil genesis. The mean grain size of all the samples from the control profile is 4.16 ϕ.

B. *Organic Matter*

All samples from the surface of the excavation units have the highest organic matter percentages, ranging from 25 to 16% (Appendix 2). These values are probably related to modern soil-forming processes and the grass vegetation maintained by the

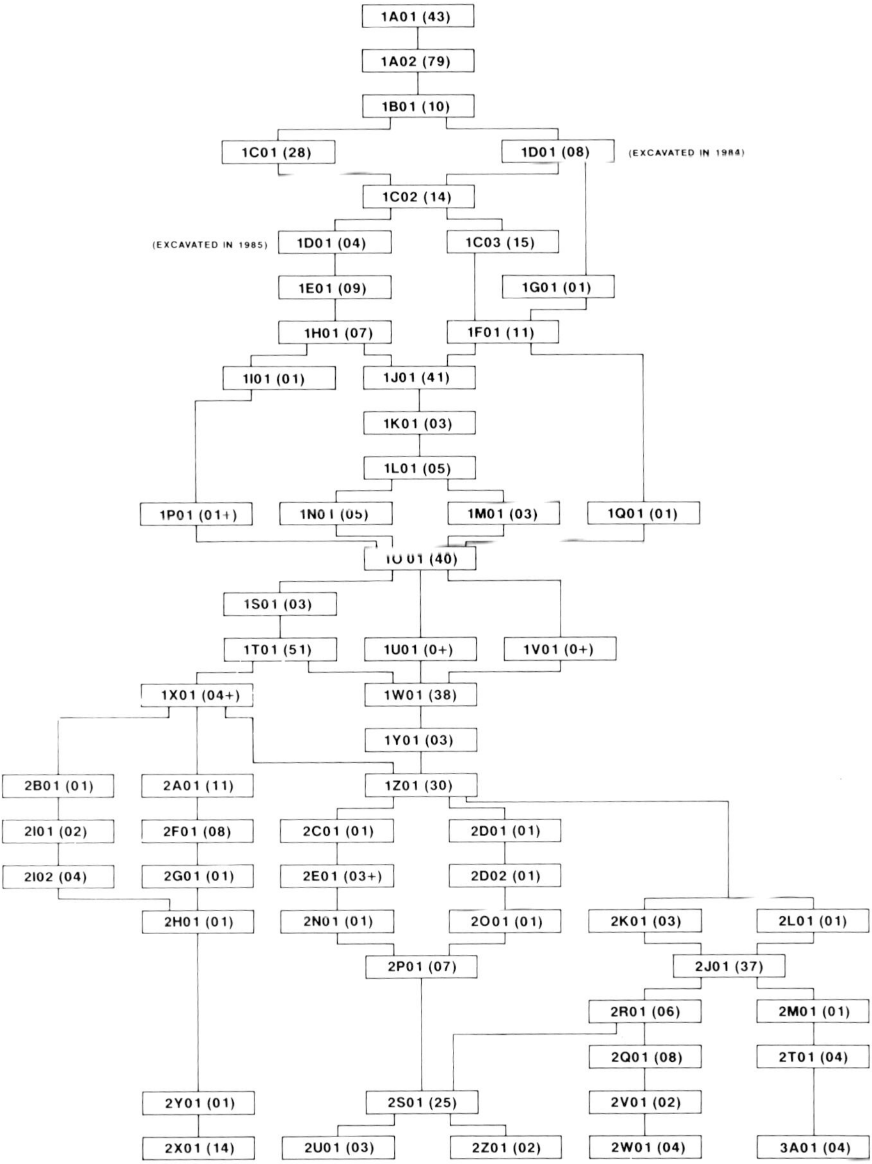

Figure 1 Harris diagram of unit 308/300. Each rectangle represents a facies, the name of which is in the box along with the volume given as "number of 8-liter buckets." For example, facies 1Z01 filled 30 8-liter buckets (a volume of 240 liters). The Harris Matrix is a graphic representation of the superpositional relationship of all the facies. Their distance from each other does not represent the size of the facies or the depth below the surface. Drawn by K. Kornbacher and drafted by J. Linse.

National Park Service. The upper 10 cm of the site, within the area historically plowed and now under grass vegetation, is the only portion of the shell midden affected by weathering processes originating at the surface. Below this 10-cm-thick zone, samples have values ranging from 22 to 9%, with tan facies (lenses of ash,

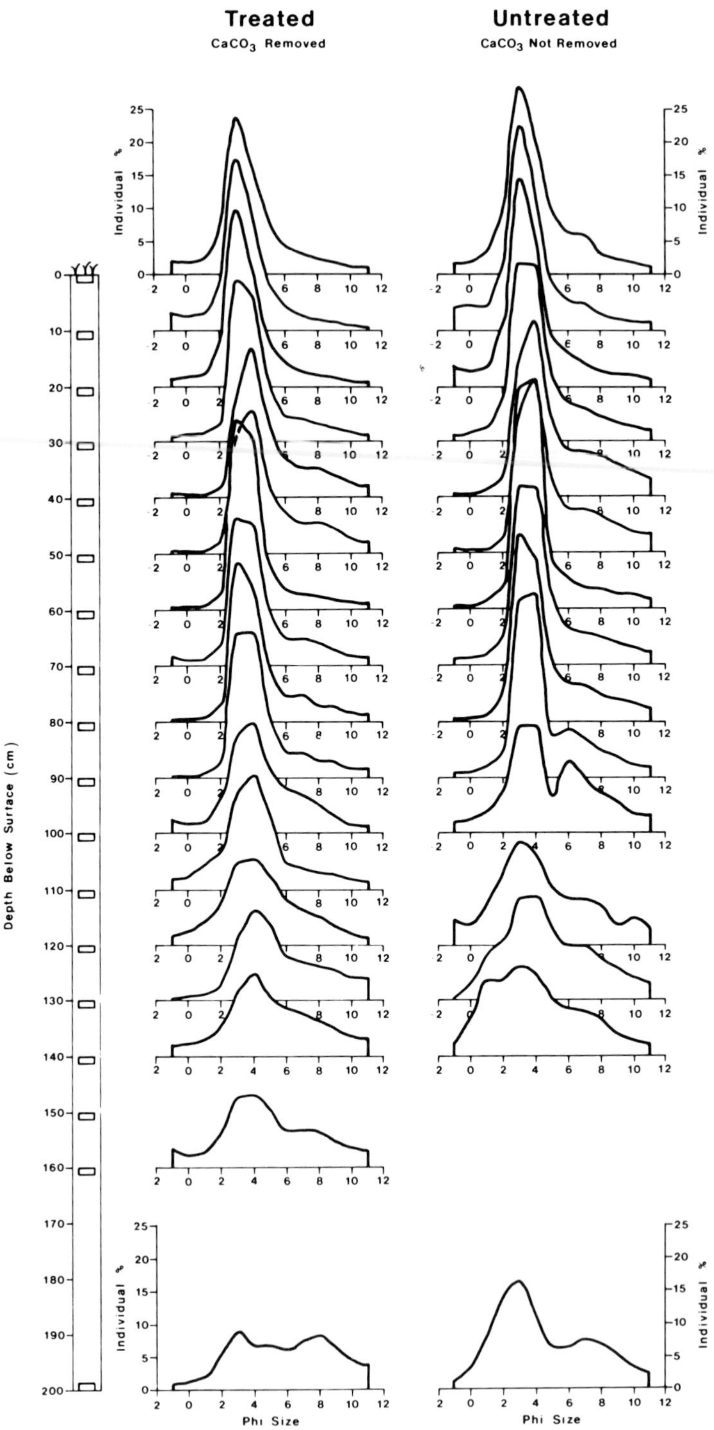

Control Profile-1983

burned shell, and small charcoal bits) having significantly lower amounts of organic matter (6 to 2%). The average amount of organic matter in the entire shell midden (including the tan facies) is 13%. Organic matter percentages seem to be roughly the same throughout the profile, and substantially higher than found in samples taken from the bay or from the control samples.

When the light/dark stratigraphic boundary was first observed at the site, one hypothesis for the difference in coloration was that the lower stratum contained more organic matter than the upper stratum (Stein 1984b). As seen in Appendix 2, facies found in the lower portions of units do not have significantly higher percentages of organic matter, especially if the values for tan facies are removed from the comparison.

C. *Carbonates*

The carbonate percentages (Appendix 2) are used to define one kind of lithostratigraphic unit which is used to detect weathering (Stein *et al.*, Ch. 6, this volume). The stratigraphic arrangement of carbonate percentages is displayed on the Harris Matrix for Unit 308/300 (Figure 3). In this figure the facies names and volumes has been removed and replaced with the percentage of carbonate found in the sample from that facies. The depth below the surface of some facies is also displayed. The Harris Matrix provides a graphic depiction of each facies's superpositional relationship without taking into account the size or depth of each facies. The percentages of carbonate in the facies at the top of the diagram (from 0 to 80 cm below the surface) are high, ranging (with three exceptions) from 43 to 66%. In the central portion of the diagram (from 80 to 95 cm) the percentages fall to values between 39 and 22%, while at the bottom of the diagram (below 95 cm) only facies with less than 29% (and predominantly 11 to 23%) are found. The two lithostratigraphic units are called Layer H and Layer L (H for high percentages of carbonate, L for low percentages of carbonate). The boundary between the two lithostratigraphic units is located just below 80 cm below the surface. The boundary is placed between the facies with samples that have high carbonate percentages and those that have low percentages. The shift from high to low is defined at 35%, the average percentage of carbonate in the site (Stein *et al.*, Ch. 6, this volume).

The same situation is observed in Unit 308/302 (Figure 4). The carbonate percentages are high from 0 to 90 cm, ranging from 41 to 70% (with three exceptions). The carbonate percentages are lower only below 90 cm (the boundary of Layer H and Layer L), ranging from 38 to 6%, with some values of 57%, 54%, and 41% in tan facies. During excavations, the dark color of the fine-grained sediment was used to define lithostratigraphic units called Layer I. The boundary between Layer I and Layer II (found stratigraphically above Layer I) was observed at about 80 cm below the surface. In Unit 308/302 the boundary between Layer H and Layer L, defined at

Figure 2 Textural analysis of samples from the control profile. The profile was an exposure of glacial drift, collected in 1983 from a wave-cut bank on Garrison Bay. Each sample was analyzed twice, once without pretreatment to remove carbonate (untreated), and another pretreated to remove carbonate (treated). The change in the shapes of the curves indicates the loss in carbonate.

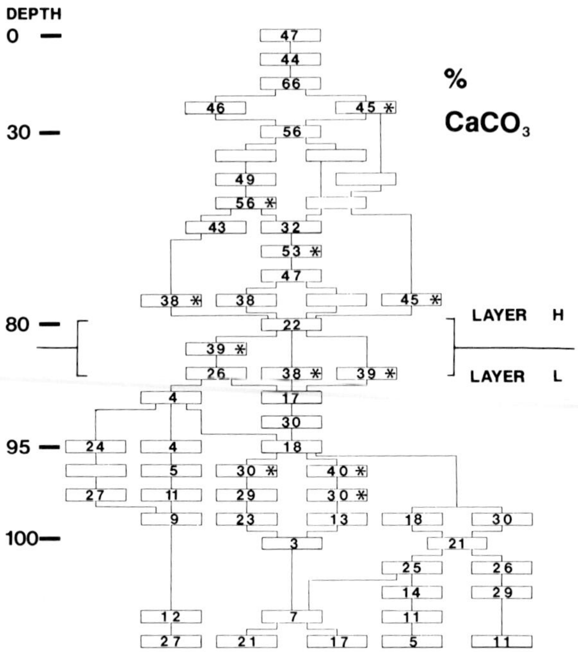

Figure 3 Carbonate percentages of facies of Unit 308/300, displayed in their correct superpositional arrangement on the Harris Matrix. Asterisks indicate facies with tan-colored fine-grained sediment. These facies often have higher carbonate percentages. Samples from blank facies were not subject to loss-on-ignition analysis. Location of the boundary of the lithostratigraphic unit based on carbonate percentages (Layer H and Layer L) is indicated. The H refers to high percentages of carbonate, the L refers to low percentages. The depth below surface (cm) of select facies is indicated on left.

90 cm below the surface, and that of Layer I and II do not coincide. The discrepancy is noted in other excavation units as well.

The percentages of carbonate in Unit 308/304 (Figure 5) are high in the facies from 0 to 120 cm below the surface (ranging from 43 to 70%, with two exceptions). The percentages fall below 40% only in those facies deeper than 120 cm below the surface. The boundary of the lithostratigraphic unit defined by carbonate percentage for purposes of observing weathering is placed at 120 cm below the surface. As in Unit 308/302, the boundary of Layer I and II in Unit 308/304, as defined by field observations, was placed at 80 cm below the surface. Thus, the two boundaries do not coincide in Unit 308/304, as they did not in Unit 308/302.

The carbonate percentages of 308/302 and 308/304 have similar distributions to those in unit 310/304 (Figure 12, Stein *et al.*, Ch. 6, this volume). All three units have high percentages of carbonate to depths well below 80 cm, where the effects of groundwater capillary actions are observed as dark-colored fine-grained sediment in facies. The carbonate percentage in the facies below 80 cm is high even though groundwater has hydrated the clays and silts as well as the organic matter.

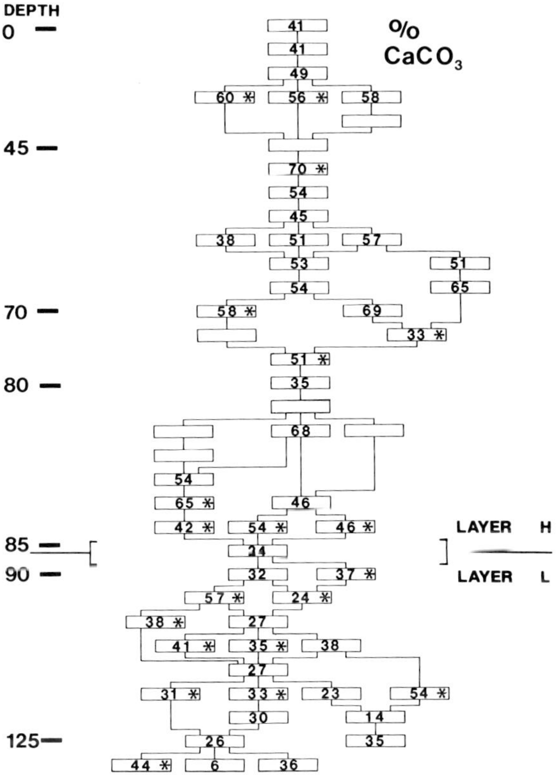

Figure 4 Carbonate percentages of facies of Unit 308/302, displayed in their correct superpositional arrangement on the Harris Matrix (see Figure 3 for further explanation). The depth below surface (cm) of select facies is indicated on the left.

D. Other Analyses

The pH of various facies is also reported in Appendix 1. These values are consistently alkaline and discussed more fully in Chapter 14 by Linse (this volume). Total phosphorus is measured for samples from Unit 310/300 and 310/306 (Table 1), as well as for samples from the control profile and Garrison Bay. No pattern is discerned in the phosphorus distribution in the midden or in the control samples. As expected, the glacial drift has very little phosphorus (mean of .22%), and the shell midden has more (mean of 1.38%). Unexpectedly the samples from Garrison Bay (taken at the substrate surface at low tide) have high percentages of phosphorus, equal to and greater than various samples from the midden. The source of the phosphorus in modern Garrison Bay sediments may be runoff from a nearby cattle ranch which drains into Garrison Bay.

VI. Discussion

The British Camp shell midden displays the two-tone stratification of dark-colored fine-grained sediment in the lower portions of the midden, and light-colored fine-grained

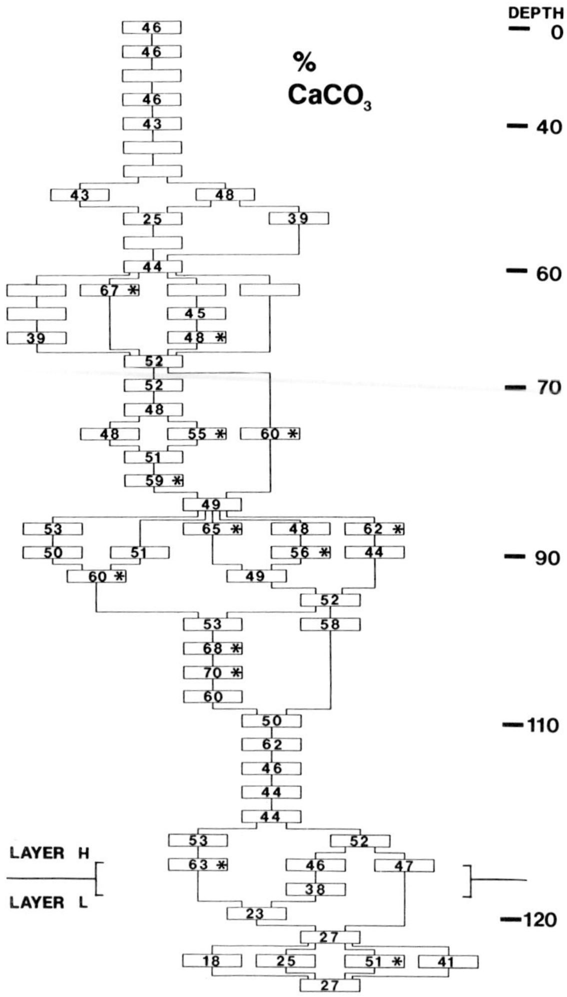

Figure 5 Carbonate percentages of facies of Unit 308/304, displayed in their correct superpositional arrangement on the Harris Matrix (see Figure 3 for further explanation). The depth below surface (cm) of select facies is indicated on the right.

sediment in the upper portions of the midden. This stratification was divided into lithostratigraphic units on the basis of field observations (Layer I, II, III, and IV, Ch. 6, this volume) with the dark/light fine-grained sediment defining the boundary between Layer I and Layer II. To test the proposal that weathering was responsible for the two-toned stratigraphy, another lithostratigraphic arrangement was designed using carbonate percentages. The carbonate data collected at the site indicate that criteria necessary for groundwater saturation to produce the stratification are present, and that the lithostratigraphic units arranged on the basis of carbonate percentages differ from the lithostratigraphic units arranged on the basis of field observations. The difference in the location of the boundaries of these two stratigraphic units is the observation that determines the site-formation processes.

Table 1 Phosphorus Data for Samples from British Camp

Unit 310/300		Unit 310/306 (1983) (cm below surface)	
Facies	%P	cm	%P
A1	2.90	0	1.15
A2	3.24	10	1.15
B	1.96	20	1.03
C	0.60	30	1.09
D	3.43	40	0.87
J	0.29	50	0.83
K	1.04	60	0.49
L	0.29	70	0.44
M1	0.86	80	1.12
M2	1.67	90	1.18
M3	3.07	100	1.18
N	1.78	110	0.41
		120	0.90
		130	1.21
		140	1.12
		150	1.12
		160	1.21
		170	1.15
		180	1.18
		190	1.18
		200	1.15

Control Profile (cm below surface)		Garrison Bay (substrate, low tide)	
cm	%P	Sample number	%P
0	0.53	1	2.23
10	0.53	2	1.04
20	0.35	3	2.11
30	0.18	4	2.11
40	0.35	5	3.29
50	0.70	6	2.23
60	0.35	7	2.11
70	0.01		
80	0.01		
90	0.35		
100	0.70		
110	0.01		
120	0.70		
130	0.53		
140	0.01		
150	0.01		
200	0.35		

A. *Attributes of the Shell Midden That Affect Weathering*

1. The shell midden is sufficiently thick to have only its lower portion intersect the water table (see Whittaker and Stein, Ch. 2, this volume), which produces the differential weathering in the zone of saturation and the zone of aeration. The top of the groundwater table was observed in 1983 while excavating a profile at the wave-cut bank, in 1984 when excavation of Unit 310/300 reached the depth of 80 cm, and again in every year of excavation when the backfill of the previous year was removed. Each observation indicates that the top of the zone of infiltration for the groundwater in this vicinity appeared at a depth close to 80 cm below the surface. Above this depth the sediment is dry, below this depth it is moist.
2. The groundwater is not supersaturated in carbonates (V. Gallucci, personal communication) and discharges and recharges rapidly as tidal levels vary.
3. The percentage of organic matter is high throughout the shell midden, and does not differ significantly from the top to the bottom of the midden. The percentages are much higher than those found in glacial drift or soil horizons (control samples), indicating that the prehistoric occupation accumulated large amounts of organic matter, which in turn contributed to the leaching of carbonate. Although some of the organic matter found in the site came from vegetation accumulating on the surface through soil-forming processes, the majority (as indicated by faunal and floral remains, and extremely high values) seems to have been deposited as part of prehistoric subsistence activities.
4. The percentages of silt and clay particles are high throughout the midden. The source of the silt and clay is most likely Garrison Bay or the glacial drift nearby. It was transported to the site by people, either intentionally as part of subsistence activities or unintentionally as residue adhering to items such as shellfish, feet, or canoes. The silt and clay absorbs water molecules and reacts with organic matter.
5. The percentage of carbonate is lower in the samples from the bottom of the shell midden than that in the samples from the top. In Unit 308/300 (Figure 3) and 310/300 (Figure 11, Ch. 6, this volume) the boundary of lithostratigraphic units based on weathering coincides with the lithostratigraphic units based on field observations of color change in the fine-grained sediment from light to dark. The boundaries of both types of lithostratigraphic units occur about 80 cm below the surface. In Unit 308/302 (Figure 4) the boundaries of the two types of lithostratigraphic units are close. The change in carbonate percentage (about 90 cm below the surface) and color (80 cm below the surface) is not exact, but is only 10 cm apart.

   In units 308/304 (Figure 5) and 310/304 (Figure 12, Ch. 6, this volume), both located in the SE corner of the excavation unit and close to the shoreline, the boundary of the lithostratigraphic unit as defined by the percent of carbonate in the fine-grained matrix does not coincide with the boundary of the lithostratigraphic unit as defined by the change in the color of the fine-grained sediment from light to dark. The boundary of Layer I and II, between moist and dry sediment, was observed in these units at about 80 cm below the surface, while the boundary of Layer L and Layer H (defined by carbonate percentages) is found at a depth of 120 cm below the surface in Unit 308/304 and 110 cm below the surface in Unit 310/304.

The explanation for the lack of coincidence in the location of the carbonate drop and the location of the observed moist fine-grained sediment in Units 308/304 and 310/304 rests with the amount of sand- and silt-sized shell found in the samples tested from these two units. As mentioned previously (Figure 18, Ch. 6, this volume), strata found at the surface to about 100 cm below the surface in these two units dip toward the shoreline and contain abundant whole and broken shell in all sizes.

The sand fractions of samples from the facies of these units were point counted and the compositional percentages of the sand fraction were calculated. The facies in the shell-rich upper portion of the units contain around 80% shell, while the sand fractions of samples from facies deeper than 110 cm below the surface have only 60 to 20% shell. In tan facies (marked with asterisks on Appendix 2 and Figure 5), in which gravel-sized objects are often absent, small-sized shell (crushed and burned) often makes up 95% of the entire facies.

In Units 308/304 and 310/304 the facies found between 80 and 120 cm have high percentages of small-sized shell and high numbers of tan facies, which also have abundant small-sized shell. When there is more shell to leach, the carbonate percentages in the matrix remain higher for a longer time. Even though the carbonate percentages measured by the loss-on-ignition method have been lowered in the last few centuries by the effects of leaching, they are still higher than the percentages in any facies that originally contained less shell. The exterior of the large-sized shells are noticeably etched by leaching, and the fine-grained sediment is moist from groundwater inundation. However, there is so much shell in the small sediment fraction that carbonate percentages remain above 40% until the shelly facies disappear (deeper than 110 cm below the surface).

6. Last, the loss of carbonate seems to have occurred over at least the last 500 years, and perhaps 1500 years. As discussed in Whittaker and Stein (Ch. 2, this volume), the position of mean sea level rose an additional meter over the last 2000 years. That rise of 1 m is enough to inundate the lower portion of the midden. However, the exact timing of that saturation is not known. Enough time has elapsed to affect significantly the carbonate in the lower and middle portions of the midden in most units.

From these data, the lithostratigraphic unit as defined by percentage of carbonate in the fine-grained matrix is assumed to be an indicator of groundwater saturation. The carbonate percentages decrease from high values in samples in the upper portion to low values in samples in the lower portion of the shell midden. Thus, groundwater intrusion and alteration of the midden sediment has occurred. However, the boundary of the two-layered stratification, as observed in field observations of the shell midden, does not have to coincide with the boundaries of the lithostratigraphic units defined by carbonate percentages. These two stratigraphic arrangements are one test that can be used to determine the cause of stratification in shell middens near fluctuating water tables.

VII. Summary

Sediment analysis is used to test the hypothesis that groundwater saturation of the British Camp shell midden produced a large-scale division in the stratification. The hypothesis suggests that the lower portion of the midden is saturated as the groundwater infiltrates the pores and permeates the sediments. This infiltration results from sea level rise, documented elsewhere in the region. The saturation changes the color of the fine-grained sediments from light to dark, and weathers differentially the lower portion of the midden. Sediment analysis at British Camp indicates that for all facies, the relative proportion of carbonate percentages are lower in the lower portion of the midden, indicating that leaching has taken place and supporting the interpretation that groundwater has produced the stratification. The amount of carbonate that is lost by leaching depends on the amount of shell deposited in each facies. The amount of shell in the facies affects the definition of boundaries between lithostratigraphic units defined on the basis of field observations and those defined on the basis of carbonate percentages.

The effects of groundwater saturation on all shell midden sites is crucial because the stratification that the groundwater produces is not culturally relevant. The distinct difference in color of the fine-grained sediment is often used as the criterion for grouping all artifacts found in a site. In the case of British Camp, artifacts from the facies with light-colored fine-grained sediments would be grouped together and compared to artifacts from facies with dark-colored sediment. The artificial grouping of artifacts using the groundwater-produced strata will result in incorrect phase definitions, especially in areas where phases are defined by artifact frequencies.

Groundwater saturation of shell middens is a process that could affect shell-bearing sites all over the world. Most shell-bearing sites are located near the coastline, or near a body of water (Bailey 1983). Worldwide sea level rose rapidly in the Holocene until 5000 B.P. Important for archaeologists who study shell-bearing sites is the fact that in many areas of the world sea level has continued to rise since 5000 B.P., with between 1 and 2 m of rise recorded (Stright 1990). Although a rise of 1 m in sea level is insignificant to Quaternary scientists, it is extremely significant to archaeologists excavating shell-bearing sites. If the necessary conditions are present in a shell-bearing site, and rising sea level causes the lower portion of the site to be saturated with groundwater, then the dual stratification will appear.

This groundwater saturation is significant because the stratification of British Camp shell midden is related only partially to cultural depositional events. Cultural events dictate whether sufficient quantities of organic matter, clay particles, and shell are deposited in any one site or layer and they dictate which artifacts will be found in each layer, but the infiltration of groundwater entering the strata from below is a postdepositional event. The infiltration could coincide with ethnostratigraphic boundaries and lithostratigraphic boundaries based on carbonate percentages (as in Units 308/300 and Unit 310/300 at British Camp). However, the boundaries might not necessarily coincide over the entire site (as in Units 308/302, 308/304, and 310/304). Therefore, the stratigraphic light/dark boundary observed at this and other sites has to be evaluated at each location to determine if it is a culturally relevant boundary or if it is a creation of the water table and the sedimentology of the substrate.

Appendix 1 Grain Size Data and pH for Samples from British Camp

Unit	facies	pH	sand	silt	clay	mean	SD	SK	K
				%		ø			
UNIT 310/300									
310300	A1	7.0	53	33	3	2.7	3.1	0.1	0.8
310300	A2	7.6	31	44	17	4.5	3.7	-0.2	
310300	B	7.8	34	42	16	4.5	3.6	-0.2	0.9
310300	C	7.6	26	48	21	5.4	3.5	-0.3	0.9
310300	D	7.8	39	34	22	5.0	4.0	0.1	1.0
310300	J	7.8	40	43	13	4.2	3.2	-0.1	0.8
310300	K *	7.8	53	33	6	3.1	3.1	0.1	0.8
310300	L	7.8	41	42	15	4.0	3.1	0.0	0.9
310300	M1 *	8.0	51	35	7	3.5	3.1	0.1	0.8
310300	M2	7.8	52	32	7	3.1	3.3	0.1	0.8
310300	M3	8.0	37	39	17	4.4	3.6	-0.1	1.0
310300	N	7.7	48	22	10	2.5	3.8	0.4	0.8
310300	P	7.8	55	26	20	4.5	3.4	0.4	0.8
310300	Q	7.2	75	5	14	2.5	2.9	0.3	2.0
310300	R1	7.7	58	28	14	4.0	3.2	0.4	0.9
310300	R2	8.0							
310300	R3	7.7	39	37	16	4.3	3.6	0.0	1.0
310300	S	7.7	42	41	10	4.1	3.2	0.0	0.9
UNIT 310/302									
310302	A3	7.9	58	28	13	3.7	3.3	0.3	0.8
310302	C1	8.0	52	31	16	4.3	3.4	0.3	0.9
310302	D1	7.9	53	31	16	4.3	3.3	0.3	0.8
310302	F *	8.0	49	41	10	4.2	2.9	0.0	0.7

Continues

Unit	facies	pH	sand	silt	clay	mean	SD	SK	K
				%		ø			
310302	G	7.9	55	31	13	4.0	3.2	0.3	0.8
310302	I	7.9	63	25	12	3.5	3.1	0.4	0.8
310302	M	8.0	63	24	12	3.6	3.2	0.4	0.8
310302	N	8.0	68	21	11	3.3	2.9	0.5	0.9
310302	O1	8.1	63	24	13	3.5	3.3	0.5	0.8
310302	O2	8.5	67	23	9	3.2	3.0	0.5	0.8
310302	P	8.1	52	37	12	4.0	3.2	0.2	0.9
310302	Q *	8.2	52	40	7	3.8	2.8	0.1	0.8
310302	U1	8.2	51	36	14	4.2	3.2	0.2	0.8
310302	U2	8.3	54	31	13	3.9	3.3	0.3	0.9
310302	U3	8.0	56	28	17	4.0	3.5	0.4	0.8
310302	V1 *	8.0	50	37	12	4.1	3.1	0.2	0.8
310302	W	8.2	57	28	15	3.8	3.2	0.4	0.8
310302	Y	8.3	45	36	19	4.8	3.4	0.1	0.8
310302	Z1	8.2							
310302	2A	7.2	50	29	21	4.6	3.8	0.3	0.9
310302	2B	7.2	51	30	19	4.6	3.7	0.3	0.9
310302	2C *	8.1	55	38	8	3.7	2.9	0.2	0.8
310302	2D *		45	38	17	4.7	3.3	0.1	0.9
310302	2E	7.8	47	27	26	5.3	4.0	0.3	0.8
310302	2F *	7.8	47	33	20	4.8	3.6	0.2	0.9
310302	2G *	7.6	44	45	11	4.2	2.8	0.1	0.9
310302	2H	7.6	50	28	22	4.8	3.8	0.4	0.9
310302	2I *	7.4	52	38	10	4.0	2.9	0.2	0.8

Continues

Unit	facies	pH	sand	silt	clay	mean	SD	SK	K
				%		ø			
310302	2J	7.3	47	35	18	4.7	3.5	0.2	0.9
310302	2K *	7.4	61	28	11	3.5	3.1	0.4	0.9
310302	2L *	7.8	57	32	11	3.8	3.2	0.4	0.9
310302	2M	7.8	77	4	19	4.3	4.0	0.7	2.1
310302	2N *	7.4	56	31	13	4.0	3.0	0.3	0.7
310302	2O	7.4	60	31	9	3.7	2.9	0.3	0.8
310302	2P	7.6	61	19	21	4.5	4.0	0.6	1.0
310302	2Q *	7.9	51	42	7	4.1	3.0	0.2	0.8
310302	2R	7.8							
310302	2S *	7.9	51	38	11	4.1	3.1	0.2	0.8
310302	2T *	7.8	51	40	9	4.1	2.8	0.1	0.8
310302	2U	7.3	62	18	20	4.5	4.1	0.6	0.9
UNIT 308/300									
308300	A1	7.4	47	35	18	4.5	3.2	0.1	0.8
308300	A2	7.4							
308300	B	7.8	60	28	12	3.6	3.1	0.4	0.8
308300	C1	7.3	51	28	21	4.5	3.6	0.3	0.8
308300	C2	7.5	41	25	34	6.0	4.5	0.2	0.7
308300	D	7.6	47	29	24	5.1	3.9	0.3	0.8
308300	E	7.5	54	27	19	4.2	3.5	0.3	0.8
308300	H *		58	34	7	3.6	2.8	0.2	0.8
308300	I	7.6							
308300	J		82	7	10	2.8	2.6	0.4	1.6
308300	K	7.7							

Continues

Unit	facies	pH	sand	silt	clay	mean	SD	SK	K
				%		ø			
308300	L	7.7							
308300	N		71	17	12	3.4	3.1	0.5	1.2
308300	P *		57	30	13	4.0	3.0	0.3	0.8
308300	Q	7.4	58	29	13	3.9	3.2	0.3	0.8
308300	O		58	29	13	4.1	3.0	0.3	0.8
308300	R *		59	28	13	4.0	3.0	0.3	0.8
308300	S *		48	36	15	4.4	3.1	0.1	0.8
308300	T	7.2	54	27	18	4.3	3.4	0.4	0.9
308300	U *	7.3	57	34	9	3.8	2.8	0.3	0.8
308300	V *	8.0	60	30	11	3.6	3.0	0.3	0.9
308300	W	7.7	51	28	21	4.7	3.5	0.4	0.9
308300	X		37	41	23	5.5	3.0	0.1	0.8
308300	Y	7.8	56	27	17	4.2	3.4	0.4	0.8
308300	Z	7.2	60	19	21	4.6	3.5	0.5	0.9
308300	2A		38	38	25	5.5	3.3	0.2	0.9
308300	2B		52	25	23	4.7	3.8	0.4	0.8
308300	2C *	7.5	57	29	14	3.9	3.2	0.4	1.0
308300	2D1 *		64	31	5	3.5	2.6	0.3	0.8
308300	2D2 *	7.3	58	31	12	4.0	2.9	0.3	0.8
308300	2E	7.4							
308300	2F		62	25	13	3.9	3.2	0.4	0.7
308300	2G	7.2	50	32	18	4.6	3.2	0.3	0.8
308300	2H	7.4							
308300	2I		48	34	17	4.6	3.3	0.2	0.9

Continues

Unit	facies	pH	sand	silt	clay	mean	SD	SK	K
			%			ø			
308300	2J		65	18	17	3.9	3.2	0.4	0.9
308300	2K	7.3	54	27	19	4.5	3.4	0.4	0.8
308300	2L		45	41	14	4.4	3.1	0.2	1.0
308300	2M	7.1	59	24	17	4.1	3.4	0.4	0.9
308300	2N		55	31	14	4.2	3.0	0.3	0.8
308300	2O		30	63	7	4.3	2.3	0.1	2.1
308300	2P		43	38	19	5.2	2.8	0.2	0.8
308300	2Q	7.6	72	19	9	3.4	2.7	0.5	1.2
CONTROL SAMPLE (cm below surface)									
0		6.3	65	26	7	3.7	2.3	0.4	1.4
10		6.6	71	20	5	3.2	2.3	0.2	1.9
20		6.6	71	20	6	3.3	2.2	0.3	1.9
30		6.8	64	26	9	4.0	2.3	0.4	1.4
40		6.4	51	32	16	4.9	2.7	0.5	1.0
50		6.3	51	31	17	4.9	2.9	0.6	1.0
60		6.4	75	18	7	3.6	1.8	0.4	2.0
70		6.7	62	25	11	4.2	2.6	0.5	1.5
80		7.2	60	25	14	4.5	2.8	0.6	1.3
90		7.3	63	24	12	4.3	2.7	0.5	1.3
100		7.6	46	32	20	5.0	3.5	0.3	1.1
110		7.6	56	30	13	4.1	3.2	0.3	1.1
120		7.9	51	29	16	4.3	3.5	0.3	1.1
130		7.4	50	33	16	4.6	3.3	0.3	1.1
140		7.7	57	29	12	3.8	3.3	0.3	1.0

Appendix 1 *(Continued)*

Unit	facies	pH	sand	silt	clay	mean	SD	SK	K
				%		ø			
160		7.2	41	33	22	5.2	3.8	0.3	1.1
200		7.7	54	28	16	4.3	3.4	0.4	0.9
Garrison Bay (substrate at low tide)									
		8.1	41	26	34	5.9	4.5	0.3	0.7

*** TAN FACIES**

Appendix 2 Facies, Organic Matter, and Carbonate Percentages for Samples from British Camp

UNIT 306/300

f	om	carb	f	om	carb	f	om	carb	f	om	carb	f	om	carb
A2	16	49	P *	5	66	2A	13	50	2H1 *	15	38	2P	12	37
B	14	52	R	13	62	2C 86	24	18	2H2 *	12	17	2Q	20	28
C	8	58	T1	10	53	2D 86	24	55	2I	8	6	2R	15	4
D	12	59	T2	16	42	2E 86	30	39	2J a	21	33	2S	13	26
F	18	47	U	12	59	2F 86	13	36	2J b	14	34	2T	14	28
G	17	51	V *	4	57	2C 87	16	41	2J2	16	31	2U	14	33
I	12	60	Wa	16	53	2D 87	14	3	2K	19	9	2V	12	14
L	20	56	Wb	19	50	2E 87	13	40	2L *	9	5	2W1	20	10
M	15	63	X	9	59	2F 87	22	23	2M	18	27	2W2	31	6
N	10	59	Y *	6	50	2F2	18	33	2N	15	22	2Y	14	21
O	11	61	Z *	6	42	2G	13	7	2O	22	8	3A	15	23

Continues

Appendix 2 *(Continued)*

UNIT 308/300

f	om	carb
A1	20	47
A2	18	44
B	9	66
C1	20	46
C2	15	56
D	16	45
E	15	49
H *	5	56
I	15	43

f	om	carb
J	14	32
K	19	53
L	12	47
N	13	38
P *	10	38
Q	8	45
O	16	22
R *	8	24
S *	0	39

f	om	carb
T	18	26
U *	8	38
V *	6	39
W	16	17
X	13	4
Y	11	30
Z	15	18
2A	16	4
2B	14	24

f	om	carb
2C *	11	30
2D1 *	5	40
2D2 *	7	30
2E	14	29
2F	21	5
2G	16	11
2H	21	9
2I	14	27
2J	18	21

f	om	carb
2K	14	18
2L	16	30
2M	16	26
2N	14	23
2O	24	13
2P	29	3
2Q	10	14

UNIT 308/302

f	om	carb
A1	20	41
A2	17	41
B1	15	49
B2	12	58
D *	5	56
F	8	60
H *	2	70
I	11	54
J	13	45
K	10	38
L	16	51

f	om	carb
M	13	57
N	13	53
O	12	54
P	12	51
Q	12	58
R	8	69
S	9	65
T	12	33
V	15	51
W	9	35
2A	10	68

f	om	carb
2B	14	54
2C	7	65
2D *	7	42
2E	13	46
2F *	2	54
2G *	5	46
2H	17	24
2I *	7	37
2J	15	32
2K *	4	57
2L	16	24

f	om	carb
2M	16	27
2N	16	38
2O	13	38
2P	9	35
2Q	11	41
2R	16	27
2S	13	31
2T	12	33
2U	13	30
2V	6	23
2W	16	14

f	om	carb
2X	13	54
2Y	17	26
2Z	20	35
3A *	8	44
3B	33	6
3C	18	36

Continues

Appendix 2 *(Continued)*

UNIT 308/304

f	om	carb	f	om	carb	f	om	carb	f	om	carb	f	om	carb
A1	20	46	S	14	39	2E	11	53	2P2 *	1	70	3A	19	38
A2	17	46	T	11	52	2F	16	48	2Q	6	60	3B	16	23
B	15	46	U	12	52	2G *	6	65	2R	15	50	3C	30	27
C	25	43	V	14	48	2H *	9	56	2S	11	62	3D	37	25
F	25	43	W *	8	55	2I	12	49	2T1	14	46	3E	19	41
G	18	48	X *	7	60	2J	14	50	2T2	15	44	3F *	14	51
H	19	25	Y	12	48	2K	22	51	2U	16	43	3G	26	18
J	17	39	Z	13	51	2L	15	52	2V	11	52	3H	16	27
K	15	44	2A *	8	59	2M	10	58	2W	13	53			
M	7	67	2B	14	49	2N *	6	60	2X *	6	63			
O	17	45	2C *	5	62	2O	17	53	2Y	13	47			
P *	7	48	2D	16	44	2P1 *	3	68	2Z	17	46			

UNIT 310/300

f	om	carb	f	om	carb	f	om	carb	f	om	carb	f	om	carb
A1	24	47	D	16	45	M1 *	4	70	N	10	63	R2	11	29
A2	17	46	J	10	64	M2	10	53	P	16	24	R3	11	25
B	12	61	K *	6	65	M3	11	40	Q	13	29	S	13	90
C	13	65	L	6	63	M4	9	34	R1	13	33			

Continues

Appendix 2 *(Continued)*

UNIT 310/302

f	om	carb
A3	17	45
C1	16	41
D1	19	43
F *	9	49
G	19	49
I	16	50
M	16	42
N	19	48
O1	15	45

f	om	carb
O2	16	43
P	15	57
Q *	9	59
U1	17	53
U2	19	45
U3	17	54
V1 *	8	59
W		
Y	14	38

f	om	carb
Z1	22	53
2A	14	48
2B	13	40
2C *	3	57
2D *		
2E	16	29
2F *	11	46
2G *	5	41
2H	15	31

f	om	carb
2I *	6	44
2J	15	38
2K *	5	48
2L *	5	20
2M	11	16
2N *	5	41
2O	7	40
2P	14	29
2Q *	6	43

f	om	carb
2R	13	40
2S *	7	50
2T *	7	47
2U	9	36

UNIT 310/304

f	om	carb
A1	19	49
A2	11	56
B1	11	63
B2	12	56
C	19	46
D1	13	64
D3	8	68
E	16	56
F	23	28
G	12	61

f	om	carb
I	13	53
K	7	70
L	15	48
M	11	54
O	18	55
P	17	49
R	12	55
S *	6	68
T	15	49
U	12	41

f	om	carb
V	11	60
W	17	43
X	24	40
Y	13	65
Z	8	58
2A	14	51
2B	18	50
2C	18	44
2D	15	45
2E	13	42

f	om	carb
2F	12	42
2G	12	44
2H*	13	33
2I	16	41
2J	9	67
2K1	15	40
2K2	14	33
2L	20	51
2M	15	37
2N	22	39

f	om	carb
2O	24	33
2P	29	25
2R	28	34
2S	41	18
2T *	12	52
2U	23	26

Continues

Appendix 2 *(Continued)*

UNIT 310/306

f	om	carb	f	om	carb	f	om	carb	f	om	carb	f	om	carb
0	25	52	50	7	71	100	7	57	150	11	35	200	14	18
10	9	59	60	7	78	110	0	89	160	12	23			
20	6	77	70	4	80	120	7	67	170	17				
30	13	82	80	5	80	130	16	43	180	23	21			
40	9	75	90	8	53	140	9	55	190	17	17			

CONTROL PROFILE

f	om	carb	f	om	carb	f	om	carb	f	om	carb	f	om	carb
0	7	2	**40**	2	2	**80**	2	2	**120**	3	2	**200**	3	2
10	3	2	**50**	2	2	**90**	1	2	**130**	3	2			
20	2	2	**60**	1	2	**100**	2	4	**140**	2	2			
30	2	1	**70**	1	2	**110**	2	5	**160**	2	3			

Garrison Bay - substrate at low tide

f	om	carb
	7	48

* = Tan facies.

Acknowledgments

The data reported here were analyzed in the University of Washington Archaeological Sediment Laboratory. Samples were curated and prepared for analysis by Mary Parr and Mark Madsen. Samples for textural analysis were pretreated by Melinda Allen, Lisa Nagaoka, and Kris Wilhelmsen. Loss-on-ignition analysis was completed with the help of Carl Harrington, and phosphorus analysis by Brigid Henderson. This manuscript has benefitted greatly from the comments of Vance T. Holliday.

References

Bailey, G. N.

1983 Problems of site formation and the interpretation of spatial and temporal discontinuities in the distribution of coastal middens. In *Quaternary coastlines and marine archaeology*, edited by P. M. Masters and N. C. Flemming. New York: Academic Press. Pp.559–582.

Birkeland, P. W.
1984 *Soils and geomorphology*. New York: Oxford University Press.
Brandon, M. T., D. S. Cowan, and J. A. Vance
1988 The Late Cretaceous San Juan thrust system, San Juan Islands, Washington. Geological Society of America, Special Paper 221, Boulder, Colorado.
Buol, S. W., F. D. Hole, and R. J. McCracken.
1989 *Soil genesis and classification, 3rd ed.* Ames, Iowa: Iowa State University Press.
Carlson, R. L.
1960 Chronology and culture change in the San Juan Islands, Washington. *American Antiquity* **25**, 562–586.
Claassen, C.
1991 Normative thinking and shell-bearing sites. In *Archaeological method and theory, vol. 3*, edited by M. B. Schiffer. Tucson, Arizona: University of Arizona Press. Pp.249–298.
Dean, W. E., Jr.
1974 Determination of carbonate and organic matter in calcareous sediments and sedimentary rocks by loss-on-ignition: comparison with other methods. *Journal of Sedimentary Petrology* **44**, 242–248.
Dunnell, R. C., and J. K. Stein
1989 Theoretical issues in the interpretation of microartifacts. *Geoarchaeology* **4**, 31–42.
Fladmark, K. R.
1982 Microdebitage analysis: initial considerations. *Journal of Archaeological Science* **9**, 205–220.
Folk, R. L.
1980 *Petrology of sedimentary rocks*. Austin, Texas: Hemphill Publishing Company.
Gorski, L.
n.d. Microstratigraphic analysis at the Carlston Annis site, Kentucky. In *Archaeology of the middle Green River area, Kentucky*, edited by W. H. Marquardt, M. C. Kennedy, and P. J. Watson. Mid-Continental Journal of Archaeology, Special Publication (in press).
Hole, F. D.
1961 A classification of pedoturbations and some other processes and factors of soil formation in relation to isotropism and anisotropism. *Soil Science* **91**, 375–377.
Holliday, V. T.
1990 Pedology in archaeology. In *Archaeological geology of North America*, edited by N. P. Lasca and J. Donahue. Geological Society of America, Centennial Special Volume 4, Boulder, Colorado. Pp.525–540.
Holliday, V. T., and J. K. Stein
1989 Variability of laboratory procedures and results in geoarchaeology. *Geoarchaeology* **4**, 347–358.
Jackson, M. L.
1969 *Soil chemical analysis-advanced course, 2nd Ed.* Department of Soils, University of Wisconsin, Madison.
Rosen, A. M.
1986 *Cities of clay: the geoarchaeology of tells*. Chicago, Illinois: University of Chicago Press.
Sharer, R. J., and W. Ashmore
1987 Archaeology: Discovering our past. Palo Alto, California: Mayfield.
Sjoberg, A.
1976 Phosphate analysis of anthropic soils. *Journal of Field Archaeology* **3**, 447–454.
Stein, J. K.
1983 Earthworm activity: A source of potential disturbance of archaeological sediment. *American Antiquity* **48**, 277–289.
1984a Organic matter and carbonates in archaeological sites. *Journal of Field Archaeology* **11**, 239–246.
1984b Interpreting the stratigraphy of Northwest shell middens. *Tebiwa* **21**, 26–34.
1985 Interpreting sediment in cultural settings. In *Archaeological sediments in context*, edited by J. K. Stein and W. R. Farrand. Center for the Study of Early Man, Institute for Quaternary Studies, University of Maine, Orono. Pp.5–19.

1987 Deposits for archaeologist. In *Advances in archaeological method and theory, vol. 11*, edited by M. B. Schiffer. Orlando, Florida: Academic Press. Pp.337–393.

1992 Formation processes of coastal sites: View from a Northwest Coast shell midden. In *Proceedings of the Circum-Pacific Prehistory Conference: IIIC. Maritime Societies in Western North America*, edited by A. R. Blukis Onat. Washington State University Press (in press).

1992 Organic matter in archaeological contexts. In *Soils, landscape evolution, and human occupation*, edited by V. T. Holliday. Washington D.C.: Smithsonian Institution Press. Pp.193–216.

Stein, J. K., and G. Rapp, Jr.

1985 Archaeological sediments: a largely untapped reservoir of information. In *Contributions to Aegean archaeology*, edited by N. C. Wilkie and W. D. E. Coulson. Center for Ancient Studies, University of Minnesota, Publications in Ancient Studies 1, Minneapolis. Pp.143–159.

Stein, J. K., and P. A. Teltser

1989 Size distributions of artifact classes: combining macro- and micro-fractions. *Geoarchaeology* **4**, 1–30.

Stright, M. J.

1990 Archaeological sites on the North American continental shelf. In *Archaeological geology of North America*, edited by N. Lasca and J. Donahue. Geological Society of America, Centennial Special Volume 4, Boulder, Colorado. Pp.439–465.

Waselkov, G. A.

1987 Shellfish gathering and shell midden archaeology. In *Advances in archaeological method and theory, vol. 10*, edited by M. B. Schiffer. Orlando, Florida: Academic Press. Pp.93–210.

Widmer, R.

1989 Archaeological research strategies in the investigation of shell-bearing sites, a Florida perspective. Paper presented at the annual meeting of the Society for American Archaeology, Atlanta, Georgia.

Wood, W. R., and D. L. Johnson

1978 A survey of disturbance processes in archaeological site formation. In *Advances in archaeological method and theory, vol 1*, edited by M. B. Schiffer. Orlando, Florida: Academic Press. Pp.315–381.

8

Shell Midden Lithic Technology: Analysis of Stone Artifacts from British Camp

Kimberly D. Kornbacher

I. Introduction

Analyses of lithic assemblages from southern Northwest Coast shell middens have seldom included debitage or expedient tools. One reason for this omission is the predominance of culture history in the region (Borden 1970; Carlson 1960, 1970, 1983; Mitchell 1971, 1990). This approach requires artifacts that exhibit stylistic attributes useful for chronology building (Dunnell 1978). Thus, culture historians have traditionally had little interest in debitage or unformed tools. Unfortunately, this narrow focus has not resulted in a secure, undisputed culture historical sequence for the Northwest Coast. With the exception of water-saturated or other anaerobic deposits (e.g., Bernick 1983; Croes 1976; Daugherty and Friedman 1983), the artifact assemblages of most excavated sites on the southern Northwest Coast consist mainly of bone and stone tools that exhibit little stylistic variation (but see Borden 1983). Archaeologists have implicitly relied upon *ad hoc* combinations of functional, technological, and stylistic attributes of stone and bone tools to develop regional chronologies.

A more productive approach to lithic analysis focuses on all constituents of the lithic assemblage, including debitage and expedient tools, and emphasizes functional and technological data. Unlike culture history, this approach can help us understand the articulation of prehistoric cultures and their environment (Binford 1968, 1977). It can also provide an empirical basis for understanding the nature of culture historical "phases" on the Northwest Coast, the ways in which they differ and how they affect the interpretation of change over time and space. In this chapter, I demonstrate the potential of such an approach using data from the British Camp site.

Radiocarbon age estimates (Stein *et al.* Ch. 6, this volume) ranging from at least 1500 B.P. to historic times indicate that the British Camp facies can be grouped within the chronostratigraphic units designated "Marpole Phase" (2400–750 B.P.) and "San Juan Phase" (750 B.P. to present) by local culture historians (Borden 1951, 1970; Burley 1980; Carlson 1954, 1960, 1970; Kidd 1964; Mitchell 1971, 1990). According to these authors and other archaeologists currently working on the Northwest Coast, a decrease in frequency of chipped stone materials is one of the distinguishing characteristics of the transition from the Marpole Phase to the San Juan Phase. The British Camp deposits are consistent with this trend. An abrupt change in the relative abundance of chipped stone artifacts occurs in midden deposits dated to approximately 1000 B.P. This difference in the abundance of flaked stone allows a division of the deposits into two "ethnozones" (Stein *et al.*, Ch. 6, this volume). Ethnozone I, defined by the abundant occurrence of lithics, refers to the lower facies (the older part of the midden). Ethnozone II is characterized by a paucity of lithics and refers to the upper, more recent facies.

Despite a general awareness of the decline in chipped stone in southern Northwest Coast middens over time, and the various attempts to explain why such a change may have occurred (e.g., Ames 1981; Borden 1970; Burley 1980; Carlson 1970; Croes and Hackenburger 1988; Thompson 1978), certain aspects of lithic technology that have the potential to increase our knowledge of this change have not been investigated. In particular, debitage and expedient tools from shell midden excavations (if collected at all) have been collected in an unsystematic fashion, resulting in loss of potentially important technological and functional information. Campbell (1981) and Clark (1991) are notable exceptions.

The goals of this chapter are to describe and compare the stone tool assemblages of Ethnozones I and II in a manner that facilitates comparison with lithic assemblages of other Northwest Coast shell middens and to provide technological and functional information about lithic industries at British Camp. This information will be used to test the hypothesis that the decline in abundance of chipped stone tools over time is related to a change in lithic technology. Specifically, manufacturing techniques, range of lithic reduction activities represented (e.g., core reduction only or core reduction, tool manufacture, and tool refurbishment, etc.), and the kinds of tools produced are all inferred from tool and debitage analyses. Based on these inferences, lithic technologies of Ethnozones I and II are compared. Results are used to determine if the decline in abundance of chipped stone tools over time corresponds to a change in lithic technology. If no identifiable difference between the two ethnozones is indicated, the hypothesis that the decline in chipped stone tools is related to a change in the technological processes involved in their production can be rejected.

II. Methods

This analysis draws upon research originally conducted by the author (Kornbacher 1989) to examine change in the lithostratigraphic units referred to as Layers I, II, III, and IV (Stein *et al.*, Ch. 6, this volume). Lithic data are analyzed here within the framework of the two ethnostratigraphic units, Ethnozones I and II (Stein *et al.*, Ch.

6, this volume). The reader is referred to Kornbacher (1989) for information about smaller scale changes in the lithic assemblage of the British Camp midden.

A. *The Data Set*

This study analyzes lithic tools and debitage excavated from the deposits of a 6 × 4-meter area of Operation A at British Camp. The units selected for study are 306/300, 308/300, 310/300, 306/302, 308/302, and 310/302 (Figure 1). Three of the 2 × 2-meter units are located contiguously along the north profile of the excavation and the remainder are adjacent units to the south. The lithic materials from 310/302 were omitted from the final analysis because of disturbance observed in the unit profile and the anomalous radiocarbon date derived from this unit (Stein *et al.*, Ch. 6, this volume).

Chapter 6 of this volume demonstrates that ethnostratigraphic and lithostratigraphic boundaries do not always correspond. However, with the exception of Unit 310/302 (in which the remains of a large pit are visible), the units chosen for this analysis (Figure 1) are located away from the shoreline and share common ethnostratigraphic and lithostratigraphic boundaries. In these units, the top of the large lithostratigraphic unit (Layer I) coincides with the change in abundance of lithic artifacts that forms the boundary of Ethnozone I and Ethnozone II. Thus for purposes of this analysis, Ethnozone I incorporates the lithic materials from Layer I, while the lithics from Layers II, III, and IV comprise Ethnozone II.

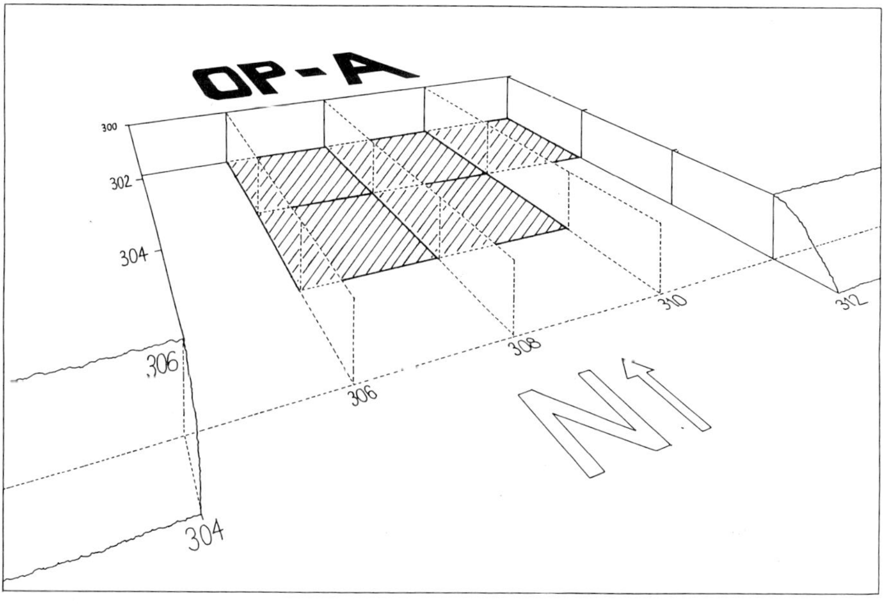

Figure 1 Analyzed portion of Operation A, British Camp.

The analysis includes all lithic materials excavated during the 1984 and 1986 field seasons that are larger than 12 mm (retained on the 1-inch and 1/2-inch screens), as well as significantly modified lithic artifacts for which three-point provenience was taken. A sample of smaller sized lithics retained on the 1/4-inch and 1/8-inch screens (3–12 mm) is also included in the analysis.

B. *Classification of the Lithic Materials*

Two different taxonomic classifications were used to meet the goals of the analysis. One is morphological and is constructed primarily for deriving descriptive and functional information about the tools; the other is specifically designed for debitage and is technological.

Prior to the measurement of attributes, all tool by-products (debitage) must be separated from the products of tool manufacture (shaped, retouched, or used objects). The technological classification is only applicable to the by-products of chipping manufacturing techniques, referred to as debitage. Although grinding, pecking, and incising are also reductive techniques, the by-products are usually not identifiable macroscopically and are not considered here.

The criterion for separating tools from debitage is whether or not the object displays any evidence of postdetachment modification attributable to human activity. For purposes of this study, the alteration must be detectable with a 10-power hand lens. Objects determined to be modified are grouped and analyzed as tools rather than debitage. If there is no evidence of postdetachment modification, artifacts are considered lithic debitage and are grouped and analyzed as such. Modified objects that exhibit only microscopic use-wear may be overlooked in this analytic process (Young and Bamforth 1990) and grouped as debitage instead of tools. However, the macroscopic identification technique is used here because it is less costly and provides a baseline estimate of tool frequency.

The necessary and sufficient conditions for membership in each artifact class (Dunnell 1971), as well as all variables measured, are discussed at length elsewhere (Kornbacher 1989) and are not repeated here.

1. The Morphological Classification

The morphological classification applied to modified objects or tools is based on artifact shape. Modified objects may be either shaped or unshaped. The morphological classification is applicable only to shaped objects. One of the goals of this classification is to provide a measure of comparability with other sites in the region. Hence, culture historical classes are used despite the fact that they derive meaning from ethnographic analogy. Artifacts that are morphologically similar to those observed in use at the time of European contact or recognized by native informants are assigned contemporary names that imply function, such as "adze," "wedge," "knife" (cf Barnett 1955; Boas 1890; Curtis 1913; Gunther 1927; Jenness n.d.; Stern 1934; Suttles 1951). Functional changes are obscured as a result. Related shortcomings of ethnographic analogy in this context concern the tendency to conflate attributes of style, function, and technology (Campbell 1981; Dunnell 1971, 1978, 1986; Thompson 1978) and the intuitive determination of class membership.

Attempts have been made to isolate the functional aspects of artifact morphology (e.g., Campbell 1981; Lewarch 1982; Thompson 1978). For example, Campbell (1981) created a "Functional Shape Classification." I considered this classification and found it most useful for unusual items that do not have ethnographic counterparts and for which a function is not normally inferred (e.g., "slate rod"). However, the traditional nature of most classes (adzes, bifaces, etc.) resulting from the application of the Functional Shape Classification indicates that a taxonomy similar to Mitchell (1971) could be applied more efficiently to the British Camp assemblage.

Despite the shortcomings of most culture historical classifications, they are widely accepted and are the typical means of comparing sites within a region (Mitchell 1971, 1990). Classifications based on morphology may also be replicable if explicit criteria for inclusion of materials in each class are provided (e.g., Campbell 1981; Kornbacher 1989). This study uses the morphological classification primarily as a descriptive, comparative device. The main purpose of the classification is to allow comparison of the British Camp materials with lithic assemblages from other southern Northwest Coast shell middens. However, based on the assumption that there is a relationship between morphology and function, some functional inferences are also drawn from the different artifact classes present in Ethnozones I and II. Additional functional information of a general nature is derived from an analysis of the edges of shaped and unshaped tools. Information is presented on the presence/absence of wear on bifaces and the angle of the edges of unshaped tools or expedient tools.

2. The Technological Classification

The main purpose of the technological classification is to provide information about lithic reduction activities represented by debitage from each ethnozone. If artifacts are debitage (chipped and not modified after detachment from the parent piece), they are grouped on the basis of Sullivan and Rozen's (1985:759) "Technological Attribute Key" (Figure 2). In selecting Sullivan and Rozen's (1985) scheme for debitage classification, I assumed that the majority of British Camp lithics represent a bifacial reduction strategy.

Archaeologists have criticized Sullivan and Rozen's (1985; Rozen and Sullivan 1989; Sullivan 1987) application of their "attribute key" (e.g., Amick and Mauldin 1989; Ensor and Roemer 1989) on a number of levels. The main concern involves Sullivan and Rozen's inferences and how they are derived. Critics caution that variable reduction techniques, raw materials, and taphonomic processes may significantly impact the debitage category proportions (Prentiss and Romanski 1989). These factors must be taken into account before inferences are made.

When decoupled from the interpretive baggage, the Sullivan and Rozen classification is a useful tool. It is a simple taxonomy based on binary oppositions (Dunnell 1971) and is a quick, efficient means of separating debitage of differing information potential (e.g., those with platforms from those without). The resulting classes are explicit, mutually exclusive, and replicable. They are used here to form

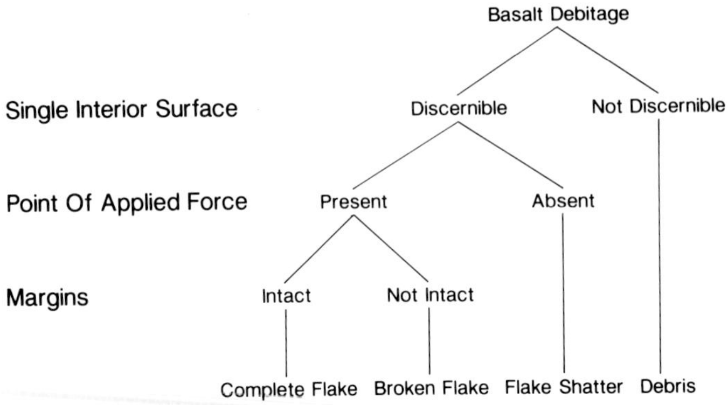

Figure 2 Taxonomic classification of debitage. Adapted from Sullivan and Rozen 1985.

initial descriptive groups from which technological attributes can be selected according to specific research questions. Used in this way, the debitage classification provides a sound basis for further research (Ensor and Roemer 1989:176).

The relative proportions of debitage categories in each ethnozone are compared for this analysis, but the distributions are not evaluated according to Sullivan and Rozen's (1985) interpretive model. Instead, consistency or variation in the proportions of debitage classes are considered indicative of differences or similarities in technological activity between the two ethnozones. Variables of weight, cortex cover, platform characteristics, and dorsal scar count provide information about the range of manufacturing activities represented by the materials from each ethnozone. Sullivan and Rozen's debitage classes are useful for separating the assemblage into groups that do or do not display the measurable attributes necessary for gathering these data.

III. Results

A. *Raw Material*

Most of the lithics (89.2%) recovered from the analyzed portion of Operation A at British Camp are made of a dense black rock of volcanic origin. This raw material, typically referred to as basalt, is the most common material type found in lithic assemblages from the southern Gulf of Georgia and Fraser Delta region (Ham 1982; Kornbacher 1989; Latas 1987; Matson 1976; Mitchell 1971; Murray 1982). Recent petrographic studies (Bakewell 1990) indicate that the material is actually dacite, a type of igneous rock quite similar to basalt but with more quartz and less calcium plagioclase.

The exact location of the source of the raw material is unknown at this time, although samples from a dacite outcrop at Watt's Point north of Vancouver, British Columbia are reportedly similar in chemistry and petrography (E. Bakewell, personal communication). Identified on the basis of macroscopic characteristics alone, cobbles of the raw material have been found in glacial deposits exposed by modern quarrying on San Juan Island and probably also occur on beaches in the archipelago. Latas (Ch. 10, this volume) did not find any nonartifactual nodules or cobbles of the material in his study of sources of fire-cracked rock; but the study was conducted within the bounds of the National Park and his results may not reflect the availability of dacite in the region.

Other raw materials used in the manufacture of stone tools deposited at British Camp include slate, sandstone, mudstone, schist, chert, and obsidian. Together these raw materials comprise less than ten percent of the assemblage and thus are not considered separately in the following discussion.

B. *The Tools*

1. Tool Types, Distribution, and Sample Size

The stone tools analyzed for this study are listed by ethnozone and industry in Table 1. Class richness (Beck and Jones 1989; Jones *et al.* 1989; Leonard and Jones 1989) is indicated at the bottom of the table and shows that Ethnozone II exhibits the greatest diversity in terms of numbers of classes (11 tool classes). The two tool classes not represented in Ethnozone I have only one member in Ethnozone II.

The relationship of class richness to sample size is contrary to that expected if the results were due to the effects of sample size alone (cf Grayson 1984; Jones *et al.* 1989). There is an inverse relationship between the total number of tools and the number of different tool types in each ethnozone.

The frequency of each tool class is also displayed in Table 1. Of the 128 stone tools, 85 are from Ethnozone I and 43 are from Ethnozone II. Sample sizes are very small for most tools, many types are only represented by one member. Utilized flake is the most prevalent tool type in both ethnozones; 36 occur in Ethnozone I and 13 are from Ethnozone II. Small bifaces, retouched flakes, and unshaped abraders are the other classes with eight or more members. All of these are chipped stone classes except unshaped abrader, the members of which are tabular objects made of abrasive sedimentary material, usually sandstone, that exhibit evidence of grinding on one or more surfaces (Figure 3).

2. Expedient Tools

Retouched and utilized flakes together can be considered expedient flake tools (Binford 1979; Parry and Kelly 1987) because they require little or no time and energy to produce. Expedient tools are distinguished from other debitage by the character of the modified edge or edges. Utilized flakes exhibit attrition or microflaking, crushing, polish, or abrasion modification thought to be attributable to use. Retouched flakes exhibit evidence of systematic detachment of at least three flakes from a portion of the perimeter or face of the object. Retouch is assumed to have been done for "rather specific purposes such as altering the shape of the edge, increasing its tensile strength, or resharpening it" (Chapman 1977:378). When dealing with

Table 1 Frequency of Artifact Types by Ethnozone

	Ethnozone I		Ethnozone II	
	N	%	*N*	%
Chipped stone				
Small biface	22	25.9	6	14.0
Large biface	4	4.7	1	2.3
Microblade	1	1.2	0	0
Preform	1	1.2	1	2.3
Retouched flake	16	18.8	6	14.0
Utilized flake	36	42.4	13	30.2
Total	80	94.1	27	62.8
Ground stone				
Knife or fragment	2	2.4	4	9.4
Unshaped abrader	1	1.2	8	18.6
Shaped abrader	0	0	1	2.3
Adze/celt	2	2.4	1	2.3
Total	5	5.9	14	32.6
Pecked stone				
Net weight	0	0	1	2.3
Pendant	0	0	1	2.3
Total	0	0	2	4.6
TOTAL	85	100	43	100
Class Richness (*N* of types)	9		11	

course-grained raw material, microflaking wear on utilized flakes may be difficult or impossible to distinguish from systematic retouch (Campbell 1981; Matson 1976; Pokotylo 1978). Specific guidelines were established for this study; if the source of the edge damage (retouch or use) is unclear, the object is considered utilized. If an object exhibits use-wear and retouch, it is classified as a retouched flake. Regardless of the classification, the number of worn and retouched edges is determined for all expedient tools, since any given discrete object may have been used for a number of different functions.

Edge angles of utilized and retouched flakes were measured to obtain information about the general function or functions of the object. Three categories were recognized: less than 45 degrees (acute), greater than or equal to 45 degrees (steep), and artifacts with an equal number of steep- and acute-angled edges. All utilized and retouched edges were measured and objects were grouped together if the majority of the edges were steep-angled, acute-angled, or if the same number of worn edges of an object were acute and steep (e.g., four worn edges, two acute-angled, two steep-angled). This information is presented in Table 2. The proportions of utilized flakes for which the majority of edges are acute and steep are very similar for the two ethnozones. In Ethnozone I, 52.8% of the utilized flakes have a majority of acute edges, 13.9% have equal numbers of acute and steep edges, and 33.3% have a majority of steep edges. In Ethnozone II, the percentages are 53.8, 7.7, and 38.5, respectively.

The situation differs somewhat for retouched flakes. In Ethnozone I, a slightly larger proportion (43.8%) of the retouched flakes have a majority of acute-angled edges. Nearly 20% (18.8%) have an equal number of acute and steep-angled edges, and 37.5% have a majority of steep-angled edges. Almost all of the retouched flakes of Ethnozone II (5 of 6 or 83.3%) have a majority of steep-angled edges. Only one retouched flake of the six (16.7%) has a majority of acute edges. Sample sizes in both ethnozones, especially Ethnozone II, are small and may be affecting these freqencies and percentages.

The relative proportions of utilized to retouched flake tools (Table 2) are very similar in the two ethnozones at British Camp. Retouched flakes comprise about 31.6% (6 of 19) of the expedient tools from Ethnozone II, and 30.8% (16 of 52) of those from Ethnozone I. The same pattern appears in the ratio of expedient tools (retouched and utilized flakes) to small bifaces in each ethnozone. If expedient tools and small bifaces are for a moment considered the total tool assemblage, then Ethnozone I has 70.3% expedient tools and 29.7% small bifaces and Ethnozone II, 76% expedient tools and 24% bifaces.

3. Morphology of Small Bifaces

Bifaces are "formed by bifacially removing flakes from the complete periphery of the object, so that both ventral and dorsal sides show extensive flake scars originating from the peripheries" (Matson 1976:105–106). These tools usually exhibit bilateral

Figure 3 Unshaped sandstone abraders.

Table 2 Expedient Tool Frequency and Edge Angle

	Ethnozone I		Ethnozone II	
	N	%	*N*	%
Expedient tools				
Retouched flakes	16	30.8	6	31.6
Utilized flakes	36	69.2	13	68.4
Total	52	100	19	100
Angle of retouched edges				
Majority acute	7	43.8	1	16.7
Majority steep	6	37.5	5	83.3
Equal number acute and steep	3	18.8	0	0
Total	16	100	6	100
Angle of worn edges				
Majority acute	19	52.8	7	53.8
Majority steep	12	33.3	6	38.5
Equal number acute and steep	5	13.9	1	7.7
Total	36	100	13	100

symmetry and have one pointed and one nonpointed end. The nonpointed end or base is often modified to facilitate hafting. Small bifaces are distinguished from large bifaces (Figure 4) because they are smaller (60 mm or less in the largest dimension), thinner in cross section (less than 12 mm), and differ in inferred function and method of manufacture. Small biface is a class that has a relatively large number of members (28 in the two ethnozones) while only five large bifaces were found. The discussion below is confined to small bifaces.

Archaeologists studying the Northwest Coast (e.g., Borden 1970; Carlson 1954, 1960, 1970; Kidd 1964; King 1950; Mitchell 1971, 1990) frequently use morphological attributes of small bifaces (usually called projectile points) for classification. Often the goal is to determine which forms persist over time and which decline in frequency or disappear altogether. However, the criteria for assigning objects to a particular projectile point type is seldom consistently applied across different tool forms. For example, the shape of the margins only, or the treatment of the base, or the shape of the margins and the treatment of the base may be the grouping criteria (e.g., leaf-shaped, side-notched, triangular-unstemmed). Which attribute is used depends on intuitive criteria that are often not articulated. Thus, membership in a group, even within the same collection, may not be replicable between investigators.

To avoid such confusion and ensure comparability, the morphological description of each small biface from British Camp is accomplished through the consistent use of two aspects of form: (1) general shape which refers to the shape of the margins in plan view, from the widest point of the blade to the point; and (2) treatment of the base which refers to the shape of the object in plan view from the widest point of the blade to the proximal end (and also concerns whether or not the object extends proximally past the widest point of the blade). These data are presented in Table 3. The majority of the tools are triangular (lateral margins are straight), 54.6% in Eth-

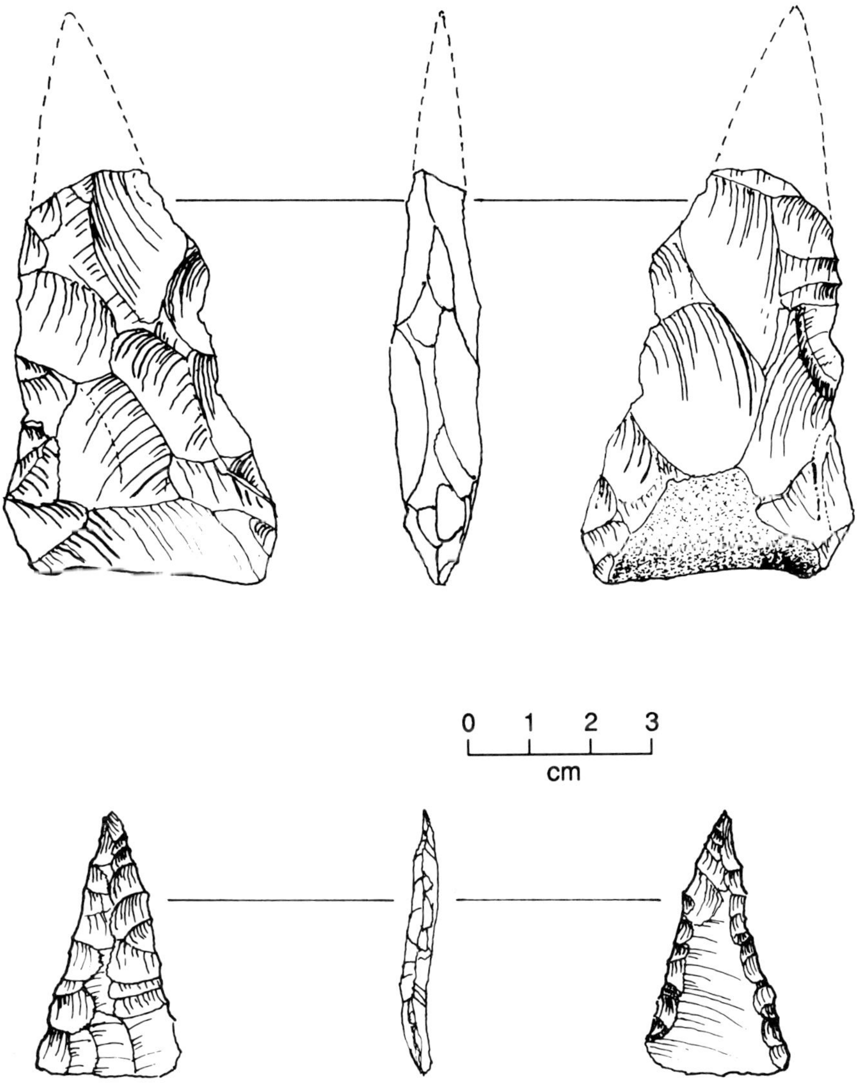

Figure 4 Large and small bifaces.

nozone I and 66.7% in Ethnozone II, with unstemmed bases. Most of the nontriangular bifaces occur in Ethnozone I facies. Two small bifaces that are not triangular, one leaf-shaped (lateral margins convex) and one leaf-triangular (one convex lateral margin, one straight) are from Ethnozone II. Examples of unstemmed and stemmed small bifaces from British Camp are shown in Figure 5.

Treatment of the base of small bifaces was originally considered in more detail than shown in Table 3. The dimensions were collapsed into stemmed and unstemmed categories to increase sample sizes. The majority of the small bifaces in both ethnozones

Table 3 Small Biface Morphological Data

Shape of margins	General shape	Ethnozone I N	Ethnozone I %	Ethnozone II N	Ethnozone II %
Indeterminate		3	13.6	0	0
Straight	Triangular	12	54.5	4	66.7
Convex	Leaf-shaped	5	22.7	1	16.7
1 straight, 1 convex	Leaf-triangular	2	9.1	1	16.7
Total		22	100	6	100
	Basal treatment				
	Indeterminate	7	31.8	0	0
	Unstemmed	12	54.5	6	100
	Stemmed	3	13.6	0	0
	Total	22	100	6	100

have unstemmed bases. The major difference in biface morphology between the two ethnozones is that all of the stemmed bifaces occur in Ethnozone I. Despite the collapsed categories, sample size may still be affecting these frequencies. An absence of stemmed forms in the recent ethnozones, however, is typical of other assemblages in the region (cf Carlson 1983).

4. Small Biface Wear and Condition

All small bifaces were grouped according to whether they were broken or complete and worn or unworn in an effort to evaluate functional differences between Ethnozone I and II. Attributes of condition and wear were collapsed from a more divisive scheme to create larger sample sizes. The results are displayed in Table 4. The data on biface condition indicate that over twice as many bifaces are broken (68.2%) as complete (31.8%) in Ethnozone I, while equal numbers are broken and complete in Ethnozone II.

Table 4 also shows the number of worn and unworn bifaces in both ethnozones. Nearly equal numbers of bifaces are worn and unworn in Ethnozone I (45.5 and 54.5%, respectively). A greater proportion of Ethnozone II bifaces exhibit no wear (66.7%) than are worn (33.3%).

Despite collapsing the original modes of biface shape, condition, and wear into dichotomous categories, the effects of sample size must be considered. Only six small bifaces are from Ethnozone II. Differences in biface condition and presence/absence of wear between the two ethnozones may be random and interpretations based on these differences spurious. Thus condition and wear attributes of Ethnozone II bifaces may be evaluated more productively on a presence/absence basis. Frequency information is provided, however, to show wear and condition attributes of the Ethnozone I bifaces.

Figure 5 Unstemmed and stemmed small bifaces from British Camp.

5. Tool Technology

Slightly different aspects of the analysis must be emphasized in order to use the tool data for addressing technological issues. Of primary concern is not the tool form but the type of manufacturing technique used and how this differs between the two ethnozones. The frequency of tools made by chipping, grinding, and pecking in each ethnozone is displayed in Table 5. The majority of the tools from the British Camp site were manufactured by chipping and grinding. Chipping is the dominant mode of manufacture in both ethnozones; in Ethnozone I, 94.1% of the tools are made by chipping.

A comparison of the relative frequency of chipped and ground stone tools in Ethnozones I and II (Table 6) indicates a decline in chipped stone over time. Over 70% (72.1%) of the chipped stone tools are from the older deposits of Ethnozone I. This difference is emphasized (92.5%) when counts are corrected for the volume of excavated material in each ethnozone (24.5 tools per 1000 liters in Ethnozone I and 1.99 tools per 1000 liters in Ethnozone II). Although the volume of Ethnozone II is much greater than that of Ethnozone I (Table 6), the frequency of lithics is greater in Ethnozone I.

Sample sizes are smaller for ground stone tools, but the majority (73.7%) of the ground stone tools recovered are from Ethnozone II. Ground stone tools are more

Table 4 Small Biface Data: Wear and Condition

	Ethnozone I		Ethnozone II	
	N	%	*N*	%
Wear				
Present	10	45.5	2	33.3
Absent	12	54.5	4	66.7
Condition				
Complete	7	31.8	3	50.0
Broken	15	68.2	3	50.0

evenly distributed when the counts are corrected for volume (1.5 tools per 1000 liters in Ethnozone I and 1.0 per 1000 liters in Ethnozone II). While the raw counts suggest that Ethnozone II deposits contain more ground stone tools, the corrected counts indicate this difference may simply be a function of the disparity in volume between the two ethnostratigraphic units.

C. *The Debitage*

1. Relative Abundance

The information presented above and in Chapter 6 of this volume documents a decline in the relative frequency of chipped stone tools over time at the British Camp site. The debitage also reflects this pattern (Table 6). Of 304 pieces of debitage greater than 12 mm (retained on the 1/2-inch or 1-inch screen), 260 (85.5%) occur in Ethnozone I. Table 6 shows the decrease in debitage between Ethnozone I and Ethnozone II is emphasized when the counts are corrected for volume of excavated material.

The decrease in abundance of chipped stone debitage is even more apparent when the small debitage is considered. Small debitage in this context refers to materials less than 12 mm but greater than 3 mm [items retained on the 1/4-inch (6 mm) or 1/8-inch (3 mm) screens]. A 25% sample of the small debitage collected from one 2 × 2-m excavation unit (308/300) was counted for this study following the protocol recommended by Ahler (1989). As Table 7 shows, nearly 95% of the small deb-

Table 5 Frequency of Stone Tools by Industry

	Ethnozone I		Ethnozone II	
Industry	*N*	%	*N*	%
Chipped	80	94.1	27	62.8
Ground	5	5.9	14	32.6
Pecked	0	0	2	4.7
Total	85	100	43	100

Table 6 Lithic Counts Corrected by Volume

	Ethnozone[a]			
	I		II	
	N	%	*N*	%
Chipped stone tools				
Uncorrected frequency	80.0	72.1	27	27.9
Corrected for volume	24.5	92.5	1.99	7.5
Ground stone tools				
Uncorrected frequency	5.0	26.3	14	73.7
Corrected for volume	1.5	59.8	1.0	40.2
Debitage				
Uncorrected frequency	260	85.5	84	14.5
Corrected for volume	79.7	92.8	6.2	7.2

[a]Volume of Ethnozone I = 3264 liters; volume of Ethnozone II = 13,624 liters; counts are lithics per 1000 liters.

itage sample is from Ethnozone I. In addition to overall abundance, the relative proportions of different size categories varies between ethnozones as well. Ethnozone I has a higher proportion of small-sized debitage from the 1/8-inch screen (83.9%) than Ethnozone II (62.8%).

The bottom portion of Table 7 shows the ratio of chipped stone tools to debitage with and without the sample of small debitage. Despite the difference in the relative abundance of debitage between each ethnozone (and the increase in proportion of debitage in both ethnozones), the ratio of debitage to tools remains very similar between Ethnozone I and II. In Ethnozone I, 99.76% of the lithics are debitage and only 0.24% are chipped stone tools. In Ethnozone II, chipped stone tools comprise 1.37% of the lithics and the debitage, 98.63%.

Table 7 Debitage Data

	Ethnozone			
	I		II	
Debitage size category	*N*	%	*N*	%
⅛ inch (sample)	28,425		1,220	62.8
¼ inch (sample)	5,195	83.9	640	32.9
>½ inch	260	15.3	84	4.3
		0.8		
Total	33,880	100	1,944	100
Ratio of debitage to tools	Deb.	Tools	Deb.	Tools
Small debitage excluded	76.5	23.5	75.7	24.3
Small debitage included	99.76	0.24	98.63	1.37

Table 8 Debitage Category Frequency

	Ethnozone			
	I		II	
Debitage category	*N*	%	*N*	%
Complete flakes[a]	94	36.2	30	35.7
Broken flakes[b]	40	15.4	12	14.3
Flake shatter	102	39.2	36	42.9
Debris	24	9.2	6	7.1
Total	260	100	84	100
Cores	6		3	

[a]Includes bipolar flakes.
[b]Includes split flakes.

2. Technological Analysis

The frequency of large debitage (greater than 13 mm) grouped using Sullivan and Rozen's (1985) taxonomy (Figure 2) is displayed in Table 8. The relative proportions of complete flakes (including bipolar flakes), broken flakes (including split flakes), flake shatter (Sullivan and Rozen's flake fragments), debris, and cores are remarkably similar between the two ethnozones. Flake shatter and complete flakes dominate both assemblages. Broken flakes are the next most numerous category (approximately 15% in each ethnozone), and debris comprises less than 10% of Ethnozone I and Ethnozone II assemblages (9.2 and 7.1% respectively). Both assemblages have very few cores; six are from Ethnozone I and only three are from Ethnozone II.

This study does not identify specific stages of manufacture represented by individual pieces of debitage. The reduction process is viewed as a continuum, not as a series of discrete, identifiable stages. However, the assemblages of Ethnozone I and II at British Camp can be compared to address the issue of whether or not they are technologically similar in terms of the range of reduction stages represented and/or the general position on the continuum indicated. For example, does one assemblage indicate only the very terminal end of the reduction process with debitage suggestive of pressure flaking, or resharpening, for example, while debitage from the other ethnozone is indicative of early stages of core reduction only? Or, are all stages of the manufacturing process represented in both assemblages from initial core reduction to refinishing?

Magne (1983), Magne and Pokotylo (1981), and Pokotylo (1978) (and others) have conducted actualistic studies to determine which debitage attributes are the most useful indicators of reduction stage information. Weight, cortex cover, platform scarring, and dorsal scar count are regarded as most effective (Magne 1983; Magne and Pokotylo 1981). General assumptions regarding the technological significance of variations in these attributes are summarized in Table 9. As the table indicates, the basic assumption from which these generalizations are derived is that lithic man-

Table 9 Assumptions of Bifacial Reduction Model for Inferring Reduction Stage

Attribute	Early stage		Late stage
Weight	More	→	Less
Cortex cover	More	→	Less
Platform scar count	Less	→	More
Platform cortex	More	→	Less
Dorsal scar count	Few	→	Many

ufacture is a reductive process in which each successive step results in the removal of additional material. Thus, the later in the reduction process, the less weight is removed, the less cortex or outer weathered surface remains, and the higher the frequency of platform scars and dorsal scars (indicative of prior platform preparation and flake removals, respectively).

Statistics on the weight (in grams) of complete flakes, broken flakes, flake shatter, and debris greater than 12 mm are displayed in Table 10. Mean debitage weight in Ethnozone I is less than that of Ethnozone II for all debitage categories, suggesting that Ethnozone II debitage represents earlier stage manufacturing. Sample sizes of Ethnozone II debitage are very low, however, and mean weights are affected by outliers. The median is a more resistant measure of central tendency and is more appropriate considering the disparity in sample size between the assemblages of the

Table 10 Debitage Weight in Grams

	Ethnozone I	Ethnozone II
Complete flakes		
Mean	6.58	7.66
Standard deviation	6.87	7.89
Median	4.12	4.79
N	94	30
Broken flakes		
Mean	5.90	6.57
Standard deviation	5.88	5.81
Median	4.57	5.40
N	40	12
Flake shatter		
Mean	4.15	4.73
Standard deviation	6.77	6.11
Median	2.79	2.48
N	102	36
Debris		
Mean	5.79	13.92
Standard deviation	7.44	23.16
Median	3.45	3.98
N	24	6

Table 11 Debitage Attributes (Cortex Cover, Platform Characteristics, and Dorsal Scar Count)

	Ethnozone I		Ethnozone II	
	N	%	*N*	%
Cortex Cover				
Complete cover	7	5.3	2	5.9
Partial cover	53	39.8	8	23.5
No cover	74	54.9	24	70.6
Total	134	100	34	100
Platform Characteristics				
Incomplete	30	22.4	7	19.4
Cortex present	64	47.8	18	50.0
Cortex absent	40	29.9	11	30.6
Total	134	100	36	100
Scarring absent platform incomplete	92	68.7	25	69.4
Scarring present	42	31.3	11	30.6
Total	134	100	36	100
Dorsal Scar Count				
0–3 scars	80	35.7	30	46.9
4–6 scars	60	26.8	19	29.8
More than 6 scars	84	37.2	15	23.7
Total	224	100	64	100

two ethnozones. The medians for each debitage category are strikingly similar. For complete flakes, broken flakes, flake shatter, and debris categories, the medians of Ethnozone I debitage weights are within 1.0 gram of the medians of Ethnozone II debitage weights. The data on weight suggest very similar reductive activities produced the lithic assemblages of Ethnozone I and II facies.

Dorsal surface cortex cover was measured for all debitage with platform-remnants (Table 11). While the majority of the flakes in both ethnozones lack cortex on the dorsal face (less than 6% are completely cortical), Ethnozone II contains a greater proportion of noncortical flakes (70.6%) than Ethnozone I (54.9%), suggesting a difference in the range of manufacturing stages represented by the two ethnozones. Ethnozone I flakes are indicative of middle and late stage reduction. Very little early stage manufacture is indicated by the debitage of either ethnozone. However, Ethnozone II flakes suggest more of an emphasis on late stage reduction than indicated in Ethnozone I.

Platform characteristics are also displayed in Table 11. Platforms that lack scarring, are crushed or otherwise incomplete, and/or exhibit cortex, are indicative of earlier stages in the manufacturing process. The proportions of scarred and unscarred platforms are very similar in the two ethnozones. The relative proportions of flakes with incomplete platforms, those with platforms that are complete and cortex covered, and those with complete platforms that lack cortex are also very similar between the two ethnozones. Flakes with cortex-covered platforms are most numerous in both ethnozones. These data indicate more of an emphasis on early than late manufacturing stages in both stratigraphic units.

Dorsal scar data are broken into ordinal level categories that are inferred to be roughly indicative of early (zero to three scars), middle (four to six scars), and late (more than six scars) stages in the reduction process (Table 11). Although a range of reduction stages is represented by the flakes of both ethnozones, the dorsal scar count data of Ethnozone I indicate nearly equal representation of early and late stages. By contrast, early stage reduction is emphasized in Ethnozone II.

IV. Discussion

The discussion below summarizes the results of the lithic analysis in terms of functional and technological information and compares the findings for Ethnozones I and II of the British Camp midden.

A. *Functional Inferences*

Data on class richness (Table 1) indicate that the tool assemblage of Ethnozone II is more diverse than that of Ethnozone I. However the difficulties with sample size cast doubt on the significance of this difference. Inferences must take into account this possibility. In addition, the difference in time represented between Ethnozone I and Ethnozone II must be considered. The facies of Ethnozone II represent almost 1000 years, while those of the portion of Ethnozone I analyzed for this study represent only about 500 years. The generalization that the number of classes represented increases with time could account for the apparently more diverse tool assemblage of Ethnozone II.

Based on tool morphology and ethnographic analogy, the tools of both Ethnozone I and II indicate a wide range of activities such as woodworking (adzes or celts), resource processing (expedient flake tools and ground stone knives), tool manufacturing and refurbishment (abraders and expedient tools), and a variety of resource procurement activities (bifaces). In general, the same kinds of tools are represented in both ethnozones. Microblade is the only tool from Ethnozone I not represented in Ethnozone II. Net weight, shaped abrader, and decorative object are the three tool types of Ethnozone II that do not occur in Ethnozone I.

Interpretation of differences in the relative frequencies of tool types between Ethnozones I and II is subject to the deleterious effects of small sample sizes. Nevertheless, proportional changes indicate a decrease in chipped stone tools (bifaces, expedient tools) and an increase in ground stone tools, especially ground stone cutting tools [generally associated with food processing (cf Burley 1980; Hayden 1981)] and abraders (probably used in the manufacture and refurbishment of ground stone and bone tools). Although utilized flakes and retouched flakes maintain an important role in Ethnozone II (as determined by the relative proportion within each ethnozone), they are more frequent in the earlier ethnozone and decline in frequency over time.

Due to their low production costs, expedient tools are typically discarded in the context of use (Binford 1979). The presence of expedient tools (retouched and utilized flakes) in both ethnozones at British Camp indicates that expedient tool use occurred throughout the occupation of the area. Inferring specific tasks for which flakes were used, however, is difficult in the absence of detailed microwear analyses or residue studies. Equating acute angles with cutting activities and steep angles with scraping activities (Wilmsen 1970) oversimplifies complex manifestations of use on individual stone tools. However, archaeologists have demonstrated (Hayden 1979) that variation in edge angle provides an indication of tool function. The gross distinction made here between acute- and steep-angled edges is assumed to be adequate for evaluating general function of an expedient tool assemblage as a whole. Based on this assumption and considering utilized flakes only, the data indicate that the variability of tasks using flakes in Ethnozone I is greater than that of Ethnozone II. The older ethnozone has a larger number of utilized flakes with acute- and steep-angled edges on the same object. The angle of the worn edges suggests the otherwise unaltered flakes from both Ethnozone I and II were most frequently used for cutting tasks. Again, sample size may be affecting this pattern.

Acknowledging that sample size may be inadequate, tentative inferences about the function of retouched flakes are also advanced. The angle of retouched edges is considered indicative of intended or potential purpose rather than a product of use. Ethnozone I flakes were retouched for an apparent variety of tasks, reflected in the number of single objects with edges modified to form acute- and steep-angled edges, perhaps suitable for cutting and scraping activities. In Ethnozone II, less potential task variability is indicated by the retouched flake edge angles (steep-angled edges predominate). Again, a 1:1 correspondence between the angle of a particular edge and function is not suggested. One can imagine a situation, for example, in which a sharp flake edge is steeply retouched to make it safer to hold in the hand while the opposite acute-angled edge is used for cutting.

Small bifaces are among the chipped stone tools that appear to decrease in frequency over time relative to other tool types. Although it is tempting to suggest that tasks for which they were used (e.g., hunting) declined in importance over time, small sample sizes preclude a conclusive statement to this effect. Biface form also differs between Ethnozone I and II. Stemmed forms are absent in Ethnozone II and stemless triangular forms dominate.

Nearly equal numbers of worn and unworn bifaces in Ethnozone I (Table 4) suggest that these tools were made and used at the British Camp location. Although wear is not necessarily indicative of use at a specific location (the tool may be curated), the large proportion of worn bifaces (45.5%) suggests the tools were used at British Camp. The number of unworn bifaces indicates that they were manufactured at this location and not used or not used intensively before they were deposited. The relative proportions of complete and broken bifaces in Ethnozone I support these inferences. Over 30% (31.8%) of the Ethnozone I small biface assemblage is complete, indicating that the tools may have been made at British Camp. Broken bifaces

(68.2%) are more problematic. They may represent breakage during use and subsequent discard, curation for recycling, or postdepositional breakage. Although the same inferences might be drawn for the small bifaces from Ethnozone II, the sample is so small that relative frequencies are probably not very meaningful. Evaluated nominally, the presence of complete and unworn bifaces, as well as broken and worn bifaces, is supportive of the idea that biface manufacture and use occurred during the time represented by Ethnozone II facies.

In summary, the tool analyses have demonstrated overall similarity between the kinds of activities conducted during the deposition of the facies of Ethnozone I and Ethnozone II. Tool manufacture and use may be inferred from the data. A decline in the intensity or importance of those activities involving the production and use of chipped stone over time is also indicated.

B. *Technological Inferences*

The technological analysis of British Camp lithic materials provides additional evidence of a decline in the importance of chipped stone tools over time. The structure of the chipped stone tool assemblages of both ethnozones, however, is remarkably similar. For instance, the ratios of tools to debitage between the two units, even when the small debitage are considered, is nearly identical. The relative abundance of complete flakes, broken flakes, flake shatter, debris, and cores in Ethnozone I is also very similar to the proportions of these classes in Ethnozone II. Although the technological and functional significance of differing proportions of these categories has not been established (Amick and Mauldin 1989; Ensor and Roemer 1989), and the effects of certain postdepositional alterations (e.g., Prentiss and Romanski 1989) are not fully known at this time, the evidence suggests similar use of chipped stone in both ethnozones, perhaps conducted at a lower level of intensity during the deposition of Ethnozone II facies.

The results of the microdebitage analysis (Madsen, Ch. 9, this volume) indicate that lithic manufacturing did not take place during the deposition of Ethnozone II facies. While other aspects of the debitage analysis are indicative of a decline in the intensity of manufacturing activities, the presence of debitage argues for chipped stone tool manufacture during Ethnozone II. In addition, the structure of the chipped stone assemblage appears very similar for the two ethnozones, indicating that chipped stone tools were produced and used in a similar manner over time.

These conflicting aspects of the data suggest that the area of the midden sampled as Operation A was used in functionally distinct ways during the times represented by Ethnozone I and Ethnozone II facies. Lithics from Ethnozone I (large-sized debitage and microdebitage) indicate tool use and production occurred in this area of the site. The virtual absence of microdebitage in Ethnozone II facies and the presence of larger sized debitage in low frequencies provide evidence that during the more recent periods of occupation, the same area was used as a refuse disposal area. I am suggesting that the moveable waste products of manufacturing and related lithic reduction activities were deposited but chipped stone tool manufacture did not occur at this location. It is likely that lithic reduction activities were conducted in close proximity to this portion of the midden, however, as the larger by-products of an identical manufacturing process were deposited in the sampled portion of the British Camp midden during this time.

Measurement of weight, dorsal cortex, platform scarring, and dorsal scars demonstrates technological similarity between the two ethnozones. However, the data are contradictory in terms of what they indicate about reduction stages represented. For example, platform scar data for both ethnozones indicate early and middle stage reduction. But cortex cover data, though basically consistent between the two ethnozones, suggest later stages of reduction.

These conflicting findings may result from the application of a bifacial reduction model to a situation in which biface production was not the dominant manufacturing activity. The assumptions outlined in Table 9 are pervasive in the debitage literature (e.g., Collins 1975; Jelinek 1976; Magne 1983; Raab *et al.* 1979; Sullivan and Rozen 1985). They are based on an implicit model of the manufacturing process that begins with reduction of a quarried nodule of raw material and ends with the production of a bifacial tool. The British Camp data indicate that the majority of the debitage was not produced by biface reduction or maintenance activities; a different reduction strategy is represented.

I suggest that a more appropriate model for this particular site (and perhaps region) involves the reduction of cobbles from eroding glacial deposits for the production of usable flake tools. Systematic reduction of cobbles for the production of expedient flake tools (cf Morrow 1984; Thorne and Johnson 1979), as well as bifacial tools, may account for the anomalous debitage assemblage.

A hypothetical model of cobble reduction is depicted in Figure 6. The rounded, weathered outer surface of a cobble only covers the dorsal surface of the first (and last) flake removed. All subsequent removals lack dorsal cortex but their platforms are cortex covered or crushed. The cores that result from this reduction strategy are usually exhausted and may look more like flakes than cores. Although bifacial tools may be manufactured from flakes produced by cobble reduction, this technique is probably highly correlated with expedient tool manufacture.

The most cost-effective means of producing flakes from a cobble is bipolar reduction or the hammer and anvil technique (Crabtree 1972). This process results in flakes that lack a prominent bulb of percussion and resemble segments of an orange, frequently with crushing on the distal or supported end; flake shatter is also a typical adjunct of bipolar reduction (Campbell 1981:352; Crabtree 1972:10). The platforms of flakes produced by this method are usually cortex covered or incomplete; they are frequently crushed with the force required to break the nodule or cobble. The process of bipolar flaking may use the entire core with the exception of wedge-shaped remnants that exhibit cortex-covered dorsal surfaces. Although cobbles can be reduced by other percussion techniques, bipolar reduction is an efficient way to produce flakes from cobbles.

Considering the British Camp data within the framework of a cobble reduction model sheds some light on the seemingly anomalous results. The model predicts that debitage produced during the reduction of cobbles will have a high proportion of debitage lacking dorsal cortex with cortex-covered or incomplete platforms. The British

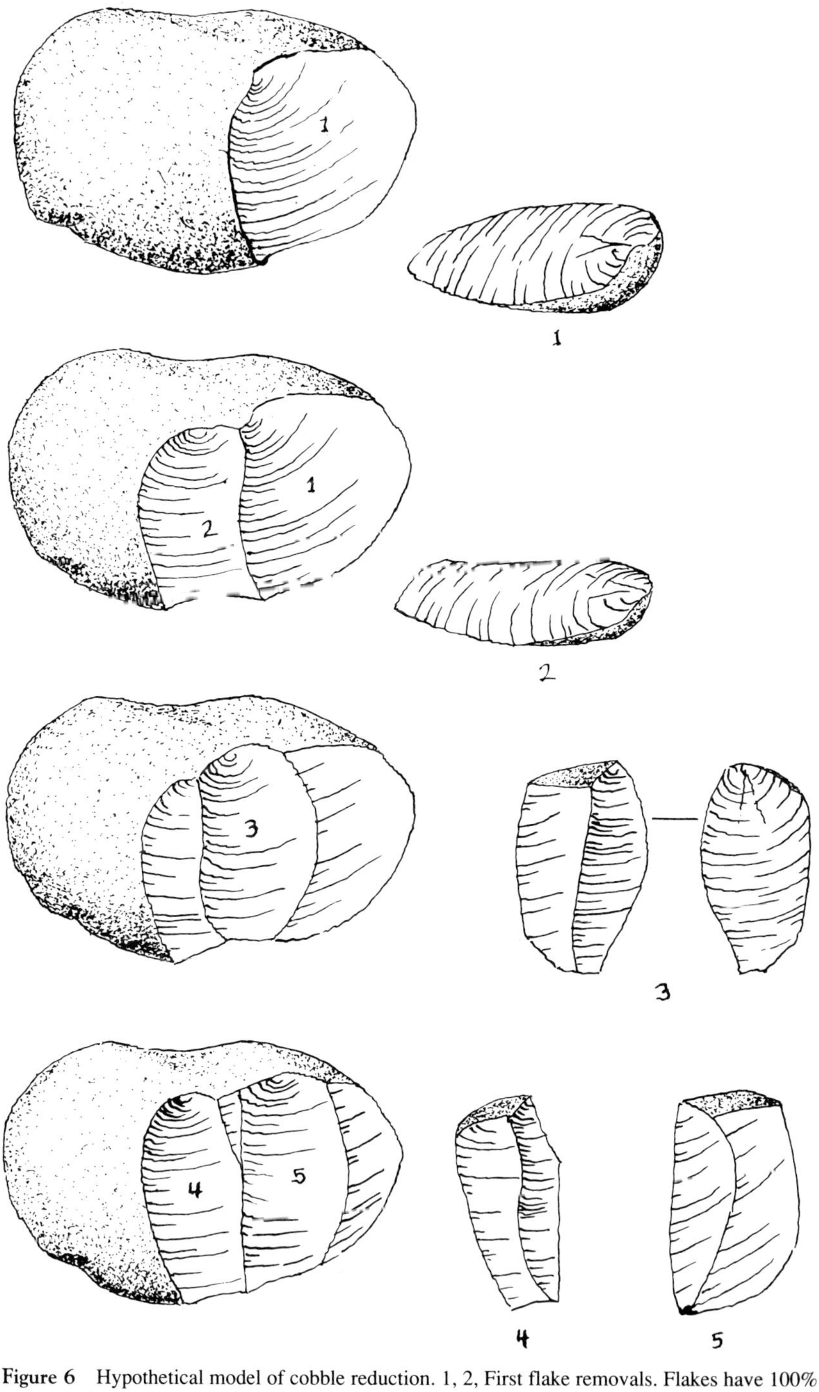

Figure 6 Hypothetical model of cobble reduction. 1, 2, First flake removals. Flakes have 100% cortex cover on the dorsal face and cortex-covered platforms. 3, 4, 5, Subsequent removals. Flakes lack dorsal cortex but have cortex-covered platforms.

Figure 7 Dacite cores from British Camp.

Camp debitage data is consistent with this prediction. The lack of cores recovered from the site is also predicted, as the reduction of small cobbles or pebbles most frequently results in exhausted, flake-like cores that may be difficult to identify (e.g., Figure 7). The Ethnozone I and Ethnozone II lithics at British Camp exhibit evidence of bipolar (or hammer and anvil) reduction, evidence again consistent with the predictions of the cobble reduction model. Finally, the most abundant tool classes in both ethnozones are expedient flake tools, an indication that usable flakes were important products of the manufacturing process. The findings are not conclusive. However, the cobble reduction model proposed here (in conjunction with a bifacial reduction model) provides a more parsimonious account of the British Camp lithic data than a traditional model of bifacial reduction.

V. Conclusion

Three substantive conclusions may be drawn from the present study: (1) the technology used to manufacture stone tools at British Camp was based on cobble reduction and focused on the production of flake tools and bifacial tools (therefore, the lithic data cannot be evaluated according to a traditional model of bifacial reduction alone); (2) the change in relative abundance of chipped stone tools indicates a decline in their importance relative to other kinds of tools and is not a result of change in the technological processes involved in their production; and (3) the function of the analyzed portion of the British Camp midden changed during the time periods represented by Ethnozone I and Ethnozone II materials. Each of these "conclusions" are advanced not as established fact but as a framework for the formulation of testable hypotheses.

In terms of the technological activities conducted, the data indicate that cobble reduction using bipolar and other percussion techniques was the dominant means of chipped stone tool production at British Camp during the period of prehistory represented by the analyzed portions of Ethnozone I and II. Actualistic research is needed to help us establish better criteria for identifying cobble reduction in the archaeological record. Studies should emphasize the use of cobbles for producing a range of tools and not be bound by traditional assumptions about a desired end product.

Regarding the decline in the relative (and absolute) abundance of chipped stone materials over time: this change has been observed at many locations in the region and interpreted as a functional shift in resource procurement activities. Some researchers (e.g., Burley 1980; Carlson 1970; Croes and Hackenburger 1988; Mitchell 1971, 1990) have hypothesized an increased emphasis on fishing between 1500 and 1000 years ago that required the production of bone and/or wood tools and deemphasized the role of chipped stone (equated with hunting) in the economy. Small bone points in particular (generally regarded as barbs for composite fish hooks) increase in relative frequency during the San Juan Phase and the outer coastal equivalent of this time period (cf Croes and Hackenberger 1988).

Although this is a logical way to account for the materials observed, it is unsubstantiated and is not an explanation of why the change occurred. We must first understand the nature of such changes before we can address possible causes. The hypothesis investigated for this paper (that the difference in abundance of chipped stone materials between Ethnozone I and Ethnozone II is related to a technological change in these materials) was advanced as a step toward understanding the nature of the change that characterizes the Marpole and San Juan Phases. This hypothesis may now be rejected. Although some features of the debitage indicate differences between the two units, this analysis has shown the technological structure of the assemblages to be very much the same. The problem may now be reformulated to hypothesize that the relative intensity of the activities involving the use of chipped stone declined over time.

Finally, I suggest that changes documented at British Camp, as well as other shell middens in the region, may be in part due to a functional shift in the portion of the site being investigated by archaeologists. For British Camp specifically, the data suggest that as the midden prograded and the location of the shoreline changed (Whittaker and Stein, Ch. 2, this volume), the use of space changed relative to the shoreline. After the deposition of Ethnozone I facies, stone tool manufacturing activities were no longer conducted in the same area. Tools were manufactured elsewhere and the refuse of that and other activities was deposited in the location of the British Camp midden sampled as Operation A.

Unlike other dense concentrations of archaeological material, shell middens reflect the dynamic interaction of people with a changing shoreline, and the contribution of people to those physical changes. The results of the lithic analyses presented here suggest that we need to alter traditional views of archaeological sites as discrete, unchangeable areas of use and consider the role of postdepositional processes in the formation of midden deposits. To increase our understanding of the processes of change that have occurred, a range of materials must be examined that may lack the stylistic information useful for telling time. As demonstrated by this study, stone tool data must be supplemented with debitage analyses if we are to understand the complexities of the aboriginal use of the region and how it may have changed over time.

Acknowledgments

This chapter relies on data generated for my thesis research. Thus I would like to thank the people who served on my advisory committee: David Pokotylo (chair) and R. G. Matson of the University of British Columbia, and Julie Stein of the University of Washington. Each contributed in important and varied ways. Eric Bangs, Claudia Vergnani Vaupel, and Connie Grafer counted the sample of 1/4-inch and 1/8-inch debitage. Mary Parr and Walt Bartholomew exercised their curatorial genius organizing and cataloging the lithic materials. Eric Rasmussen illustrated Figure 6, my conception of cobble reduction, and Figure 4. Tim Hunt drafted Figure 1 and Figure 2. Kris Wilhelmsen, Chris Pierce, Fran Whittaker, Julie Stein, and David Pokotylo reviewed earlier versions of the chapter. All made helpful comments and suggestions and contributed substantially to the quality of the chapter. Finally, I would like to thank all the students of the San Juan Island Archaeological Project field school for their hard work, humor, and enthusiasm.

References

Ahler, S. A.

1989 Mass analysis of flaking debris: studying the forest rather than the tree. In *Alternative approaches to lithic analysis*, edited by D. O. Henry and G. H. Odell. Archaeological Papers of the American Anthropological Association Number 1. Pp.85–118.

Ames, K.

1981 The evolution of social ranking on the northwest coast of North America. *American Antiquity* **46**(4), 789–805.

Amick, D. S., and R. P. Mauldin

1989 Comments on Sullivan and Rozen's "Debitage Analysis and Archaeological Interpretation." *American Antiquity* **54**, 166–168.

Bakewell, Edward

1990 Petrographic and geochemical source-modeling of lithics from archaeological contexts. B.A. Honor's thesis, Department of Geology, University of Washington, Seattle.

Barnett, H. G.

1955 The Coast Salish of British Columbia. University of Oregon, Eugene.

Beck, C., and G. T. Jones

1989 Bias and archaeological classification. *American Antiquity* **54**, 244–262.

Bernick, K.

1983 A site catchment analysis of the Little Qualicum Creek Site, SiSc 1: A wet site on the east coast of Vancouver Island, B.C. *Archaeological Survey of Canada Paper* No. 118. National Museum of Man Mercury Series, Ottawa.

Binford, L. R.

1968 Archaeological perspectives. In *New perspectives in archaeology*, edited by S. R. Binford and L. R. Binford. Chicago: Aldine. Pp.5–32.

1977 General introduction. In *For theory building in archaeology*, edited by L. R. Binford. New York: Academic Press. Pp.1–10.

1979 Organization and formation processes: looking at curated technologies. *Journal of Anthropological Research* **35**, 255–273.

Boas, F.

1890 Second general report on the Indians of British Columbia I. The LkungEn. *British Association for the Advancement of Science* **60**, 563–582.

Borden, C.

1951 Facts and problems of Northwest Coast prehistory. *Anthropology in British Columbia* **2**, 35–52.

1970 Culture history of the Fraser Delta Region: an outline. *B.C. Studies* 6–7, 95–112.

1983 Prehistoric art of the Lower Fraser Region. In *Indian art traditions of the Northwest Coast*, edited by Roy L. Carlson. Archaeology Press, Simon Fraser University, Burnaby, British Columbia. Pp.131–166.

Burley, D.

1980 Marpole: Anthropological reconstructions of a prehistoric Northwest Coast culture type. Simon Fraser University, Department of Archaeology, Publication No. 8, Burnaby, British Columbia.

Campbell, S. K.

1981 The Duwamish no. 1 site: A lower Puget Sound shell midden. University of Washington, Office of Public Archaeology, Research Report 1.

Carlson, R. L.

1954 Archaeological investigations in the San Juan Islands. Unpublished Master's thesis, Department of Anthropology, University of Washington, Seattle.

1960 Chronology and culture change in the San Juan Islands, Washington. *American Antiquity* **25**(4), 562–586.

1970 Excavations at Helen Point on Mayne Island. In *Archaeology in British Columbia, new discoveries. B.C. Studies*, No. 6-7. Pp.113–125.

1983 Prehistory of the Northwest Coast. In *Indian art traditions of the Northwest Coast*, edited by R. L. Carlson. Archaeology Press, Simon Fraser University, Burnaby, British Columbia. Pp.13–32.

Chapman, R. C.

1977 Analysis of the lithic assemblages. In *Settlement and subsistence along the Lower Chaco River*, edited by Charles A. Reher. Albuquerque: The University of New Mexico Press. Pp.371–452.

Clark, L. A.

1991 Archaeology of Seal Rock (35LNC14). In *Prehistory of the Oregon Coast: The effects of excavation strategies and assemblage size on archaeological inquiry*, edited by R. L. Lyman. San Diego: Academic Press. Pp.175–240.

Collins, M. B.

1975 Lithic technology as a means of processual inference. In *Lithic technology: Making and using stone tools*, edited by Earl Swanson. The Hague: Mouton. Pp.15–34.

Crabtree, D. E.

1972 An introduction to flintworking. Occasional Papers of the Idaho State University Museum, No. 28, Pocatello.

Croes, D. R.

1976 The excavation of water-saturated sites (wet sites) on the Northwest Coast of North America (editor). *Archaeological Survey of Canada Paper* No. 50. National Museum of Man Mercury Series, Ottawa.

Croes, D. R., and S. Hackenberger

1988 Hoko River archaeological complex: modeling prehistoric northwest coast economic evolution. In *Prehistoric economies of the Pacific Northwest Coast*, edited by Barry L. Issac. Research in Economic Anthropology, Supplement 3. Pp.19–86.

Curtis, E. S.

1913 *The North American Indian, Vol. IX, Coast Salish.* Norwood, Massachusetts.

Daugherty, R., and J. Friedman

1983 An introduction to Ozette Art. In *Indian art traditions of the Northwest Coast*, edited by R. L. Carlson. Archaeology Press, Simon Fraser University, Burnaby, British Columbia. Pp.183–198.

Dunnell, R. C.

1971 *Systematics in prehistory*. New York: Free Press.

1978 Style and function: a fundamental dichotomy. *American Antiquity* **43**(2), 192–202.

1986 Methodological issues in Americanist artifact classification. In *Advances in archaeological method and theory vol. 9*. New York: Academic Press. Pp.149–207.

Ensor, H. B., and E. Roemer, Jr.

1989 Comments on Sullivan and Rozen's "Debitage Analysis and Archaeological Interpretation." *American Antiquity* **54**, 175–178.

Grayson, D. K.
1984 *Quantitative zooarchaeology: Topics in the analysis of archaeological faunas*. Orlando, Academic Press.

Gunther, Erna
1927 Klallam ethnography. University of Washington Publications in Anthropology 1(5). Pp.171–314.

Ham, L.
1982 Seasonality, shell midden layers and Coast Salish subsistence activities at the Crescent Beach Site, DgRr 1. Ph.D. dissertation, Department of Anthropology and Sociology, University of British Columbia, Vancouver.

Hayden, B.
1979 *Lithic use-wear analysis* (editor). New York: Academic Press.
1981 Research and development in the Stone Age: technological transitions among hunter-gatherers. *Current Anthropology* **22**(5), 519–548.

Jelinek, A. J.
1976 Form, function, and style in lithic analysis. In *Cultural change and continuity: Essays in honor of James Bennett Griffin*, edited by C. E. Cleland. New York: Academic Press. Pp.19–33.

Jenness, D.
n.d. The Saanich Indians of Vancouver Island. Manuscript in the Archives of the Ethnology Division, National Museum of Man, Ottawa.

Jones, G. T., C. Beck, and D. K. Grayson
1989 Measures of diversity and expedient lithic technologies. In *Quantifying diversity in archaeology*, edited by R. D. Leonard and G. T. Jones. Cambridge: Cambridge University Press. Pp.69–78.

Kidd, R. S.
1964 A synthesis of western Washington prehistory from the perspective of three occupation sites. Unpublished Master's thesis, Department of Anthropology, University of Washington, Seattle.

King, A.
1950 Cattle Point, a stratified site in the southern northwest coast region. *Memoirs of the Society for American Archaeology*, 7, Menasha.

Kornbacher, K. D.
1989 Shell midden lithic technology: an investigation of change at British Camp (45SJ24), San Juan Island. Unpublished Master's thesis, Department of Anthropology and Sociology, University of British Columbia, Vancouver.

Latas, T.
1987 A depositional history of English Camp angular rock: a test of fire-cracked rock attributes. Unpublished Master's thesis, Department of Anthropology, University of Washington, Seattle.

Leonard, R., and G. T. Jones
1989 *Quantifying diversity*. Cambridge: Cambridge University Press.

Lewarch, D.
1982 Analysis of lithic artifacts. In *The Cannon Reservoir human ecology project: An archaeological study of cultural adaptations in the southern Prairie Peninsula*, edited by M. J. O'Brien, R. E. Warren, and D. E. Lewarch. New York: Academic Press. Pp.145–169.

Magne, M.
1983 Lithics and livelihood: stone tool technologies of central and southern Interior B.C. *Archaeological Survey of Canada Paper* No. 133. National Museum of Man, Mercury Series, Ottawa.

Magne, M., and D. L. Pokotylo
1981 A pilot study in bifacial lithic reduction sequences. *Lithic Technology* **10**, 34–47.

Matson, R. G.
1976 The Glenrose Cannery site. *Archaeological Survey of Canada Paper* No. 52, National Museum of Man, Mercury Series, Ottawa.

Mitchell, D.
1971 Archaeology of the Gulf of Georgia, a natural region and its culture types. *Syesis* **4**, suppl. 1. Victoria.

1990 Prehistory of the coasts of southern British Columbia and northern Washington. In *Handbook of North American Indians, vol. 7, Northwest Coast*. Washington, D.C.: Smithsonian Institution. Pp.340–348.

Morrow, C. A.

1984 A biface production model for gravel-based stone industries. *Lithic Technology* **13**, 20–28.

Murray, R.

1982 Analysis of artifacts from four Duke Point sites, near Nanaimo, B.C.: an example of cultural continuity in the southern Gulf of Georgia region. *Archaeological Survey of Canada* No. 113. National Museum of Man, Mercury Series, Ottawa.

Parry, W. J., and R. L. Kelly

1987 Expedient core technology and sedentism. In *The organization of core technology*, edited by J. K. Johnson and C. A. Morrow. Pp.285–302. Boulder, Colorado: Westview Press.

Pokotylo, D. L.

1978 Lithic technology and settlement patterns in Upper Hat Creek Valley, B.C. Unpublished Ph.D. dissertation, Department of Anthropology and Sociology, University of British Columbia, Vancouver.

Prentiss, W., and E. J. Romanski

1989 Experimental evaluation of Sullivan and Rozen's debitage typology. In *Experiments in lithic technology*, edited by D. S. Amick and R. P. Mauldin. BAR International Series 528. Pp.89–100.

Raab, L. Mark, Robert F. Cande, and David W. Stahle

1979 Debitage graphs and Archaic settlement patterns in the Arkansas Ozarks. *Mid-Continental Journal of Archaeology* **4**, 167–182.

Rozen, K. C., and A. P. Sullivan, III

1989 Measurement, method, and meaning in lithic analysis: problems with Amick and Mauldin's middle-range approach. *American Antiquity* **54**, 169–175.

Stern, B. J.

1934 *The Lummi Indians of northwest Washington*. New York: Columbia University Press.

Sullivan, Alan P., III

1987 Probing the sources of lithic assemblage variability: a regional case study near the Homolovi Ruins Arizona. *North American Archaeology* **8**(1), 41–71.

Sullivan, Alan P., III, and K. C. Rozen

1985 Debitage analysis and archaeological interpretation. *American Antiquity* **50**, 755–779.

Suttles, Wayne

1951 Economic life of the Coast Salish of Haro and Rosario Straits. Ph.D dissertation, Department of Anthropology, University of Washington, Seattle.

Thompson, G.

1978 Prehistoric settlement changes in the southern Northwest Coast: A functional approach. University of Washington, Department of Anthropology Reports in Archaeology No. 5, Seattle.

Thorne, R. M., and J. K. Johnson

1979 Cultural resource survey of Opossum and Muddy Bayous, Tallaahatchee County, Mississippi. Report submitted to Vicksburg Office, U.S. Army Corps of Engineers, Vicksburg, Mississippi.

Wilmsen, E.

1970 Lithic analysis and cultural inference. University of Arizona Anthropological Papers No. 16.

Young, D., and D. B. Bamforth

1990 On the macroscopic identification of used flakes. *American Antiquity* **55**(2):403–409.

9

Lithic Manufacturing at British Camp: Evidence from Size Distributions and Microartifacts

Mark E. Madsen

I. Introduction

Many aspects of the archaeological record are the intentional products of human behavior, either by manufacture or by use. These products of behavior are what we usually seek to understand in terms of the activities that produce them. Because activities have an empirical as well as a behavioral component, it is possible to study them directly in the archaeological record. One cannot reconstruct the behaviors associated with a particular activity in a testable fashion; however, the results of the activity are empirical. Archaeologists can make use of this fact to interpret archaeological remains by establishing the physical linkages between specific activities and attributes which can be measured in the archaeological record. Such an approach is different than the behaviorally oriented approach which has traditionally characterized activity analysis in archaeology (e.g., Binford 1978; Kent 1987).

Establishing necessary linkages between activities and their archaeological signatures requires understanding the physical processes which govern an activity and the assemblage attributes which are diagnostic of a given process. Fladmark (1982) and Stahle and Dunn (1982, 1984) have begun exploring such an approach for lithic manufacturing. They have found that size distributions and microartifacts are particularly useful in experimental situations for differentiating kinds of manufacturing techniques as well as mapping the spatial distribution of the debris resulting from such techniques.

At the British Camp site (45SJ24), San Juan Island, Washington, we find few finished tools and large amounts of debitage. Spatial subdivisions of the overall lithic assemblage could represent areas where lithic reduction was taking place or instead could be dumps where debitage from knapping and other activities involving stone

Deciphering a Shell Midden

tools was deposited. A comparison between experimental size distributions (including microartifacts) for several different lithic reduction techniques and archaeological distributions from British Camp can help resolve this question for each spatial subdivision of the overall lithic assemblage.

At British Camp, however, there is evidence for postdepositional alterations within the midden sediments. Stein (Ch. 7, this volume) describes postdepositional processes which are creating a stratigraphic break and cross-cut facies boundaries in all excavation units. This break is referred to as the light/dark layer distinction, seen in many Northwest Coast shell middens, and described in detail by Stein (1984, Ch. 7, this volume) and Nelson *et al.* (1986). Chemical and faunal studies of this stratigraphic break reveal that the contact reflects the current average height of the groundwater table; the "dark" color of the groundwater-saturated zone is due to the hydration of organic matter and clay minerals by saturation in groundwater. In addition to chemical alterations of the midden sediments by groundwater, it is possible that water percolating downward from the surface within the highly porous midden matrix can move particles, including microdebitage, and deposit these particles at the top of the groundwater table when the flowing water enters the groundwater table and loses competence to transport large particles. Should microdebitage be translocated postdepositionally within the midden, microartifacts and size distributions would not necessarily reflect the activities responsible for the deposition of lithic debitage assemblages. Clearly this possibility must be assessed.

II. Research Questions

A. *Lithic Manufacturing*

The basic physical model for percussion flaking involves the application of force with a relatively inelastic percussor to a rock body. In order for a crack to form and propagate, tensile stresses within the rock must exceed the fracture toughness of the material, which varies according to material type (Lawn and Wilshaw 1975). Removal of material occurs when a crack forms in the rock body and propagates until the lines of force travel completely through the intervening rock body (Cotterell and Kamminga 1987). The size and shape of the resultant flake is dependent upon several variables, including the angle of force application, the nature of the raw material, the rate of loading (i.e., the interval of time over which force is applied by the percussor to the rock body), and the momentum of the percussor at the moment of impact (Cotterell and Kamminga 1987; Speth 1972).

The creation of a suite of particles of different sizes during percussion flaking has been noted by many (e.g., Fladmark 1982; Speth 1972). Fladmark (1982) particularly noted the presence of microscopic flakes resulting from the knapping process. These microflakes, or microdebitage, are produced by the same events that produce larger flakes. Their production has not been well studied but is probably the result of the force wave encountering tiny inhomogeneities within the rock body, at which point the force wave bifurcates. When multiple force waves moving through a material propagate completely through a rock body, the result can be the removal of smaller particles between and at the edges of a larger flake (Lawn and Wilshaw 1975). Shattering at the point of impact between percussor and rock body may also be a source for microdebitage (Cotterell and Kamminga 1987).

Whatever the exact nature of microdebitage production, Fladmark (1982) experimentally established that microdebitage is produced in prodigious amounts by ordinary percussion flaking. More importantly, Fladmark demonstrated the utility of microdebitage for addressing significant archaeological problems. Since Fladmark's work appeared, an expanding group of scholars have used microartifacts to delineate activity areas (Hull 1987; Metcalfe and Heath 1990; Vance 1987) and study site-formation processes (Madsen and Dunnell 1989; Rosen 1986, 1989; Stein and Teltser 1989).

Size distributions are now in frequent use for studying lithic reduction sequences from debitage assemblages (e.g., Patterson 1982, 1990; Stahle and Dunn 1982, 1984); however, all studies by lithic analysts to date focus on large debitage, ignoring the smallest size of debris. Fladmark (1982) remarks that one of the potentials of microdebitage analysis is the use of granulometry (i.e., size distributions) to study the lithic reduction techniques used in systemic contexts to produce archaeological assemblages of debris. In his initial work, Fladmark experimentally demonstrated that different kinds of percussion flaking (e.g., hard hammer, pressure flaking, etc.) exhibit distinctive size distributions of debitage, including microdebitage.

In this chapter I will test whether lithic assemblages from a sample of facies at British Camp represent the remains of lithic manufacturing loci, rather than areas where lithics were used or discarded. Assemblages which represent the *de facto* refuse of lithic manufacturing should include microdebitage and also exhibit size distributions similar to those generated experimentally by Fladmark. Furthermore, I will compare particular assemblages of debitage with size distributions for specific manufacturing techniques, in order to determine whether specific reduction techniques can be recognized within lithic assemblages at British Camp.

B. *Postdepositional Alterations*

Because this study focuses on microartifacts and size distributions, the effects of postdepositional alterations to microdebitage must be taken into account. Microartifacts, by virtue of their size, are capable of being moved by a variety of natural transport agents, such as wind and running water (Dunnell and Stein 1989; Fladmark 1982). The effect of removing microdebitage preferentially from an assemblage would be to bias the size distribution of the assemblage. This bias can interfere with attempts to interpret assemblages using size distributions.

Normally one would expect that sand-sized particles [as defined by Folk (1980), 2 mm to 0.0625 mm] would not move appreciably in a stratigraphic profile; mobility is usually restricted to the silt and clay fractions (i.e., < 0.0625 mm) (Buol *et al.* 1989). However, shell midden sediments are very porous, especially compared to the texture of most soil parent materials. Large shells and other gravel-sized objects in the midden tend to create voids, in which particles can easily move downward within the deposits. Running water follows these voids, and in combination with gravity can translocate sediment particles in the process.

If microartifacts were postdepositionally transported by the movement of water downward within the stratigraphic column at British Camp, microartifacts would be expected to be absent from the upper portion of the solum and to accumulate somewhere lower in the midden. In fact, the upper portion of the solum at British Camp is dry, with all surface water percolating through the highly porous midden to the groundwater table, seasonally located within the lower portion of the midden sediments.

Percolating water may translocate microartifacts and other large particles downward within the porous midden sediments. If such movement occurred, a zone of accumulation of microartifacts (and other sand-sized particles) should be found directly above the top of the groundwater table, where channelized flow from above loses competence and deposits its transported load.

If microartifacts are being transported, the sand-sized fractions of sediment samples from above groundwater should be impoverished in all kinds of sand-sized particles relative to samples from the groundwater boundary region and below; on the other hand, if microartifacts were not deposited into site sediments by humans in the past, microartifacts will be absent but other particle types abundant.

III. Methods

In order to test the hypotheses outlined above, two kinds of data are necessary. Data on the stratigraphic distribution of lithics from individual size fractions are required in order to control for the effects of translocation on microartifacts within the midden. Data concerning the size distribution of lithics from individual facies are necessary to differentiate manufacturing debris from refuse deposits, as well as to examine the possibility of recognizing specific reduction techniques as Fladmark suggested.

Stratigraphic units, called facies, are defined during excavation and have measurable horizontal and vertical limits (Stein *et al.*, Ch. 2, this volume). Facies are not necessarily stacked up in layer-cake fashion; rather, the vertical and horizontal relationships of facies are controlled by use of the Harris Matrix method (Harris 1989; Kornbacher 1989).

All facies used in this analysis were derived from the same 2 × 2-m excavation unit (Unit 308/300). Twelve facies of the thirty-one excavated as of 1987 were sampled for this study. Facies 1A is the first facies excavated in the unit, followed by 1B, and so on. The current average position of the groundwater table, identified by the postdepositionally altered zone, occurs at the top surface of Facies 1O (in Unit 308/300) (Stein, Ch. 7, this volume). Thus Facies 1A, 1B, 1C, 84D, 85D, 1E, and 1J are above the groundwater table, and Facies 1O, 1T, 1W, 1Y, and 1Z have been within the groundwater table for varying periods of time.

Size distributions were derived by measuring a sample of all lithics from each facies chosen, using size classes including the largest lithics down to those approximately 0.25 mm in size (Table 1). Since the size distribution of lithics includes both macroscopic lithics and their microscopic counterparts, two different recovery methods are necessary. Macroscopic lithics (from 2 mm to the largest lithics found) require large samples to obtain statistically significant estimates of abundance.

Table 1 Size Classes Used in Analysis

Size class	Interval
−6 ϕ[a]	>64 mm
−5 ϕ	64 mm 32 mm
−4 ϕ	32 mm–16 mm
−3 ϕ	16 mm–8 mm
−2 ϕ	8 mm–4 mm
−1 ϕ	4 mm–2 mm
0 ϕ	2 mm–1 mm
1 ϕ	1 mm–0.5 mm
2 ϕ	0.5 mm–0.25 mm

[a]The ϕ number for each size class is the negative logarithm (base 2) of the lower limit in millimeters for each size class. The ϕ scale is a standard scale for particle size analysis (Folk 1980).

Shackley (1975) advocates a sample size of approximately 50 kg for objects 64 mm (−6 ϕ) in size, which corresponds to the largest lithics found in excavation as yet. Macroscopic lithics were taken from excavated material screened in the field. A sample of material from the excavation of each facies weighing 50 kg was used for this analysis. All lithics from that 50 kg were used in the grain-size analysis.

Microdebitage must be sampled differently from macroscopic lithics. For this study, microdebitage was sampled from an 8-kg sample taken from each facies. The 8-kg sample was first screened through a 2-mm (−1 ϕ) screen, and a subsample of approximately 80 grams of sediment was taken for later analysis of microartifact content.

In the laboratory, all macroscopic lithics were cleaned and dried before sizing and weighing. Large lithics were assigned to ϕ-size classes (Folk 1980) by sizing them with standard geological sieves in whole phi intervals (−6 to −1 ϕ). Sizing was accomplished by hand-passing each lithic through sieves until the intermediate axis of the lithic would not pass any further sieve mesh (Figure 1). The lithics in each ϕ-size fraction were then weighed to 0.0001 grams on an analytic balance and the weights totaled for each ϕ-size fraction.

The subsample of sediment taken for microartifact analysis was soaked in a solution of peptizing agent (sodium hexametaphosphate) to prevent flocculation of clays. The subsample was then wet-sieved to separate the silt and clay fractions from the sand fraction which contains the identifiable microartifacts (Folk 1980). After drying, the sand fraction was shaken for 15 minutes through a stack of geological sieves in whole ϕ-size intervals (−1 ϕ to 4 ϕ, see Table 1). The material retained on each sieve was weighed to 0.0001 grams on the analytic balance and stored for point counting of microartifacts.

In order to quantify the microdebitage content of each size fraction derived from dry-sieving, the point-counting method described in Stein and Teltser (1989) and

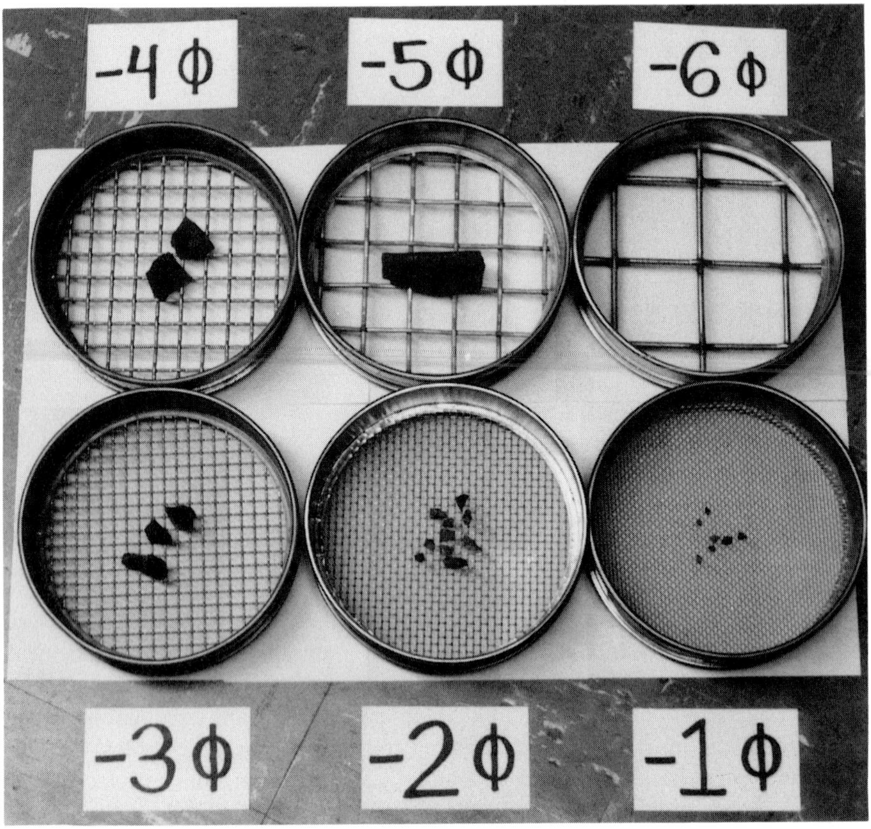

Figure 1 Geological sieves used to determine the size of large lithics. Sieves range from 64 mm (−6 ϕ) to 2 mm (−1 ϕ). Lithics with intermediate axis smaller than the mesh size are retained on the sieve.

modified from Chayes (1956) and Galehouse (1971) was used. In this method the grains to be counted were spread out on a grid surface and successive lots of 100 particles are counted, and compositional types tallied within each lot of particles. For each successive 100 particles, the cumulative percentage of each compositional type was calculated. When the cumulative percentage of all compositional types was within 0.1% of the previous cumulative percent, statistical redundancy was reached and counting ceased. Table 2 shows an example of point counting and redundancy for one facies.

A problem with the point-counting method is the asymmetry between presence and absence in the sampling procedure. One particle is enough to establish the presence of a material type, but the number of particles needed to constitute the absence of a material type cannot be known. The convention used here to establish absence is a 1000-particle limit: if a material type is not present in the counting of 1000 particles, it is considered absent from the sample. This means that material types present

Table 2 Example of Point Counting and Redundancy Calculations for Unit 308/300, Facies 10, Size Fraction 0 ϕ

N[a]	Lithics	% Lithics	Cumulative % lithics	% Change
100	3	3.0	3.0	—
200	2	2.0	2.5	0.5
300	0	0.0	1.67	0.83
400	1	1.0	1.5	0.17
500	1	1.0	1.4	0.1
600	2	2.0	1.5	0.1
700	3	3.0	1.71	0.21
800	1	1.0	1.63	0.08[b]
900	1	1.0	1.56	0.07
1000	1	1.0	1.50	0.06 Redundant

[a]Explanation of columns: *N* are successive lots of 100 particles; Lithics are the count of microdebitage particles in 100 particles; % Lithics is the percentage lithics are of 100 particles; Cumulative % Lithics is the cumulative percentage of lithics in all lots yet counted; % Change is the change in cumulative percent with the addition of 100 more particles.

[b]Although the change in cumulative percent is less than 0.1, I often counted several more lots of 100 particles to ensure that redundacy had in fact been achieved.

in a sample will have frequencies above 0.1%, which is the tolerance limit for the redundancy method of point counting used here.

Macroscopic and microscopic lithics were sampled from different sample sizes, and thus data derived as above cannot simply be combined to form size distribution. Macroscopic lithics (> 2 mm) are derived from a 50-kg sample of excavated material. Microscopic lithics are derived from subsamples of varying sizes. For each size fraction of lithics, the total weight of material removed from a facies was divided by the sample size to obtain a splitting factor (e.g., 345 kg/50 kg for macroscopic lithics from Facies 1J, Unit 308/300). The splitting factor for each size fraction was then multiplied by the lithic weight in each size fraction to obtain corrected weights. These corrected weights were then used to build size distributions. Table 3 shows an example of correcting weights and computing the size distribution for one facies.

Once corrected weights for each size fraction in a facies were known, the derivation of size distributions for lithics within a facies was accomplished by calculating the relative percentage of lithics from the facies which belong in each size fraction. The stratigraphic distribution of lithics was determined by arranging the relative percentage of lithics from each facies, calculated as above, from a given size fraction. In practice this simply meant reproducing the frequency bars of a given particle size class from all facies onto a stratigraphic axis.

Fladmark's (1982) experimental size distributions for lithic manufacturing techniques were compared to the archaeological data gathered here by calculating the rank-order correlation (Spearman's ρ) between archaeological and experimental size distributions. A nonparametric statistic was chosen for two reasons: first, size distributions are closed-array data and thus are not appropriate for Pearson's correlation coefficient (Cowgill 1990) and second, size distributions often violate assumptions of normality required for parametric statistics. Size distribution data from each facies were compared to Fladmark's distributions for hard-hammer, soft-hammer, and pressure flaking. A significant positive correlation indicates that two distributions are very similar, supporting an inference that a given manufacturing technique, as defined by Fladmark, was used in the production of a particular assemblage of lithics from the British Camp site.

IV. Results

Figure 2 presents the size distribution of lithics from the twelve facies under consideration; Facies 85D has no lithics in the sample and thus is not represented by a

Table 3 Example of Weight Correction and Calculation of Size Distribution for Unit 308/300, Facies 1J

Calculation of splitting factors

Size class (phi)	Total facies weight (g)	Subsample (g)	S.F.
−5[a]	345,425	49,200	7.0208
−4	345,425	49,200	7.0208
−3	345,425	49,200	7.0208
−2	345,425	49,200	7.0208
−1	345,425	1013.3	340.89
0	345,425	79.8	4328.6
1	345,425	79.8	4328.6
2	345,425	79.8	4328.6

Corrected weights and relative frequencies of lithics

Size class (ϕ)	Lithic Wt.	S.F.	Corrected Wt. (g)	Relative %
−6	0.0000	7.0208	0.0000	00.00
−5	84.9501	7.0208	596.4205	29.53
−4	59.3286	7.0208	416.5362	20.62
−3	8.3140	7.0208	58.3712	2.89
−2	5.0044	7.0208	35.1351	1.74
−1	0.2540	340.89	83.5203	4.14
0	0.0817	4328.6	353.6466	17.51
1	0.0528	4328.6	228.5501	11.32
2	0.0572	4328.6	247.5959	12.26

[a]No lithics were assigned to the −6 ϕ size class for Facies 1J, so no splitting factor is calculated.

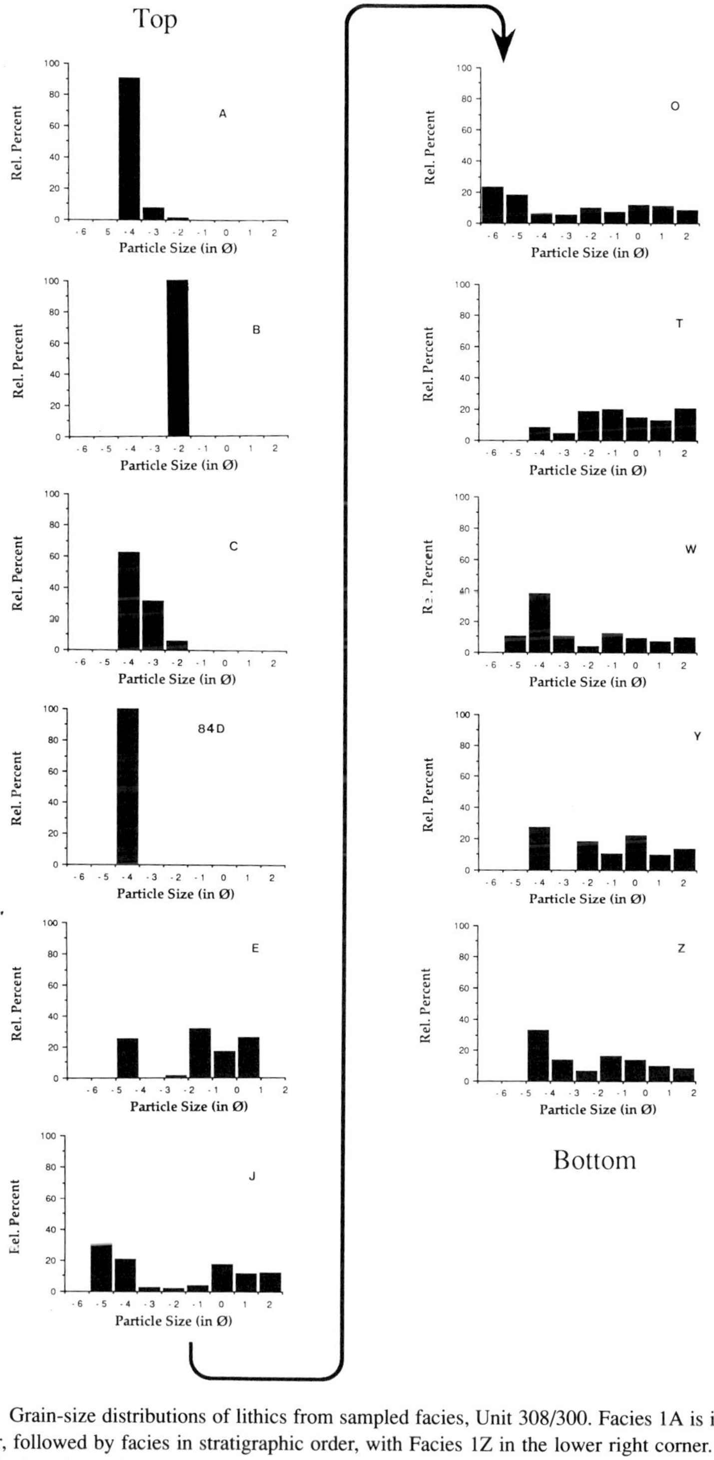

Figure 2 Grain-size distributions of lithics from sampled facies, Unit 308/300. Facies 1A is in the top left corner, followed by facies in stratigraphic order, with Facies 1Z in the lower right corner. On each size distribution, the vertical axis indicates relative percentage of lithics from the facies that occur within each size class; horizontal axis is size in ϕ-scale classes.

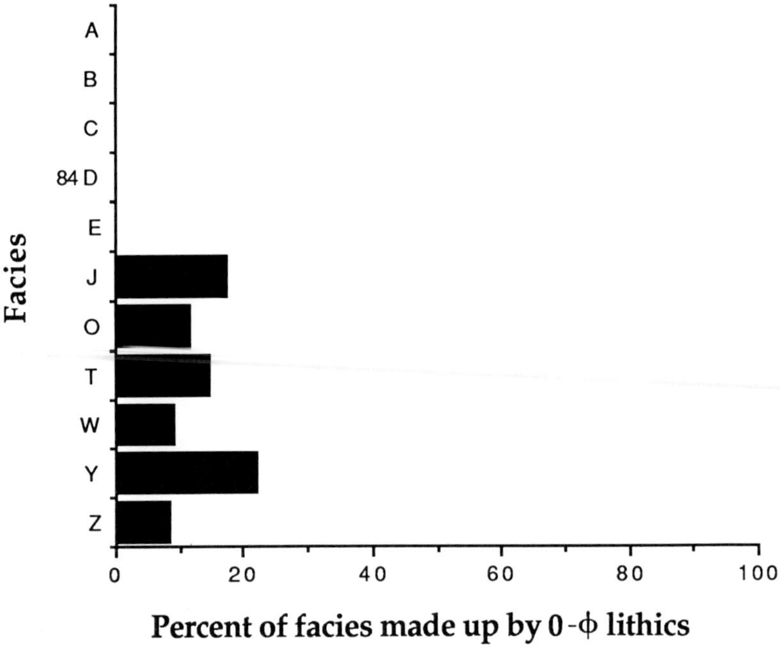

Figure 3 Stratigraphic distribution of 0-ϕ lithics across facies, Unit 308/300. Vertical axis indicates facies (A, *top*, Z, *bottom*); horizontal axis is the percentage of size fraction that is lithics for each facies.

graph. Two distinct types of size distributions are discernible. The first kind of distribution is represented by Facies 1A. No microdebitage is found, a fact which is indicated by the empty cells in the smaller size ranges on the graph. Facies 1A, 1B, 1C, 84D, and 85D all have this kind of distribution. The second kind of distribution is typified by Facies 1O. Abundant microdebitage is found, a fact which is indicated by the presence of frequency bars in the smaller size ranges of the graph. Facies 1E, 1J, 1O, 1T, 1Y and 1Z have size distributions of this kind. In fact, in Facies 1T microdebitage is the major component of the lithic sample by weight.

The distribution of the smallest size fractions (0, 1, and 2 ϕ) of lithics across stratigraphic units is presented in Figures 3–5. Because each graph depicts the relative density of lithics in each facies in relation to other facies, one fact is immediately apparent. The maximum density of lithics differs in stratigraphic location for each of the depicted size fractions. Zero-ϕ lithics attain their highest density in Facies 1Y, relatively low in the sampled profile, whereas 1-ϕ lithics show a peak at Facies 1E, the highest facies with any microdebitage at all. While it is true that the lower facies in this section of the midden have relatively more lithics than upper facies, this is not manifested by a monotonic increase.

When Fladmark's experimental data (Figures 6–8) are compared against the size distributions from the eleven facies that contained lithics, several significant correlations result (Tables 4–6). Correlations with soft-hammer data occur in Facies 1A,

1C and 1Z (Table 4). No significant correlations exist between any facies and the hard-hammer distribution (Table 5). One significant correlation exists between pressure-flaking data and Facies 1T (Table 6).

V. Interpretation

A. *Postdepositional Alterations*

The present average level of the groundwater table, marked by the postdepositionally altered layer, occurs above Facies 1O (in Unit 308/300). Data on the stratigraphic distribution of lithics from the microartifact fractions show that such a zone of accumulation does not exist above the groundwater-affected zone (Figures 3–5). The maximum amounts of microdebitage present in the sampled facies vary in stratigraphic location; the smallest lithics sampled (2-ϕ) are most abundant in samples from Facies 1T, well below the depth a zone of accumulation would occur.

In order for lithics to accumulate in an illuviation layer above the groundwater table, upper facies must be depleted; facies above act as a source for concentrations of lithics below. Figure 2 shows that this is clearly not the case. While the uppermost facies do not have any microdebitage, Facies 1E has a higher density of 1-ϕ lithics than lower facies. Facies 1E and 1J have roughly the same density of 0-ϕ lithics, and both have a higher density of 1-ϕ lithics than Facies 1O.

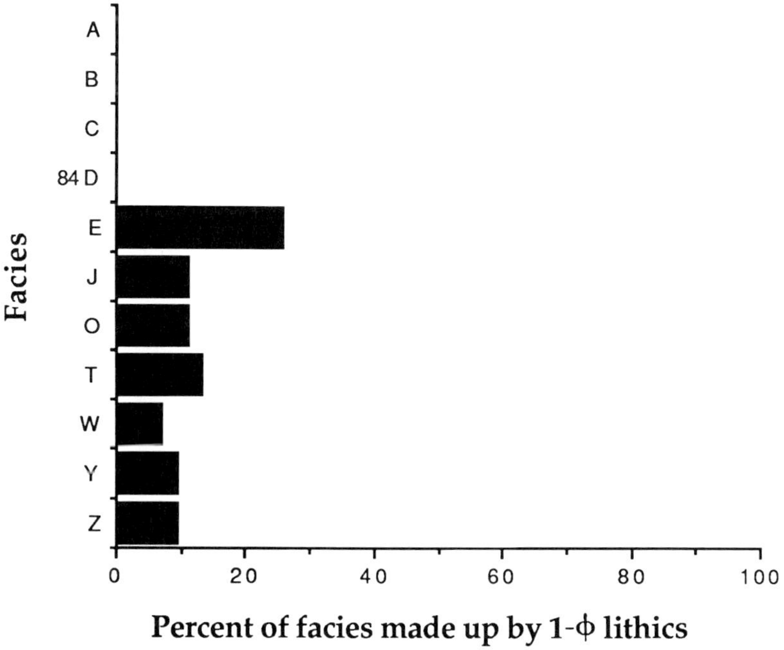

Figure 4 Stratigraphic distribution of 1-ϕ lithics across facies, Unit 308/300. Vertical axis indicates facies (A, *top*, Z, *bottom*); horizontal axis is the percentage of size fraction that is lithics for each facies.

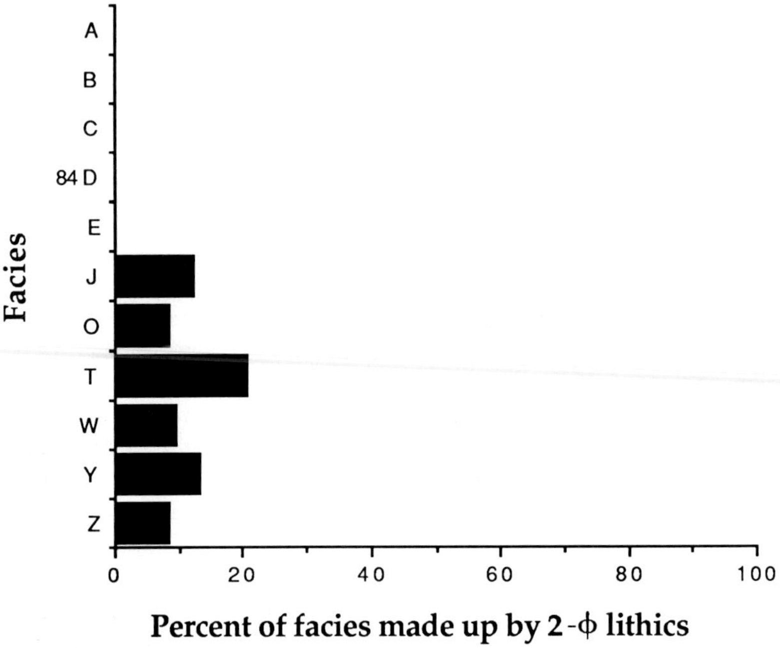

Figure 5 Stratigraphic distribution of 2-ϕ lithics across facies, Unit 308/300. Vertical axis indicates facies (A, *top*, Z, *bottom*); horizontal axis is the percentage of size fraction that is lithics for each facies.

Additionally, since movement by water is a physical process, compositional types of particles are not singled out for translocation (although particle morphology will affect entrainment by percolating groundwater). If translocation is occurring in upper facies (e.g., Facies 1A, Facies 1E, etc.), all particles of a given size would be transported, not simply lithics. Analysis of sediments from the sampled facies revealed abundant sand-sized particles and microartifacts other than lithics (e.g., sea urchin spines and bone fragments). Significantly, many of the compositional types present in upper facies (e.g., sea urchin spine fragments) are tabular in shape and would therefore exhibit the same sort of behavior as microflakes under groundwater percolation. The abundance of these compositional types in upper facies argues against lithic translocation within the midden.

B. *Lithic Manufacturing*

Because microartifact distributions have not been affected by postdepositional alterations at British Camp, lithic size distributions can be compared to Fladmark's experimental data from different manufacturing techniques. The results of comparing Fladmark's experimental data to each of the archaeological size distributions reveal only four significant correlations (at $p = 0.05$). Three of these occurred with the soft-hammer percussion data (Table 4). Facies 1A and 1C both show significant correla-

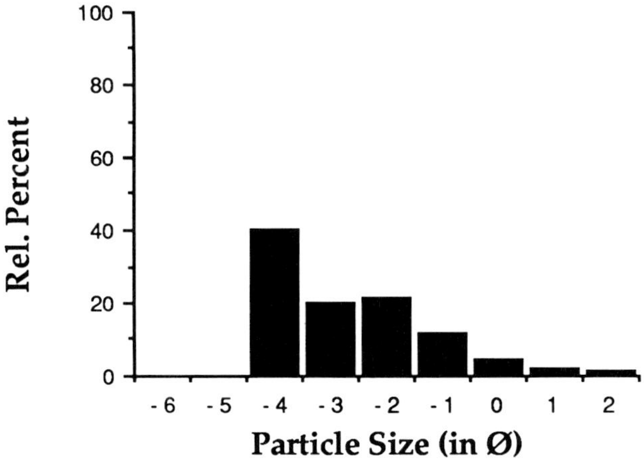

Figure 6 Size distribution of soft-hammer percussion flaking debris. Vertical axis indicates relative percentage of lithics from the facies that occur within each size class; horizontal axis is size in ϕ-scale classes. Data modified from Fladmark (1982).

tions with the soft-hammer data, but neither facies includes microdebitage in its assemblage (Figure 2). Neither assemblage can be considered the remains of a knapping episode. The size distribution indicates that only large lithics are present; moreover, large lithics are present in the same proportion as in the experimental flaking data.

One possible interpretation for this situation is that Facies 1A and 1C represent locations where lithic debris (excluding microdebitage) were dumped from another location. The fact that British Camp is a midden, and therefore in part a refuse dump,

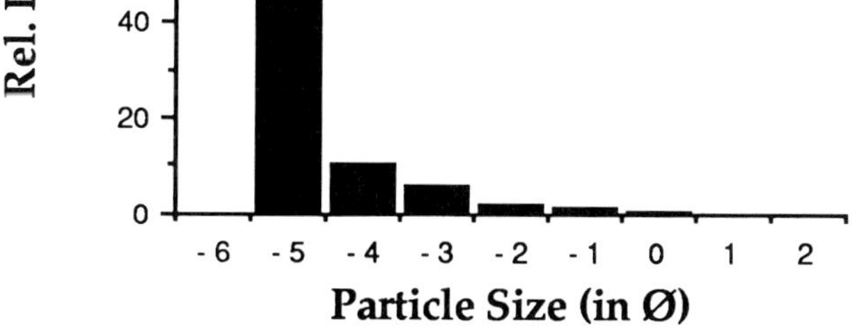

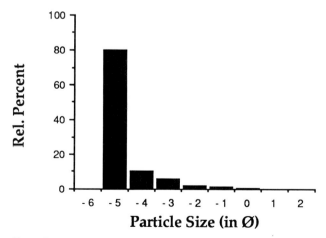

Figure 7 Size distribution of hard-hammer percussion flaking debris. Vertical axis indicates relative percentage of lithics from the facies that occur within each size class; horizontal axis is size in ϕ-scale classes. Data modified from Fladmark (1982).

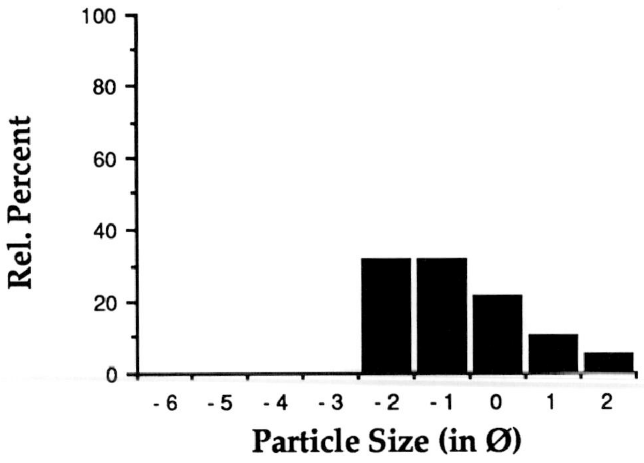

Figure 8 Size distribution of pressure-flaking debris. Vertical axis indicates relative percentage of lithics from the facies that occur within each size class; horizontal axis is size in ϕ-scale classes. Data modified from Fladmark (1982).

supports this notion. One would suspect, a priori, that in a midden most lithic assemblages would be of this variety.

Facies 1Z is the other significant correlation with the soft-hammer percussion data. Figure 2 shows that microdebitage is present in the assemblage, and visual comparison of this distribution to the soft-hammer size distribution (Figure 6) indicates great structural similarity. The only major difference between the two distributions is the relative paucity of large material in the Facies 1Z distribution. Why such is the case is not known. On the basis of these data, however, it seems likely that the lithic assemblage of Facies 1Z represents debris from the soft-hammer reduction of stone tools.

Table 4 Spearman's ρ Correlation between Facies and Soft-Hammer Percussion Data

Facies	ρ	*p*
1A	0.825	0.019
1B	0.413	0.243
1C	0.825	0.019
84D	0.550	0.119
1E	0.463	0.190
1J	−0.067	0.849
1O	−0.762	0.031
1T	0.294	0.209
1W	0.446	0.106
1Y	0.613	0.083
1Z	0.731	0.039

Table 5 Spearman's ρ Correlation between Facies and Hard-Hammer Percussion Data

Facies	ρ	*p*
1A	0.577	0.103
1B	0.138	0.697
1C	0.577	0.103
84D	0.413	0.243
1E	−0.022	0.951
1J	0.494	0.163
1O	−0.385	0.276
1T	−0.269	0.447
1W	0.653	0.065
1Y	0.085	0.809
1Z	0.261	0.461

One significant correlation exists between archaeological data and the pressure-flaking distribution. The lithic assemblage from facies 1T consists mainly of small lithics, including an abundance of microdebitage (Figure 2). Comparison of the 1T size distribution with that of the pressure-flaking data reveals some differences in structure, however. Rather than steadily decreasing toward the fine fractions, the 1T distribution is polymodal. Simple correlation coefficients, especially nonparametric statistics, are not measures of the structure of a distribution but rather the overall amount of variance two distributions share with respect to two or more variables. On the basis of these data it is not possible to unequivocally attribute the lithic assemblage from facies 1T to pressure flaking of stone tools.

Rather, Facies 1T may represent an example of a location where several different kinds of lithic deposition occurred, resulting in an assemblage characterized by a polymodal size distribution. Stein (1987) discusses lithic size distributions in terms

Table 6 Spearman's ρ Correlation between Facies and Pressure-Flaking Data

Facies	ρ	*p*
1A	−0.265	0.454
1B	0.503	0.155
1C	0.265	0.454
84D	−0.359	0.309
1E	0.598	0.091
1J	−0.219	0.536
1O	−0.035	0.921
1T	0.781	0.027
1W	−0.201	0.569
1Y	0.436	0.218
1Z	0.140	0.691

of the position of the mode; different types of lithic reduction techniques produce size distributions with modes at different particle sizes. Hard-hammer percussion produces a mode in the gravel fraction, soft-hammer percussion yields a mode in the coarse sand fraction, and pressure-flaking percussion produces a mode in the fine sand and silt fraction. Given the position of the three modes in the size distribution of Facies 1T, a mixture of hard-hammer and soft-hammer reduction techniques may be responsible for producing the sampled assemblage.

As noted above, no significant correlations between hard-hammer percussion and the archaeological data exist. This suggests that hard-hammer percussion is not in evidence in the lithic assemblages examined in this study. This seems unlikely, however. Detailed examination of flake attributes by Kornbacher (1989) reveals abundant hard-hammer percussion at British Camp. However, Kornbacher's analysis also shows that the vast majority of initial core reduction was accomplished at British Camp by cobble reduction (including bipolar techniques) rather than bifacial reduction. Fladmark's size distributions were all derived from bifacial reduction experiments, as are all other published lithic size distribution studies (e.g., Patterson 1990; Stahle and Dunn 1984). This fact alone probably accounts for the apparent discrepancy between Kornbacher's results and the results obtained in this study.

Additionally, Fladmark's size distributions are from obsidian core reduction, rather than dacite, from which chipped stone tools at British Camp are made (Bakewell 1990). Because of this difference, size distributions from Fladmark's experiments were most usefully considered as ordinal-level data; that is, across material types one expects the rank ordering of frequencies to be invariant while the absolute values are expected to be different. While hard-hammer percussion is certainly present in lithic assemblages at British Camp, the study reported in this chapter was limited to examining hard-hammer bifacial reduction.

Apart from positive evidence of lithic manufacturing, most lithic assemblages sampled for this study did not correlate with any experimental data. Of course, these conclusions do not indicate that these assemblages are uninterpretable but simply that within the limited scope of the experimental research, models for activities other than manufacturing activity were not generated and tested. Most of the lithic assemblages in the midden possibly represent use or refuse disposal activities.

VI. Conclusion

The size distribution and stratigraphic data from British Camp indicate that no microdebitage was found in the upper facies of the stratigraphic profile sampled. The absence of microdebitage in upper facies may indicate the decrease in frequency of lithic manufacturing at British Camp, an interpretation supported by Kornbacher's (1989) attribute study of macroscopic lithics. Alternatively, it may indicate a spatial shift in manufacturing activity away from the area examined in this study and to a different location, while the present location (Unit 308/300) became an area used for activities like resharpening or refuse disposal. Such an interpretation requires independent confirmation with stratigraphic and chronological evidence. There is evidence that soft-hammer percussion is a primary lithic reduction technique at British Camp, appearing following the initial reduction by cobble-reduction techniques.

The British Camp data show that once postdepositional alterations are understood, it is possible to use size distributions and microartifacts to interpret artifact assemblages in terms of the activities which created them. In contrast to previous uses of size distributions, the smallest sizes of debitage (microartifacts) were included in analysis and the combination of both approaches proved more successful than either alone. The archaeological record of past human activities is patterned both by human behavior and postdepositional processes; the approach taken in this paper allowed the contribution of each to be isolated and understood for British Camp lithics.

Acknowledgments

Many people assisted me in the course of my research and the writing of this chapter. In particular I would like to recognize the intellectual contributions of J. K. Stein and P. T. McCutcheon. Others who contributed in important ways to the research include R. C. Dunnell, K. Kornbacher, and K. Wilhelmsen. Additionally, J. K. Stein, K. R. Fladmark, P. A. Teltser, and C. P. Lipo read and made valuable comments on versions of this manuscript. The author was supported during much of the writing for this chapter by a National Science Foundation graduate fellowship. Aspects of this research were presented in papers at the 1988 meetings of the Northwest Anthropological Conference and the Society for American Archaeology.

References

Bakewell, E.
1990 Petrographic and geochemical source-modeling of lithics from archaeological contexts. B.A. Honors thesis, University of Washington, Seattle.

Binford, L. R.
1978 *Nunamiut ethnoarchaeology.* New York: Academic Press.

Buol, S. W., F. D. Hole, and R. J. McCracken
1989 *Soil genesis and classification, 3rd ed.* Ames, Iowa: Iowa State University Press.

Chayes, F.
1956 *Petrographic modal analysis.* New York: Wiley.

Cotterell, B., and J. Kamminga
1987 The formation of flakes. *American Antiquity* **52**, 675–708.

Cowgill, G. L.
1990 Why Pearson's r is not a good similarity coefficient for comparing collections. *American Antiquity* **55**, 512–520.

Dunnell, R. C., and J. K. Stein
1989 Theoretical issues in the interpretation of microartifacts. *Geoarchaeology* **4**, 31–42.

Fladmark, K. R.
1982 Microdebitage analysis: initial considerations. *Journal of Archaeological Science* **9**, 205–220.

Folk, R. L.
1980 *Petrology of sedimentary rocks.* Austin, Texas: Hemphill.

Galehouse, J. S.
1971 Point-counting. In *Procedures in sedimentary petrology*, edited by R. Carver. New York: Wiley. Pp.385–407.

Harris, E.
1989 *Principles of archaeological stratigraphy, 2nd ed.* Orlando, Florida: Academic Press.

Hull, K. L.
1987 Identification of cultural site formation processes through microdebitage analysis. *American Antiquity* **52**, 772–783.

Kent, S.
1987 *Method and theory in activity area research: An ethnoarchaeological approach.* New York: Columbia University Press.

Kornbacher, K. D.
1989 Shell midden lithic technology: An investigation of change at British Camp (45SJ24), San Juan Island. Unpublished Master's thesis, Department of Anthropology and Sociology, University of British Columbia, Vancouver, B.C.

Lawn, B. R., and T. R. Wilshaw
1975 *The fracture of brittle solids.* Cambridge: Cambridge University Press.

Madsen, M. E., and R. C. Dunnell
1989 The role of microartifacts in interpreting low-density plowzone records. Paper delivered at the 53rd Annual Meeting of the Society for American Archaeology, Atlanta, Georgia.

Metcalfe, D., and K. M. Heath
1990 Microrefuse and site structure: the hearths and floors of the Heartbreak Hotel. *American Antiquity* **55**, 781–796.

Nelson, M. A., P. J. Ford, and J. K. Stein
1986 Turning a midden into mush: evidence of acidic conditions in a shell midden. Paper delivered at the 51st Annual Meeting of the Society for American Archaeology, New Orleans, Louisiana.

Patterson, L. W.
1982 The importance of flake size distribution. *Contract Abstracts and CRM Archaeology* **3**, 70–72.
1990 Characteristics of bifacial-reduction flake-size distribution. *American Antiquity* **55**, 550–558.
1986 *Cities of clay: the geoarchaeology of tells.* Chicago: University of Chicago Press.

Rosen, A. M.
1989 Ancient town and city sites: a view from the microscope. *American Antiquity* **54**, 564–578.

Shackley, M. L.
1975 *Archaeological sediments.* New York: Wiley.

Speth, J. D.
1972 Mechanical basis of percussion flaking. *American Antiquity* **37**, 34–60.

Stahle, D. W., and J. E. Dunn
1982 An analysis and application of the size distribution of waste flakes from the manufacture of bifacial stone tools. *World Archaeology* **14**, 84–97.
1984 An experimental analysis of the size distribution of waste flakes from biface reduction. Arkansas Archaeological Survey Technical Paper, No. 2.

Stein, J. K.
1984 Interpreting the stratigraphy of northwest coast shell middens. *Tebiwa* **21**, 26–34.
1987 Deposits for archaeologists. In *Advances in archaeological method and theory, vol. 11*, edited by M. Schiffer. Orlando, Florida: Academic Press. Pp.337–395.

Stein, J. K., and P. A. Teltser
1989 Size distributions of artifact classes: combining macro- and micro-fractions. *Geoarchaeology* **4**, 1–30.

Vance, E. D.
1987 Microdebitage and archaeological activity analyses. *Archaeology* **40**, 58–59.

10

An Analysis of Fire-Cracked Rock: A Sedimentological Approach

Timothy W. Latas

I. Introduction

Fire-cracked rocks are probably the most common artifact in Pacific Northwest sites. In fact, many sites appear to be composed of four primary constituents: fire-cracked rock, shell, bone, and lithic artifacts (Benson 1975; Campbell 1981; Carlson 1960, 1979; Chatters and Thompson 1979; Croes and Blinman 1980; Dunnell and Beck 1979; Dunnell and Campbell 1977; Gaston and Jermann 1975; Greengo and Houston 1970; Kennedy *et al.* 1976; King 1950; Larson and Jermann 1978; Lewarch and Reynolds 1975; Lorenz *et al.* 1976). Many archaeologists simply report the presence of the ubiquitous fire-cracked rocks in archaeological sites and, occasionally, they may count and weigh them. However, the physical and chemical properties that are used to identify this artifact are not explicitly defined or are recognized from a set of criteria developed from experiments.

By using an artifact definition based in sedimentological terms, fire-cracked rocks are considered to have a complex depositional history of being transported from a host of sources, deposited in a variety of environments exposed to fires or high temperatures, and altered by those environments. Certain physical and chemical properties of the rock are influenced and changed according to the type and magnitude of factors encountered at each stage of this depositional history. Those properties which are expected to be altered during this unique depositional history could then be used as attributes for the identification of these artifacts. In particular, fire-cracked rock would reflect attributes which indicate that these artifacts are derived from local sources, transported by humans, deposited in environments that were exposed to fire or extreme temperatures, and fractured and oxidized in these hostile environments.

This sedimentological approach of artifact analysis was used to assess whether large, angular cobbles of the British Camp shell midden (45SJ24) were fire-cracked

Deciphering a Shell Midden

rock. From 1983 to 1989, the San Juan Island Archaeological Project excavated and screened shell midden deposits at site 45SJ24, located in the San Juan Island National Historical Park, British Camp, Washington. The excavated materials were sorted by size and material types, including angular rock, rounded rock, shell, charcoal, and bone. Other artifacts such as flaked and ground stone items were also separated during this screening operation. The large angular rock from this excavation was initially considered to be fire-cracked rock. However, a review of literature on fire-cracked rock found that traditional attributes appeared to be unreliable and that a reconstruction of the depositional history of the angular cobbles was required in order to determine whether this rock was fire-cracked rock.

II. Traditional Attributes of Fire-Cracked Rock

Fire-cracked rock is generally considered "to reflect cooking practices involving the heating of stones for use in earth ovens or for stone boiling" (House and Smith 1975:75), but may also result from a variety of other social and cultural activities. In the northwest United States, these activities include (1) sweat lodge activities (Catlin 1913; Gilmore 1925; Haerberlin and Gunther 1930; Hough 1926); (2) heating of small shelters (Bowers 1948); (3) breakdown of crystalline materials such as granite for ceramic temper (Gilmore 1925; Wedel 1940; Will and Hecker 1944; Wilson 1910); (4) hearth enclosure (Lovick 1983); and (5) structure burning (Lovick 1983). Thoms (1989) presents a compendium of ethnographic references on the variety of functions for which earth ovens were employed in Asia, Europe, and North America. These ethnographic accounts may describe many of the activities leading to the postdepositional fracturing and oxidizing of fire-cracked rock; however, they do not provide a set of attributes that can be used to identify fire-cracked rock.

Very few attempts have been made to explicitly define attributes of fire-cracked rock. House and Smith (1975) placed chert, quartzite, and sandstone cobbles from the Crowley's Ridge gravel of the Cache River Basin, Arkansas in hot wood fires that were hot enough to partially vitrify the cortex of the samples. After firing, they found that all of their rock types displayed either red or black oxidation. These rock types also fractured when allowed to cool overnight, but did not fracture when used as boiling stones. Based on these observations, House and Smith suggested that air cooling produces irregular and jagged fractures with some of the fragments resembling primary flakes with remnants of the cortex, but lacking the typical striking platforms or bulbs of percussion of flakes.

Based on a series of field experiments, Thoms (1984), with the assistance of R. Schalk (personal communication), conducted a series of thermal fracturing experiments on rocks of the Libby Reservoir area of Montana. They hypothesized that rocks used as linings in fire pits could be distinguished from boiling stones on the basis of fracture pattern. To test this hypothesis, Thoms and Schalk conducted an experiment that consisted of heating rocks to a known temperature and either allowing these rocks to slowly cool in the air or rapidly cooling these rocks in a bucket of water. Using a variety of rock types, Thoms and Schalk found that mudstone, granite, and granodiorite tend to fracture in curvilinear planes when air cooled, frequently producing large and small spalls with remnants of the cortex. Quartzite, when rapidly cooled by submersing the rocks in water, exhibited planar fractures and produced "blocky chunks" (Thoms 1984:182).

In addition to attributes of fire-cracked rock, Thoms has also described typical attributes for earth ovens, hearths, and firepits from the Williamette Valley in Oregon and Calispell Valley in northeastern Washington (Thoms 1989). The majority of ovens used primarily for camas processing in these areas were circular basins, ranging from 0.5 to 3.0 m in diameter and surrounded by a low earthern berm. Some of the fist-sized rocks in the ovens appeared to be fractured *in situ* but most were angular rocks that lacked articulated fragments. Thoms suggests that the lack of articulated fragments indicates most of the angular rock had been used several times (Thoms 1989:403). In fact, excavation of an experimental camas oven found that the majority of rocks had cracked or fragmented *in situ*, and remained articulated (Thoms 1988, 1989). In addition to the angular rock, the Calispell archaeological ovens contained a variety of chipped stone artifacts, milling artifacts, ash, charcoal, carbon-stained sediments, and remains of bone, shell, seeds, nuts, and roots. Oxidized (reddened) and hardened sediments were usually found on the rims, sides, and bottoms of these ovens. Other Calispell features also contained these fire-cracked rocks, including fire pits and hearths which are smaller in diameter (50 to 70 cm) than an oven and lacked well-defined bottoms or sides.

Thoms's research also suggests that access to a nearby, abundant rock supply was a major constraint on the spatial distribution of camas ovens in Calispell Valley (Thoms 1989). In a field experiment, Thoms found that a 2.0-m diameter oven required nearly 295 kg of rock and that gathering of this rock was labor intensive. Ethnographic descriptions indicate that intensive rock gathering was minimized whenever possible by using rocks from old ovens (Downing and Furniss 1968; Thoms 1989:255).

Both House and Smith (1975) and Thoms (1984) compared their experimental results with archaeological collections. House and Smith (1975) concluded that archaeological samples from the Cache River Basin were fractured during air cooling after being exposed to high temperatures. Thoms (1984) suggested that blocky chunks of rocks in the Libby Reservoir sites were boiling stones and that chunks of rocks with curvilinear fracture planes were steaming stones.

These previous investigations suggest that the traditionally recognized attributes of fire-cracked rock are: spalls or cobble-sized rocks with red or black discolorations, planar, curvilinear, irregular, jagged, or potlid fractures, and the absence of flaking attributes, such as bulbs of percussion, ripples, or striking platforms. These rocks generally have remnants of the cortex and are usually associated with a variety of artifacts in archaeological deposits. These deposits may also contain carbon-stained sediments, charcoal, and oxidized sediment. These traditionally recognized attributes were, however, difficult to apply to British Camp angular rock. These angular rock exhibited very little or no cortex remnants, nearly all lacked flaking attributes, appeared to be discolored, and were associated with charcoal in almost all deposits of the midden.

III. Fire-Cracked Rock as Sediment

By considering fire-cracked rock as sediment, a specific set of unique influences on physical and chemical properties of the rocks is expected to occur during the depositional history of these artifacts. The concept of depositional history can be simply explained as reconstructing the "life history" of a sediment (Blatt *et al.* 1972; Hassan 1978; Krumbein and Sloss 1963; Reineck and Singh 1980; Stein 1985, 1987; Stein and Rapp 1985; Twenhofel 1950). A depositional history consists of four stages: (1) source, (2) transport agent, (3) depositional environment, and (4) postdepositional alterations. Each of these stages is interpreted using a variety of geological principles, chemical and physical laws, and attributes of sediments and deposits.

The concepts of deposit and sediment are of primary importance to reconstructing a depositional history. A deposit, in the archaeological literature, is a three-dimensional unit that is distinguishable in the field on the basis of observable changes in some physical properties (Schiffer 1983). The deposit can therefore be considered as the fundamental sedimentary unit. Geologists recognize two types of deposits: those deposits which are formed from chemical precipitation of minerals, such as limestones or evaporites (Folk 1980); and those deposits consisting of an aggregate of sedimentary particles, or sediments (Folk 1980; Stein 1987). Sediments are particulate matter that are transported by some process from one location to another (Blatt *et al.* 1972; Hassan 1978; Krumbein and Sloss 1963; Stein 1985, 1987). No restriction is made on the kind or size of materials, types of transport agents, areas of deposition, or influence of postdepositional processes. Therefore, most particles in archaeological deposits are sediments (Stein 1985).

Attributes of sediments and deposits that are used to interpret depositional histories have been defined by sedimentologists (Krumbein and Sloss 1963; Reineck and Singh 1980; Stein 1985, 1987; Stein and Rapp 1985). The attributes of roundness, lithology, size, sphericity, color, oxidation, and fracture may be described for sediments. For deposits, the vertical and horizontal relations, geomorphic expressions, content, and grain size distribution are likely to be described (Stein 1987).

A hypothetical depositional history of fire-cracked rock would consist of: (1) deriving cobble-size rocks from local sources; (2) humans transporting these rocks to sites of deposition; (3) depositing the rocks in environments which are subjected to high temperatures produced by fires and possibly accompanied by very low humidity, boiling water, or steam; and (4) postdepositional fracturing and oxidizing the rocks in these environments. Local sources of rock are expected to be exploited because the effort to collect a sufficient quantity of cobble-sized rocks to line fire pits, form ovens, or other heated-rock features is very labor intensive (Thoms 1989). Transport of the rocks is considered as a human function because humans conduct the actual collection and placement of the rocks though domestic pack animals (e.g., dogs or horses) may be used to actually carry the cobbles. This type of transport is not expected to alter the shape of the cobbles during transport. As shown by ethnographic references and archaeological investigations, deposits related to fire-cracking activities have particular shapes and contain a variety of other materials including charcoal, carbon-stained sediment, flaked and ground stone artifacts, and remains of plants, shells, and bone (Thoms 1989). These deposit attributes indicate that the use of fire was controlled and was associated with a variety of activities. Traditionally, archaeologists refer to these deposits as features which may be described as fire pits, hearths, and ovens. As demonstrated by experiments, postdepositional alterations of cobbles in a fire-cracking environment include fracturing and oxidation of the rocks. Each of these stages can be interpreted from attributes of the sediments and deposits. This reconstruction of the "life history" of rocks found in a site will evaluate whether the rocks are fire-cracked rock or artifacts of other functions of the site.

In order to test this sedimentological approach to artifact analysis, angular cobbles and their deposits from the British Camp site were closely examined and described, subjected to a variety of fire-cracking experiments, and compared with geologic information from British Camp to determine whether the depositional history of these angular cobbles resembles the complex depositional history of fire-cracked rock. The field and laboratory work was conducted in 1985 and 1986, and initial results were presented in 1987 (Latas 1987).

A. *Sources*

The sources of the British Camp angular cobbles are interpreted from a comparison of the lithology of midden rock with the lithology of potential sources from the area surrounding the site (e.g., outcrops and rocky beaches). The lithology of the midden and potential source rock was determined by identifying rock types from hand specimens and thin sections of rock from the midden and potential sources.

1. Determining Sources

Determining rock types is based on descriptions of hand specimens and microscopic examinations of rock thin sections. Percentages of minerals, mineral grain size, orientation of minerals, linear or planar features, microstructures, color, and weathering characteristics are commonly described during examination of hand specimens of rocks. Additional information concerning lithology can be developed by examining thin sections of rock under a polarizing microscope following petrographic procedures similar to those outlined in Turner and Verhoogen (1951) and Williams *et al.* (1954). Calculating relative frequencies of various rock types in a depositional unit, such as a midden, may further characterize the contribution of multiple sources on the deposit.

a. Lithologies of the Midden Rocks. To determine the lithologies of midden rock, a representative sample of angular cobbles was needed from the deposits excavated from the units at British Camp. During the 1985 field season at British Camp, screening operations were required to separate, count, and weigh the angular rock, rounded rock, shells, charcoal, flaked artifacts, and ground stone artifacts that were caught on the 0.5-inch (12.5 mm) and 1.0-inch (25 mm) wire mesh screens. These screening

Table 1 Weight of Angular Rock Found in Midden

Week	Weight of total screened materials (kg)	Weight of total 25-mm angular rock (kg)	Weight of total 12.5-mm angular rock (kg)
Second week	2.210	225	90
Fourth week	1,800	197	83
Seventh week	2,510	330	125
Totals	6,520	752	298

operations were conducted for nearly eight weeks. In order to acquire a representative sample of angular rock from the upper, middle, and lower portions of the midden, the 12.5-mm and 25-mm angular rock captured during the second, fourth, and seventh weeks of screening were set aside for futher consideration. Table 1 shows that during the second, fourth, and seventh weeks, a total of 6530 kg of excavated materials were processed through the 12.5-mm and 25-mm screens. Of this total, 4.6% (300 kg) was 12.5-mm angular rock and 11.9% (780 kg) was 25-mm angular rock. There appears to be little variation between these weekly sample weights. Therefore, as a generalization, the percentage of angular rock in the midden seems to be relatively constant for the portion of the midden represented by these three weekly samples. No calculations were made to determine the relative frequencies of the rounded rock, shells, charcoal, and artifacts for all eight weeks of screening.

Although finer size fractions (less than 12.5 mm) probably contain small fire-cracked rock, this investigation focused on the larger fragments because the small fractions were not sorted into angular and rounded rock classes. To perform this task, an inordinate amount of time and effort would be required. In addition, description of lithologies for rocks smaller than 12.5 mm is also extremely tedious.

To simplify rock classification of the 1180 kg of 12.5-mm and 25-mm angular rock, a subsample of the angular rock was obtained from the three weekly samples by saving every tenth shovelful during removal of the 12.5-mm and 25-mm piles. These second-, fourth-, and seventh-week subsamples were then sorted by rock type, counted, and weighed (Tables 2, 3, 4, and 5). The total subsample includes 4135 angular rocks caught on the 12.5-mm and 25-mm screens, weighing a total of 92 kg. This total subsample represents 8.5% of the total angular rock sample and 1.4% of the total amount of materials processed through the screens during the second, fourth, and seventh weeks. Though "cobbles" are usually classified in the United States as ranging in size from 64 to 256 mm (Folk 1980), this term is used informally here as a descriptor of British Camp midden rocks greater than 12.5 mm in diameter.

In the midden sample, eight rock types are recognized, including three kinds of igneous rock (diorite, granodiorite, and basalt), two kinds of sedimentary rocks (chert and micaceous sandstone; both show indications of low-grade metamorphism), and three types of metamorphic rock (green schist, slate, and gneissic schist) (Table 2). The relative percentages of each rock type for the second, fourth, and seventh week show chert, green schist, basalt, and granodiorite to be the principal constituents of

Table 2 Descriptions of Angular Rock Types from the British Camp Midden

Rock type	Physical description
Diorite	Fine grained (<0.5 mm), equigranular, olive green, mafic, plagioclase and epidote, planar fractures, slightly rounded cortex.
Granodiorite	Fine to medium grained (0.1–0.5 mm), equigranular, whitish gray, biotite plagioclase quartz granodiorite, randomly oriented planar fractures, grainy rounded cortex, thin (1.0 mm) red oxidation rind.
Basalt	Aphanitic [very fine grained (<0.1 mm)], equigranular, dark green to black, plagioclase and quartz, planar to slight conchodial (curvilinear) fracture, rounded cortex. Some 1.0-inch samples resemble spalls or flakes, red oxidation on some fragments.
Schist	Well-developed schistosity, dark olive green to black quartz-mica schist, some fracturing normal to schistosity, brownish-red oxidation.
Chert	Cryptocrystalline, dark gray chert with cross-cutting quartz veins, irregularly shaped fractures, some are slightly conchodial (curvilinear), some fragments with rounded cortex, red oxidation.
Sandstone	Micaceous, red, fine-grained quartz sandstone, friable, planar fractures parallel schistosity.
Gneissic schist	Banded, fine grained (0.1 mm), gneissic schist, curvilinear to planar fracture along schistosity, fractures also perpendicular to schistosity.
Slate	Gray micaceous slate found as thin platy fragments, no fractures or oxidation observed.

the 25-mm samples, while green schist is the dominant rock type in the 12.5-mm samples (Tables 3, 4, and 5; Figure 1).

The occurence of basalt cobbles in the midden angular sample requires additional attention because most of the flaked artifacts from the midden are composed of aphanitic volcanic rock, commonly refered to as basalt. Based on an examination of hand specimens of the flake artifacts and the angular basalt cobbles, differences were noted in the color and texture of these two artifacts. A recent thesis on the petrology and geochemistry of black, aphanitic "basalt" lithics from the British Camp site has found that these lithics are actually a dacite rather then a basalt which may originate from sources at Mt. Garibaldi (Bakewell 1990). The basalt angular cobbles, however, had a dark green to greenish-black aphanitic appearance and had a strong resemblance to basalt samples from exposures of the Deadman Bay Volcanic pillow basalts on the west side of San Juan Island which have been previously described by Brandon *et al.* (1988) and Vance (1968, 1975, 1977). Detailed petrographic and geochemical analyses of these basalt cobbles would more accurately classify these cobbles. For the purposes of this investigation, cobbles that are dark green to black aphanitic (crystalline texture too small to be distinguished with the unaided eye) were classified as basalt. A more complete discussion of the flake technology is presented in Chapter 8, this volume.

Comparing the relative frequencies of each rock type between the weeks indicates some changes in relative abundance of the rock types recovered over the seven

Table 3 Angular Rock Count for the Second Week of Screening

	25-mm angular rock				12.5-mm angular rock			
Rock type	Count	(%)	Wt (g)	(%)	Count	(%)	Wt (g)	(%)
Diorite	45	(9)	1600	(7)	46	(3)	500	(6)
Granodiorite	53	(11)	3200	(13)	158	(12)	1200	(14)
Basalt	109	(22)	5400	(22)	103	(7)	900	(10)
Schist	145	(29)	7100	(30)	770	(57)	5200	(59)
Chert	121	(24)	4825	(20)	261	(3)	800	(9)
Sandstone	23	(5)	1700	(7)	23	(2)	225	(3)
Totals	496	(100)	23,825	(100)	1361	(100)	8825	(100)

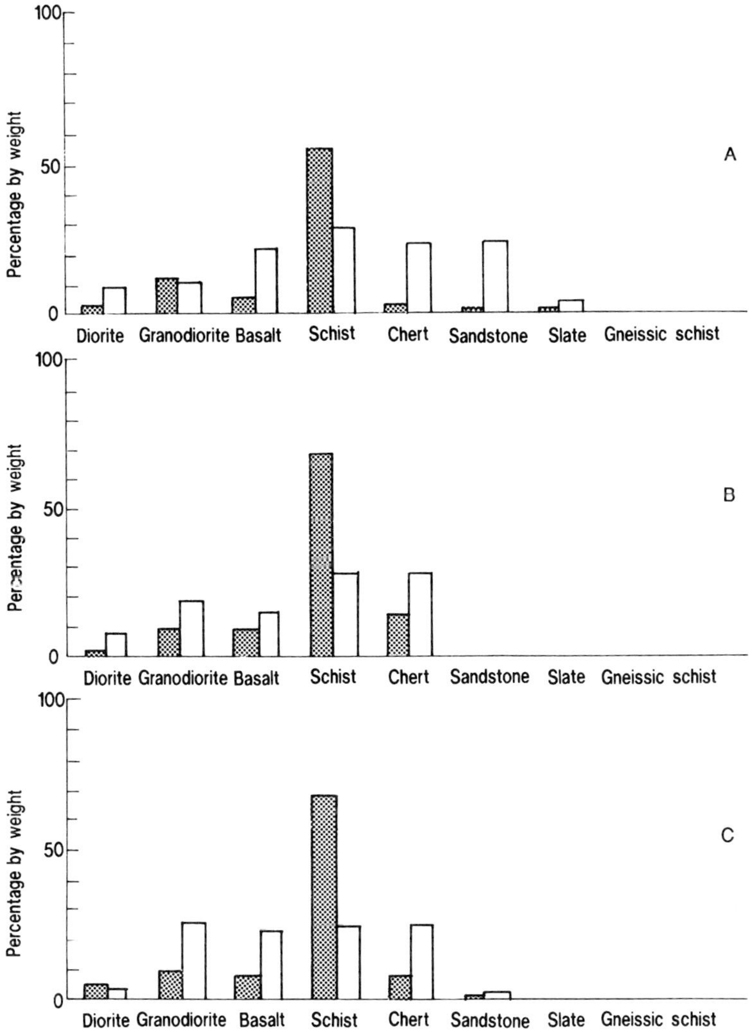

Figure 1 Relative percentages by weight of angular midden rock types as determined from rock identification of subsamples of weekly screening at British Camp Operation A, with gray bars representing percentages of 12.5-mm angular rock and white bars representing percentages of 25-mm angular rock: A, Second week of screening relative percentages by weight of rock types; B, fourth week of screening relative percentages by weight of rock types; C, seventh week of screening relative percentages by weight of rock types.

weeks of excavation but these changes are probably dependent on sample size though some changes may be the result of cultural selection of particular rock types. Another aspect of the angular rock counts is relatively low percentages of micaceous sandstone, slate, and gneissic schist throughout the samples (Tables 3, 4, and 5; Figure 1). Perhaps these rock types are primarily artifacts of a tool technology rather than use of rocks in a firing environment.

Table 4 Angular Rock Count for the Fourth Week of Screening

	25-mm angular rock				12.5-mm angular rock			
Rock type	Count	(%)	Wt (g)	(%)	Count	(%)	Wt (g)	(%)
Diorite	17	(7)	850	(7)	10	(2)	175	(3)
Granodiorite	46	(18)	2750	(22)	56	(9)	600	(12)
Basalt	41	(15)	1825	(15)	29	(9)	375	(7)
Schist	72	(28)	4025	(31)	437	(69)	3100	(61)
Chert	72	(28)	2875	(23)	93	(15)	800	(16)
Sandstone	—	—	—	—	5	(1)	75	(1)
Rhyolite	10	(4)	250	(8)	—	—	—	—
Totals	258	(100)	12,575	(100)	630	(100)	5125	(100)

Table 5 Angular Rock Count for the Seventh Week of Screening

	25-mm angular rock				12.5-mm angular rock			
Rock type	Count	(%)	Wt (g)	(%)	Count	(%)	Wt (g)	(%)
Diorite	18	(4)	1075	(3)	61	(5)	675	(7)
Granodiorite	228	(26)	8325	(26)	114	(10)	1175	(13)
Basalt	110	(24)	10,375	(32)	91	(8)	775	(8)
Schist	116	(26)	7650	(23)	790	(69)	5775	(61)
Chert	75	(26)	7650	(23)	91	(8)	775	(8)
Sandstone	11	(2)	1975	(6)	12	(1)	325	(3)
Slate	—	—	—	—	6	(1)	100	(1)
Gneissic schist	5	(1)	225	(1)	—	—	—	—
Totals	453	(100)	32,575	(100)	1137	(100)	9450	(100)

b. Lithology of Potential Sources. Lithology of nearby, potential sources is based on the rock descriptions of local outcrops, talus slopes, and beaches. To locate potential sources, a geologic map of the British Camp park was produced. The results of this mapping effort found three geologic accreted terranes overlain by Pleistocene glacial deposits containing cobbles (Figure 2). Accreted terranes are lithologically distinctive units of oceanic crust sliced by shear faults and drawn out into thin strips as the oceanic plate was obliquely subducted under the continental margin, resulting in a collage of relatively narrow, but long, blocks of oceanic rocks (Jones *et al.* 1982). These terranes on the west side of San Juan Island range in age from the Early Permian through Early Cretaceous and include the Deadman Bay terrane, Garrison terrane, and Constitution Formation which are composed of oceanic rocks and are structurally juxtaposed by thrust faulting (Brandon *et al.* 1988; Vance 1968, 1975, 1977). Late Quaternary glacial sediments blanket these terranes (Armstrong *et al.* 1965; Easterbrook 1968; Whetten *et al.* 1977).

During mapping, the lithology, color (fresh and weathered surfaces), size of weathered material, roundness of edges, and fracture patterns of outcrops and talus were described at many outcrops. Usually a small rock sample was also collected at the outcrop for thin-section analyses. At the conclusion of the mapping exercise,

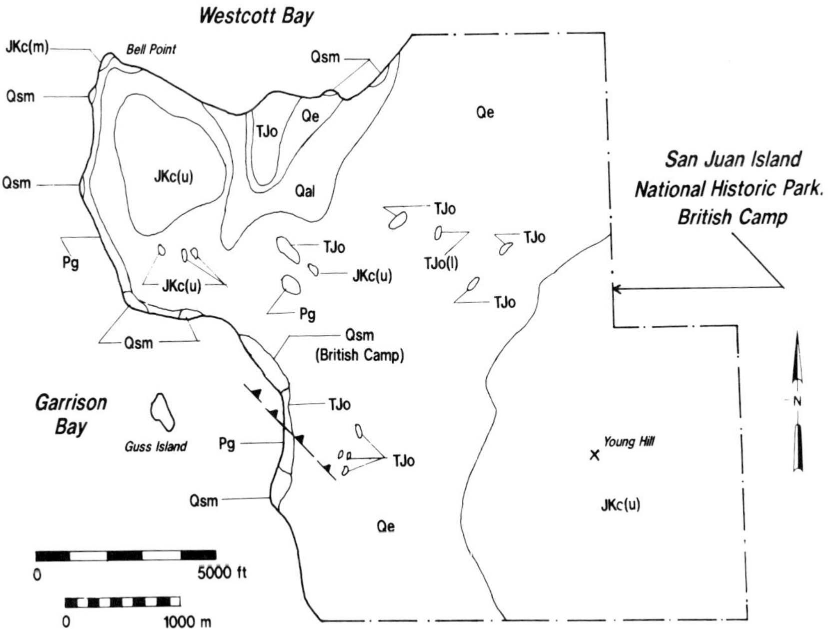

Figure 2 Geologic map of British Camp, San Juan Island National Historic Park, San Juan Island, Washington. Shown on this map is the areal distribution of deposits within the park: shell middens (Qsm), alluvium (Qal), Everson Glaciomarine Drift (Qe), Constitution Formation upper sandstone member [JKc(u)], Constitution Formation middle conglomerate member [JKc(m)], Orcas Chert ribbon chert (TJo), Orcas Chert limestone [TJo(l)], and Garrison Schist (Pg). Also shown is a low-angle thrust fault (line with sawtooth edge).

formation names were assigned based on mapped lithologies and comparison of these lithologies with previously described geologic units.

The rocks from three different accreted terranes were found in outcrops within the park. The Orcas Chert of the Deadman Bay terrane is exposed in small outcrops scattered throughout the park (Figure 2). This formation consists of dark gray, highly recrystallized cryptocrystalline chert (J. Garver, personal communication) and forms lenses 5 cm thick and 5–100 cm long, banded with silt-sized quartz partings, and weathering into gray, angular cobbles with oxidized cortices. Outcrops of the Garrison terrane were found along wave-cut banks along Garrison Bay (Figure 2) and consist of a highly recrystallized, dark olive green to blackish-green quartz-mica (quartz tectonite) weathering into dark gray to black, pebble- and cobble-sized rectangular plates with an oxidized cortex. In thin section, very fine sand-sized quartz grains appear in a matrix of reddish brown mica (biotite?) along with lawsonite and prehnite (J. Garver, personal communication). Two of the three informal members of the Constitution Formation were recognized in the park. The lower member crops out at the western tip of Bell Point (Figure 2) and consists of tectonized green recrystallized chert cobble conglomerate with silt-sized chert matrix weathering into grayish-green, angular cobble-sized blocks. The upper member of the Constitution Formation found in the park forms the topographic highs at Bell Point and Young Hill (Figure 2). It is a dark gray detrital plagioclase, volcaniclastic, chert sandstone, rich in prehnite and veins of lawsonite, which weathers into angular pebbles and cobbles with a mottled brown and gray color.

A veneer of Late Quaternary glacial and marine sediments blankets the Paleozoic and Mesozoic rocks of the park (Armstrong *et al.* 1965; Clague 1981). These Late Quaternary formations include the Everson Glaciomarine Drift and alluvium. The Everson Glaciomarine Drift is light brown, pebbly mud with rounded pebbles and cobbles, which blankets about three-fourths of the park's surface (Figure 2). The beaches are typically mud flats with cobbles derived from erosion of the shoreline deposits.

A pebble typology was completed for a drift outcrop and several locations along the beach to determine the cobble composition of these deposits. The study consisted of identifying and counting rocks greater than 12.5 mm in diameter (or cobbles) exposed in 1-m grid systems along the bank and point counts along lines oriented perpendicular to the beaches (Figure 3).

Results of these rock counts are shown on Figure 4 as bar graph representations of the relative percentages of rock types. For the drift outcrop, most of the cobbles are chert (72%) or granodiorite (17%), with minor quantities of andesite (7%) and diorite (4%). The beach samples were located at three different places along the beach (Figure 3). The easternmost beach sample, taken below a cliff exposure of Garrison schist and Orcas Chert, was dominated by schist (78%) and chert (20%) (Figure 4). The next beach sample was taken below the British Camp rose garden where a thin mantle of shell midden overlays glaciomarine drift (Figure 3), and is dominanted by chert pebbles and cobbles (82%), with minor quantities of diorite (8%), granodiorite (3%), basalt (5%), schist (1%), and sandstone (1%) (Figure 4). The westernmost beach sample below the British Camp commissary shows the dominant

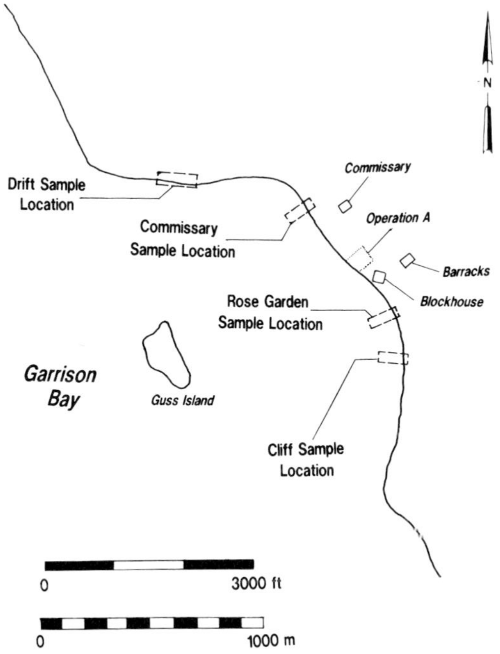

Figure 3 Locations of glacial-drift and beach samples used to construct cobble typologies as shown in Figure 4. The glacial drift sample location was located approximately 800 m northwest of Operation A on a wave-cut bank. The Commissary, Rose Garden, and Cliff beach sample locations were located in the intertidal mudflats of Garrison Bay. The relative positions of these sample locations to Operation A and British Camp buildings can be seen on this figure. The extent of the sample areas is designated by a dotted lines.

rock type to be chert (82%), with minor amounts of diorite (11%), granodiorite (4%), and basalt (3%) at the beach below the British Camp commissary (Figures 3 and 4).

2. Sources of the Midden Rock

The source for most of the angular rocks at British Camp appears to be local outcrops and beaches. Composition of the midden angular rock sample consists of diorite, granodiorite, basalt, schist, chert, sandstone, slate, and gneissic schist (Figure 1). Potential sources for the angular rock are found within a short distance of the British Camp midden, including outcrops of the Orcas Chert, Garrison Schist, and the sandstone of the middle member of the Constitution Formation (Figure 2). Diorite and granodiorite appear to be derived from either the Everson Glaciomarine Drift or beaches (Figure 3). The abundance of local sources and relative high percentages of these rock types in the midden indicate that labor-intensive rock gathering was minimized as referenced in ethnographic accounts and research conducted by Thoms (1989).

No sources of basalt, slate, micaceous sandstone, or gneissic schist were found in the outcrops surrounding the midden. Additionally, no outcrops of the basalt were

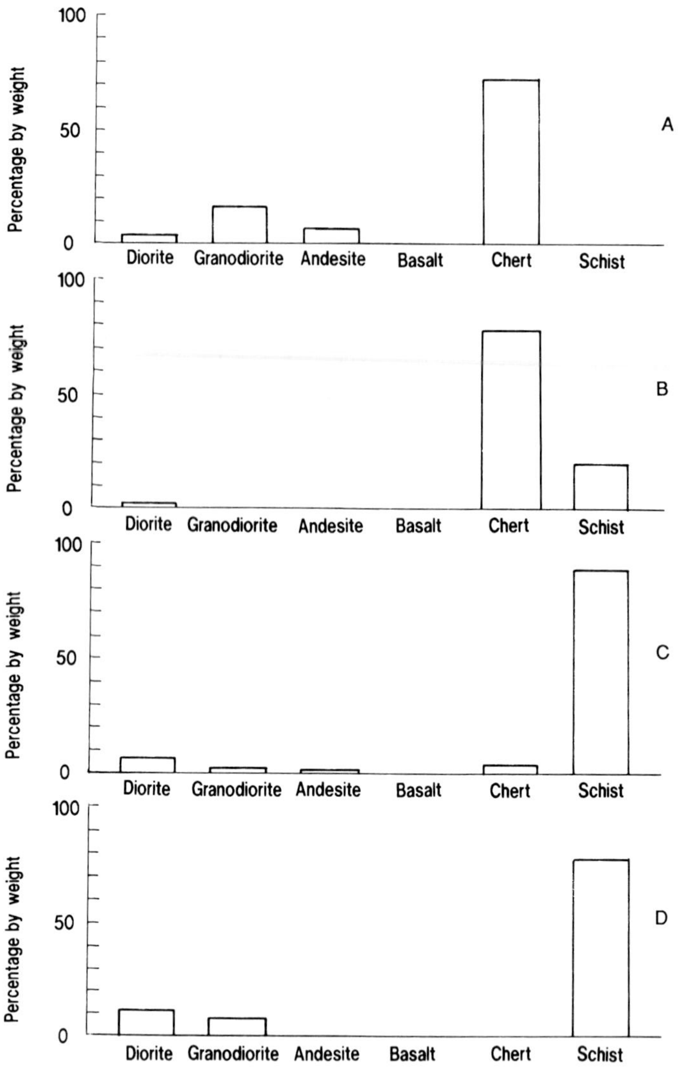

Figure 4 Relative percentages by weight of various rock found at the glacial drift and beach sample locations: A, The glacial drift sample was dominated by chert, with minor amounts of diorite, granodiorite, and andesite; B, the cliff beach sample was dominated by chert and schist; C, the Rose Garden beach sample was dominated by schist, with minor quantities of diorite, granodiorite, and andesite; D, the Commissary beach sample was dominated by schist, with minor quantities of diorite and granodiorite.

found within the park and no basalt was found in the glaciomarine drift or on the beaches, yet it comprises between 5 and 20% of the midden angular rock. This suggests that basalt is being derived from outside of the park, probably from outcrops of the Deadman Bay Volcanics described from the west coast of San Juan Island (Brandon *et al.* 1988) or possibly dacite sources on Mt. Garibaldi (Bakewell 1990).

These basalt cobbles may have originally been used at the site as lithic tools and later discarded or used as linings in fire pits.

The occurrence of slate, micaceous sandstone, and gneissic schist on San Juan Island is not documented in the literature or in outcrops within the park. These rocks are probably brought in from other islands or the mainland and should be considered as exotic rock to the midden assemblage. The occurance of such exotics suggests these rocks are probably not fire-cracked rock and are artifacts of lithic technology.

B. *Transport Agents*

Grain size, roundness, and relationships of source areas to areas of deposition are indicators of transport agents. Application of procedures as outlined in Folk (1980) and Reineck and Singh (1980) was used to describe grain size and roundness, because grain size distribution is the result of energy expended during transport of the particles, and roundness of grain (defined as the sharpness of grain edges) reflects the degree of abrasion encountered by a grain during transport. Thus, these two parameters are used to characterize energy and movement of particles by the transport agent. An examination of the relationship of source areas to areas of deposition can provide insights to the types of mechanisms that would be required to move cobbles across the intervening terrain.

1. Methods for Evaluating Transport

Since this investigation was restricted to angular rock caught on the 12.5-mm and 25-mm mesh screen, the number of transport agents which move this size of particle and avoid rounding the angular rock are limited. However, each sorted sample was closely examined in the field for rounding of the edges which would indicate transport by rolling, such as stream or wave transport.

To evaluate the types of possible transport agents capable of moving cobbles from source areas to areas of deposition, the results of the source analysis is coupled with distance and descriptions of terrain between sources and the British Camp midden.

2. Transport Agents of Angular Cobbles

All rock types examined during field sorting of the midden materials were found to have sharp edges. Though cobbles of basalt, diorite, and granodiorite exhibited remnants of rounded cortices, these cobbles had sharp, unrounded edges which have been formed by removal of the cortex. These angular edges indicate that the rocks had been transported without abrasion occurring or that the edges were produced by postdepositional alterations. The schist, chert, and sandstone samples had sharp edges and lacked evidence of a cortex. This indicates that these rocks may have been derived from a source of angular cobbles and were not subjected to abrading during transport, or that these rocks may have been altered on site.

A more rigorous explanation of transport agents is developed from a simple analysis of terrain and distance between the potential sources and the British Camp midden. Though Figure 2 does not show topography of the area, most of the terrain around the midden is relatively flat with a 2–5% slope toward the bay. Potential sources are located within a few meters to several hundred meters from the midden (Figure 2). Evaluation of these relationships indicates that the transport agent must be capable of moving cobbles across relatively flat terrain and, in some cases, upslope from the beaches over some distance (Figure 2).

Only a few geologic transport agents, such as glaciers and high-velocity water, are competent to transport cobble-sized particles (Reineck and Singh 1980). Remnants of the rounded cortices on the basalt, diorite, and granodiorite cobbles may reflect this type of transport. However, to maintain possible original angularity of the schist, chert, and sandstone cobbles, the transport agent had to move the cobbles by suspension.

By evaluating the terrain and distances between the potential sources and the midden, no geologic agent is capable of moving cobbles upslope from the source. This shows that human transport is involved with the movement of cobbles at British Camp.

C. *Depositional Environments*

Ethnographic accounts describe a variety of social and cultural activities which would expose rocks to fire-cracking environments, including steaming, stone boiling, enclosing firepits, heating of small temporary shelters, breaking down crystalline materials for ceramic temper, burning down structures, and building canoes (Bowers 1948; Catlin 1913; Gilmore 1925; Haeberlin and Gunther 1930; Hough 1926; Smith 1940). In the Northwest, cooking activities such as camas roasting, shellfish steaming, and stone boiling are probably the most common activities where rocks are exposed to firing environments (Smith 1940; Thoms 1989).

1. Attributes of a Fire-Cracking Environment

In order to evaluate if the angular cobbles of British Camp were deposited in an environment which would lead to fire-cracking, certain characteristics of the midden must be examined. For a fire-cracking environment, grain size distribution, shape, artifact contents, and contacts of the deposits were considered to be the most appropriate field indicators of this depositional environment.

To evaluate these attributes, several sources of information on the midden were consulted. Profile drawings prepared at the close of the 1985 field season presented detailed information on shape, contacts, and distribution of the materials larger than 1 cm in the profile of deposits, or facies, in the midden (Figure 5). The concept of a facies is discussed more fully in Chapter 2, this volume, and will be used in the following discussion to designate a deposit in the midden that was recognized in the field-based physical differences (e.g., color, shape, grain size) with surrounding deposits. Grain size distributions based on the weights of angular, rounded, and total rock captured on the mesh screens are used as an indicator of relative proportions of the various particle sizes (Figures 6 and 7). Consultation of field notes and other

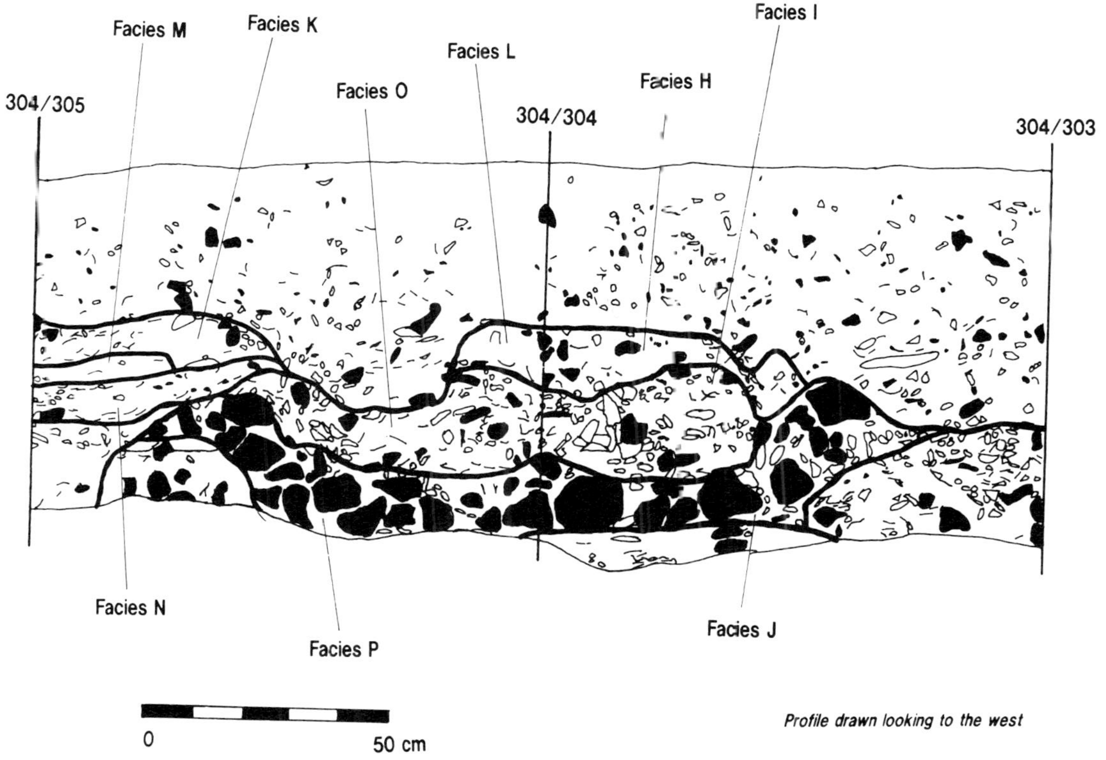

Figure 5 Profile drawing of a portion of Units 304/302 and 304/304 from 304/303 through 304/305. This profile shows the relative size of objects greater than 1 cm in the deposits and the various shapes of the deposits. Black objects are rock; white objects are bone and artifacts; shells are curved lines. Note that Facies P/Facies J is a relatively large-diameter basin which is similar to camas ovens described by Thoms (1989).

personnel involved with research on the midden provided information on the contents of the deposits, especially burned materials (Table 6).

2. Three Midden Depositional Environments

Three distinct deposits were recognized in the British Camp midden. In units 304/302 and 304/304, profile drawings of between 304/303 and 304/305 show the shape, contacts, and relative size of the larger particles of the deposits (Figure 5). Figures 6 and 7 illustrate the grain size distributions on materials greater than 3 mm for selected deposits in this profile of the midden.

The first type of deposit is tabular to lense shaped, with contacts between underlying and overlying deposits parallel to one another, and lacking cobbles (e.g., Facies I between units 304/303 and 304/304, and Facies M between Units 304/304 and 304/305) (Figure 5). Using weights of fractions caught on the 3.20-mm, 6.40-mm, 12.5-mm, and 25-mm screens, grain size distributions for Facies I of 304/302 and Facies M of 304/304 are skewed toward the finer sized fractions and few cobbles (Figures 6 and 7).

Table 6 Deposits in Three Excavation Units of the British Camp Midden with Burned Bone

Unit	Facies	Level (10 cm)
310/300	P	1
310/300	R	1
310/300	R	3
310/302	A	1
310/302	A	2
310/302	A	3
310/302	B	1
310/302	C	1
310/302	D	1
310/302	E	1
310/302	F	1
310/302	G	1
310/302	L	1
310/304	L	1
310/304	M	1
310/304	N	1
310/304	O	1
310/304	P	1
310/304	R	1
310/304	U	1
310/304	U	1
310/304	GG	1
310/304	II	1

Note: Burned bone occurs in deposits near the surface (A facies) to depths near the bottom of the midden (II facies).

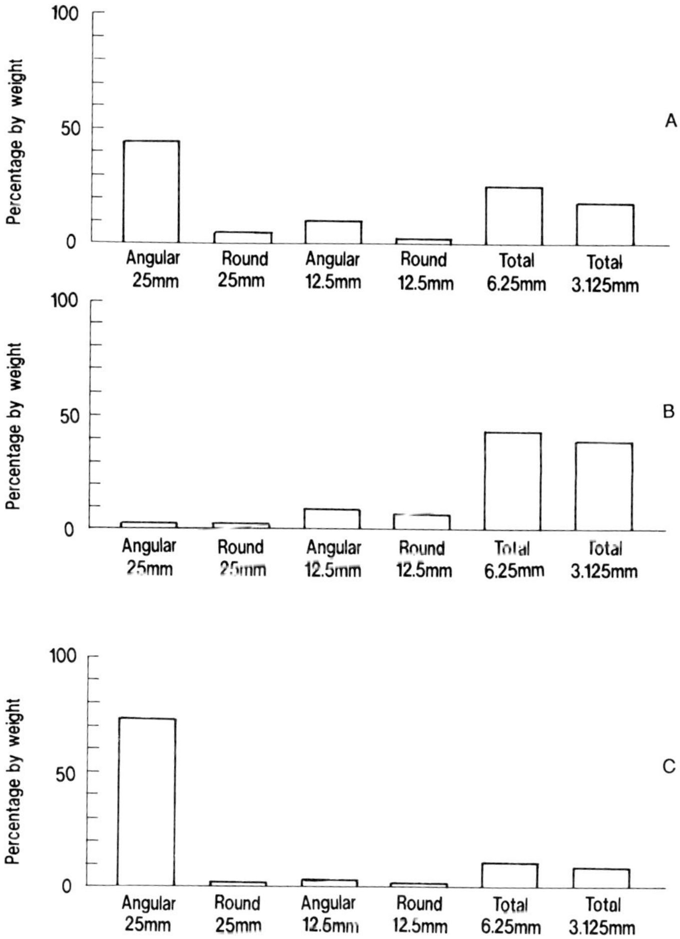

Figure 6 Grain size distributions for selected deposits as shown in the profile drawing (Figure 5). These bar graphs show the relative percentage by weight of the 25-mm angular rock, 25-mm rounded rock, 12.5-mm angular rock, 12.5-mm rounded rock, and total 6.25- and 3.125-mm rock from facies in excavation unit 304/302: A, Relative percentages of rocks from Facies H (see Figure 5 between 304/303 and 304/304); B, relative percentages of rocks from Facies I (see Figure 5 between 304/303 and 304/304); C, relative percentages of rocks from Facies J (see Figure 5 between 304/303 and 304/304).

The second type is also tabular to lense shaped with smooth contacts between underlying and overlying deposits parallel to one another, and a bimodal grain size distribution in the finer fractions and cobble-sized fraction (e.g., Facies H, between units 304/302 and 304/304, and Facies 0 between units 304/304 and 304/305) (Figures 5, 6, and 7). The presence of angular cobbles suggests that these deposits may have originated from fire-cracking activities, however, they lack the traditional shape of firepits or ovens.

The third type of deposit appears as a parabolic-shaped, cobble-lined depression which cross-cuts other deposits (Facies J between units 304/303 and 304/304, and Facies P, between Units 304/304 and 304/305) (Figure 5). The relative frequencies

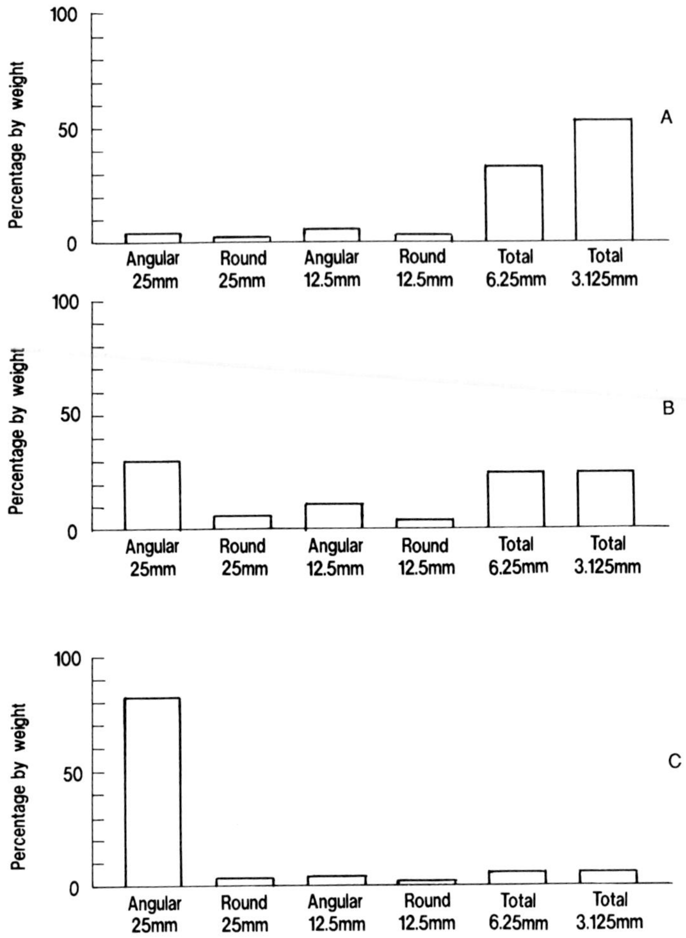

Figure 7 Grain size distributions for selected deposits as shown the profile drawing (Figure 5). These bar graphs show the relative percentage by weight of the 25-mm angular rock, 25-mm rounded rock, 12.5-mm angular rock, 12.5-mm rounded rock, and total 6.25- and 3.125-mm rock from facies in excavation unit 304/302: A, Relative percentages of rocks from Facies M (see Figure 5 between 304/304 and 304/305); B, relative percentages of rocks from Facies O (see Figure 5 between 304/304 and 304/305); C, relative percentages of rocks from Facies P (see Figure 5 between 304/304 and 304/305).

of grain sizes indicate that Facies J of 304/302 and Facies P of 304/304 are skewed toward the large-sized fractions (Figures 6 and 7). The parabolic shape, cross-cutting contacts, and cobble grain size distribution are attributes usually expected with a firepit or oven, as described by Thoms (1989).

Preliminary results of botanical and faunal analyses suggest that these three types of deposits all contain some burned bone and charcoal. Preliminary analyses of faunal remains from three excavation units in the British Camp midden (310/300,

310/302, and 310/304) found that many facies contained burned bone as shown on Table 6 (P. Ford, personal communication). Charcoal was found in all units (M. Nelson, personal communication). The presence of charcoal in all deposits and burned bone in many of them does not appear to be conclusive evidence that the deposit is the focus of a firing environment, but does suggest that the use of fire played a dominant role in activites at the site.

D. *Postdepositional Alterations*

Literature on fire-cracked rock indicates that in a fire-cracking environment the rocks may undergo a transformation of physical and chemical properties, resulting in a cobble-sized artifact with red or black discolorations, and planar, curvilinear, irregular, jagged, or potlid fractures. These cobbles lack attributes of flaking, such as bulbs of percussion, ripples, and striking platforms. This postdepositional stage of fire-cracked rock is based on comparisons made from patterns of oxidation and fracture in the midden rock and at potential sources.

1. Determining Postdepositional Alterations

The changes in size, roundness, fracture, and oxidation of an assemblage of rock samples from the British Camp park were documented during experiments and the results compared to samples from the midden for analysis. Prior to conducting firing experiments, representative rock samples from potential sources were collected from within the park. The lithology, oxidation, size, roundness, microstructures, and fracture patterns were described for rocks at outcrops, from hand specimens, and using thin sections. Three different experiments were then conducted to evaluate the effects of steaming, boiling, and long-term campsite use on the rock samples collected from potential sources. Results of the experiments were then compared to descriptions of source rock and midden rock to evaluate if firing environments produced the oxidation and fracture patterns observed on midden rocks.

a. Experiments. Cobble-sized samples of basalt, granodiorite, diorite, schist, chert, and limestone were used in these experiments to evaluate the effects of a firing environment on fracture and oxidation for each rock type (Tables 7 and 8).

The steaming experiment consisted of heating the rock samples in glowing charcoal embers for 2 hours and then pouring cold water on them to produce steam. Water was applied until the samples did not produce steam but were warm to the touch. The results of this experiment indicate that certain rock types developed small fractures and no oxidation (Tables 7 and 8). Schist fractured along foliation planes, forming planar fractures, while chert and diorite developed hairline fractures in the weathered cortex, forming slight conchoidal fractures. Limestone fractured along bedding planes, forming irregular fractures. Granodiorite did not display any fractures; however, individual mineral grains were loosened in the weathered cortex. Basalt was the only rock that did not show any evidence of fracturing from exposure to steaming conditions. None of the rocks developed any visible oxidation

Table 7 The Presence/Absence of Oxidation Noted on Rock Specimens Collected

Rock type	Outcrop	Midden	Steaming	Boiling	Long-term
Diorite	–[a]	–	–	–	–
Granodiorite	+	+	–	–	–
Basalt	–	+	–	–	–
Schist	+	+	–	–	–
Chert	+	+	–	–	–

[a] –, absent; +, present.

during the steaming experiment (Table 7). In summary, some fracturing did occur during steaming but this occurs within the weathered cortex of the rock or along planes of weaknesses (e.g., bedding or schistosity) inherit to the rock type (Table 8).

The boiling experiment consisted of heating a duplicate set of rock samples in glowing charcoal embers for 2 hours and submersing them in a bucket of cool water. None of the rock types in this experiment demonstrated any fracturing or oxidation (Tables 7 and 8). The only response noted after heating the rock samples for over 2 hours in a charcoal fire and submersing them in cool water was a slight darkening of exposed surfaces by soot. Perhaps several repetitions of this experiment would eventually cause fracturing. However, the British Camp rock types were found to be resistant to fracturing and oxidation under conditions resembling boiling.

A long-term use experiment consisted of placing rocks as a lining in a firepit which were repeatedly heated by wood fires over them throughout the summer of 1985. Fires were typically built in the pit each night for a period of 8 weeks. This experiment was used to evaluate the fracture and oxidation of rocks under conditions of a roasting pit. The pit consisted of a shallow depression lined with cobble-sized rocks collected from the tidal flat area below Operation A. The pit was excavated at the end of the field season to recover the cobbles. Fracture and oxidation patterns were described for all rocks before and after the experiment (Tables 7 and 8). From this experiment, many of the rocks had become darkened with soot and ash, but none appeared to be oxidized. Schist was the only rock type to exhibit fracturing. However, this fracturing may be due to trampling the pit while no fires were in the pit, which was noted to occur occasionally throughout the summer.

Table 8 The Presence/Absence of Fracture Noted on Rock Specimens Collected

Rock type	Outcrop	Midden	Steaming	Boiling	Long-term
Diorite	–[a]	+	+	–	–
Granodiorite	–	+	+	–	–
Basalt	–	+	–	–	–
Schist	+	+	+	–	+
Chert	+	+	+	–	–

[a] –, absent; +, present.

b. Weathering Patterns of Potential Sources. During mapping of the park, descriptions were compiled on oxidation and fracture patterns of potential source outcrops. These observations found that schist and chert were fractured in outcrop along foliation, bedding, or joint planes. Only cobbles of granodiorite and diorite were observed in the glacial drift and beach mud and these did not appear to be fractured. Most rocks exhibited a red to a brownish-red oxidation of exposed surfaces, fracture planes, or cortices. Schists tended to highly oxidize on all exposed surfaces and along fractures. Granodiorite cobbles exhibited a rind of oxidation. Cherts have oxidation only along fractures. Cobbles of diorite did not show any oxidation patterns.

c. Oxidation and Fracture of Midden Rock. Descriptions of fracture and oxidation patterns in the midden sample were prepared during sorting and rock identification (Table 2). Schist, diorite, and granodiorite have relatively planar fractures. Chert exhibited planar fractures parallel to bedding planes and conchoidal fractures across bedding planes. Most of the basalt had planar fractures, except very fine-grained basalt which tended to fracture conchoidally.

A red to brownish-red oxidation was observed on the surface of basalt, schist, chert, and granodiorite (Table 2). Red oxidation occurred only on the cortex basalt and granodiorite. Oxidation did occur on all exposed surfaces of the schist and of the chert. No gray or black discolorations were found on the midden angular rock.

2. Comparison of Experiment, Source, and Midden Attributes

Comparison of the midden, outcrop, and experiment patterns of oxidation and fracture indicates that fracture can occur from outcrop weathering or from a steaming environment. Oxidation appears to be the result of outcrop weathering and is unlikely to occur from using these rocks in boiling, steaming, or firepit environments. Long-term use of certain rocks in a campsite firepit was found to result in fracturing of schist cobbles, possibly from trampling of the pit. In summary, postdepositional fracturing of many of the midden rocks is difficult to distinguish from fracturing present in outcrops. The experiments indicate that oxidation occurs from outcrop weathering and not exposure to firing environments.

IV. Conclusions

Several conclusions regarding this sedimentological analysis of fire-cracked rock can be made. First, study of the depositional history of suspected fire-cracked rock at the British Camp site shows that several rock types found in the midden could be fire-cracked rock. This conclusion is based on the findings that: (1) most midden rocks are derived from sources found within the immediate vicinity of the site; (2) midden rocks appear to have been transported by humans to the site; (3) most angular midden rocks are found in deposits which resemble ethnographic firepits or debris strata that contain a variety of materials including burned bone and charcoal; and

(4) some midden rocks are susceptible to fracturing by fire-related (steaming) activities. Certain midden rocks were found not to conform to this scenario which may indicate that cultural selection of these particular materials was occurring based on attributes suitable for other activities, such as lithic tool production.

In addition to evaluating if the angular rocks were fire-cracked, traditional attributes of fire-cracked rock were indirectly tested. Direct observation of oxidation and fracture of midden and source rock found that oxidation, a traditional attribute of fire-cracked rock, occurred in outcrops at sources and not only within cultural contexts. Fractures present in the midden rock may have also occurred at the source as a result of weathering. Other traditional attributes such as the shape of the fracture were not evaluated. However, many of these attributes could be related to the inherent characteristics and weathering of the rock rather than to postdepositional alteration of the rock in a fire-cracking environment.

Finally, this sedimentological approach was found to be a valuable tool for artifact analysis. By evaluating the "life history" of this artifact, the reliability of traditional and expected attributes was tested, while other factors affecting the artifact through its depositional history were also documented and evaluated. This method allows archaeologists to assess the cultural and natural processes operating on artifacts and archaeological sites.

Acknowledgments

The research presented here results from the assistance of many people. Julie K. Stein offered guidance and critical editing of this work. George T. Jones, Charlette Beck, and all of the students of the 1985 San Juan Island Archaeological Project provided support throughout the field season. John Garver of the University of Washington Geology Department provided patient assistance with thin sections. Reid Ferring of the Institute of Applied Sciences at the University of North Texas offered valuable advice and comments during the preparation of the final manuscript. Finally, I would like to thank the staff of Woodward–Clyde Consultants who provided crucial support during the preparation of this chapter.

References

Armstrong, J. E., D. R. Crandall, D. J. Easterbrook, and J. N. Noble
1965 Late Pleistocene stratigraphy and chronology in southwestern British Columbia and northwestern Washington. *Geological Society of America Bulletin* **76**, 321–330.

Bakewell, E. F.
1990 Petrography and geochemical source-modeling of lithics from archaeological contexts: A case study from British Camp, San Juan Island, Washington. Manuscript on file in Archaeology Division, Burke Museum, University of Washington, Seattle.

Benson, C. L.
1975 Archaeological reconnaissance in the Clear Creek drainage, eastern Kitsap Peninsula. University of Washington, Office of Public Archaeology Reconnaissance Reports No. 3.

Blatt, H. G., G. Middleton, and R. Murray
1792 *Origin of sedimentary rocks*. Englewood Cliffs, New Jersey: Prentice-Hall.

Brandon, M. T., D. S. Cowan, and J. A. Vance
1988 The Late Cretaceous San Juan thrust system, San Juan Islands, Washington: A case history of terrane accretion of the Western Cordillera. Geological Survey of America, Special Paper 221.

Bowers, A. W.
1948 A history of the Mandan and Hidatsa. Ph.D. dissertation, University of Chicago.

Campbell, S. K.
1981 The Duwamish no. 1 site. University of Washington Office of Public Archaeology Research Reports No. 1.
1979 The early component of Bear Cave. *Canadian Journal of Archaeology* **3**, 177–194.

Carlson, R. L.
1960 Chronology and culture change in the San Juan Islands, Washington. *American Antiquity* **25**, 562–586.

Catlin, G.
1913 North American Indians, vol. 1. Philadelphia: Leary, Stuart.

Chatters, J. C., and G. Thompson
1979 Test excavating within a proposed right-a-way of SR-2, Forbes Hill to North Monroe Interchange, at 45SN29, 45SN48, and 45SN49, Snohomish County, Washington. University of Washington, Office of Public Archaeology, Reconnaissance Report No. 23.

Clague, J. J.
1981 Late Quaternary geology and geochronology of British Columbia, part 2: Summary and discussion of radiocarbon-dated Quaternary history. Geological Survey of Canada Paper 80-35.

Croes, D. R., and E. Blinman
1980 Hoko River: a 2500 year old fishing camp on the Northwest Coast of North America. Washington State University Laboratory of Anthropology Reports of Investigations No. 58.

Downing, G. R., and L. S. Furniss
1968 Some observations of camas digging and baking among present-day Nez Perce. *Tebiwa* **11**, 48–59.

Dunnell, R. C., and C. Beck
1979 The Caples site, 45-SA-5, Skamania County, Washington. University of Washington, Department of Anthropology Reports in Archaeology No. 6.

Dunnell, R. C., and S. K. Campbell
1977 Aboriginal occupation of Hamilton Island, Washington. University of Washington, Department of Anthropology Reports in Archaeology No. 4.

Easterbrook, D. J.
1968 Pleistocene stratigraphy of Island County. Washington Department of Water Resources, *Water Supply Bulletin* **25**, 1–34.

Folk, R. L.
1980 *Petrology of sedimentary rocks*. Austin, Texas: Hemphill.

Gaston, J., and J. V. Jermann
1975 Salvage excavations at Old Man House (45-KP-2) Kitsap County, Washington. University of Washington, Office of Public Archaeology Reconnaissance Reports No. 4.

Gilmore, M.
1925 Arikara uses of clay and other earth products. Museum of American Indian, Heye Foundation, Indian Notes 2(4). Pp.283–289.

Greengo, R. E., and R. Houston
1970 *Excavation at the Marymoor Site*. Seattle: The Magic Machine.

Haerberlin, H., and E. Gunther
1930 *The Indians of Puget Sound*. Seattle: University of Washington Press.

Hassan, F. A.
1978 Sediments in archaeology: Methods and implications for paleoenvironmental and cultural analysis. *Journal of Field Archaeology* **5**, 197–213.

Hough, W.
1926 Fire as an agent in human culture. U.S. National Museum, Bulletin 139.

House, J. H., and J. M. Smith
1975 Experiments in the replication of fire-cracked rock. In *The Cache River archaeological project*, assembled by M. B. Schiffer and J. H. House. Arkansas Archaeological Survey Research Series 8. Pp.75–80.

Jones, D. L., A. Cox, P. Coney, and M. Beck
1982 The growth of western North America. *Scientific American* **247**(5), 70–84.

Kennedy, H. K., R. S. Thomas, and J. V. Jermann

1976 A cultural resource assessment of the Chambers-Clover Creek Drainage Basin, Pierce County, Washington. University of Washington, Office of Public Archaeology Reconnaissance Reports No. 6.

King, A. R.

1950 Cattle Point: A stratified site in the southern Northwest Coast region. Society for American Archaeology, Memoir 7.

Krumbein, W. C., and L. S. Sloss

1963 *Stratigraphy and sedimentation.* San Francisco: W. H. Freeman.

Larson, L. L., and J. V. Jermann

1978 A cultural resources assessment of the Niqually National Wildlife Refuge. University of Washington, Office of Public Archaeology Reconnaissance Reports No. 21.

Latas, T. W.

1987 A depositional history of English Camp angular rock: A test of fire-cracked rock attributes. Unpublished Master's thesis, on file with the Department of Anthropology, University of Washington, Seattle.

Lewarch, D. E., and K. J. Reynolds

1975 Report of archaeological investigations on Hamilton Island, Skamania County, Washington. University of Washington, Office of Public Archaeology Reconnaissance Reports No. 5.

Lorenz, T. H., G. R. Spearman, and J. V. Jermann

1976 Archaeological testing at the Duwamish no. 1 site, King County, Washington. University of Washington, Office of Public Archaeology Reconnaissance Reports No. 8.

Lovick, S.

1983 Fire-cracked rock as tools: Wear-pattern analysis. *Plains Anthropologist* **28**, 41–52.

Reineck, H. E., and I. B. Singh

1980 *Depositional sedimentary environments.* New York: Springer-Verlag.

Schiffer, M. B.

1983 Toward the identification of formation processes. *American Antiquity* **48**, 675–706.

Smith, M. W.

1940 *The Puyallup-Nisqually.* New York: Columbia University Press.

Stein, J. K.

1985 Interpreting sediments in cultural settings. In *Archaeological sediments in context*, edited by J. K. Stein and W. R. Farrand. Center for the Study of Early Man, Institute for Quaternary Studies, University of Maine, Orono. Pp.5–19.

1987 Deposits for archaeologists. In *Advances in archaeological methods and theory.* New York: Academic Press.

Stein, J. K., and G. Rapp, Jr.

1985 Archaeological sediments: A largely untapped reservoir of information. In *Contributions to Aegean archaeology*, edited by N. C. Wilkie and W. D. E. Coulson. Center for Ancient Studies, University of Minnesota, Publications in Ancient Studies 1. Minneapolis, Minnesota. Pp.143–159.

Thoms, A. V.

1984 Project report number 2, cultural resources investigations for Libby Reservoir, Northwest Montana, volume 1, environment, archaeology, and land use patterns in the middle Kootenai River Valley. (editor) Center for Northwest Anthropology, Washington State University, Pullman.

1988 The structure of camas as a staple food resources: A perspective from the Calispell Valley, northeastern Washington. Paper presented at the 41st Annual Northwest Anthropological Conference, Tacoma, Washington.

1989 The northern roots of hunter-gatherer intensification: Camas and the Pacific Northwest. Ph.D. dissertation, Washington State University, Pullman.

Turner F. J., and J. Verhoogen

1951 *Igneous and metamorphic petrology.* New York: McGraw-Hill.

Twenhofel, W. H.

1950 *Principles of sedimentation.* New York: McGraw-Hill.

Vance, J. A.
1968 Metamorphic aragonite in the Prehnite-Pumpellyite facies, northwest Washington. *American Journal of Science* **266**, 299–315.
1975 Bedrock geology of San Juan County. In *Geology and water resources of the San Juan Islands*, edited by R. H. Russell. Washington Department of Ecology Water Supply Bulletin 46. Pp.3–19.
1977 The stratigraphy and structure of Orcas Island, San Juan Islands. In *Geology excursions in the Pacific Northwest*, edited by E. H. Brown and R. C. Ellis. Western Washington University, Bellingham.

Wedel, W. R.
1940 Cultural sequence in the Central Great Plains. *Smithsonian Institution, Miscellaneous Collections* **100**, 291–342.

Whetten, J. T., D. L. Jones, D. S. Cowan, and R. E. Zartman
1977 Ages of Mesozoic terranes in the San Juan Islands, Washington. In *Paleozoic paleogeography of the western United States*, edited by J. H. Stewart, C. H. Stevens, and A. E. Fritsche. Pacific Coast Symposium I, Society of Economic Paleontologists and Mineralogists, Pacific Section, Los Angeles. Pp.117–132.

Will, F. G., and T. C. Hecker
1944 Upper Missouri Valley aboriginal culture in North Dakota. *North Dakota Historical Quarterly* **11**(102).

Williams, H., F. J. Turner, and C. M. Gilbert
1954 *Petrography*. San Francisco: W. H. Freeman.

Wilson, G. L.
1910 Mandan-Hidatsa fieldwork, Fort Berthold Reservation, North Dakota. Unpublished Master's thesis, University of North Dakota.

11

Shell Midden Deposits and the Archaeobotanical Record: A Case from the Northwest Coast

Margaret A. Nelson

I. Introduction

The study of prehistoric subsistence adaptations has been a primary focus of archaeological investigation for the past quarter century. A vitally important, though still underinvestigated, aspect of the analysis of subsistence systems and landscape utilization is the study of the prehistoric utilization of plants.

There has been a particular lack of emphasis on the archaeobotanical record of the Northwest Coast, although a few recent studies have appeared (e.g., Bernick 1983; Ecklund-Johnson 1984; Gill 1983; Rhode 1982). The dearth of archaeobotanical analyses in the region, and in shell midden studies in general, is the result of several factors. The overwhelming volume of well-preserved bone and shellfish remains in most sites makes them the most obvious focus of analysis. Mesic climatic conditions compound the problem because they provide a poor preservational context for uncharred plant materials (Lopinot and Brussell 1982). In addition, on the Northwest Coast, regional ethnographies and the later ethnohistoric record have tended to emphasize high visibility, organizationally complex subsistence activities such as salmon fishing (e.g., Hewes 1973) and to deemphasize gathering activities, particularly the gathering of plants (Norton 1985). Consequently, with a few exceptions, archaeologists working in the area have neither expected to find, nor looked for, plant remains.

A reading of the early ethnohistoric literature suggests a more substantial reliance on plant foods than do later ethnographies. Nevertheless, even ethnographic studies list a wide range of plant resources, despite the fact that they were undertaken well after population sizes had been drastically reduced by disease (Boyd 1985), nonnative foods had been introduced (e.g., Suttles 1951a), and Euroamerican settlement had substantially limited access to prime resource areas (Norton 1979; White 1980).

Plants were utilized for food, clothing, tools, housing, and a variety of other purposes (e.g., Boas 1925; Drucker 1955; Gunther 1945; Haeberlin and Gunther 1930; Stern 1934; Suttles 1951b; Turner 1979). Recently, the contribution of plant foods to the precontact diet for the southern Northwest Coast was estimated to have been about half the total caloric intake (Keely 1980).

Archaeobotanical analyses of the Hoko River and Ozette sites on the outer coast of Washington (Ecklund-Johnson 1984; Gill 1983), of the Black River sites in Puget Sound (Rhode 1982), and of fortuitous finds at several other Northwest Coast sites (Bernick 1983; Croes 1976) confirm that plant remains are preserved in archaeological contexts on the Northwest Coast. However, these sites all exhibit unusually good preservation conditions and are not particularly representative of either the preservational or depositional context of most Northwest Coast sites, or of most shell middens. Archaeobotanical analyses of shell middens, which comprise the majority of known Northwest Coast sites, must be undertaken if we are to have a more complete understanding of prehistoric economies and patterns of land use than we currently do.

Plant remains from archaeological sites do not, of course, provide a straightforward record of plant resource use even under the most favorable conditions. Depositional and postdepositional processes, including those that affect the preservation and distribution of plant remains, have a fundamental impact on our ability to interpret and make statements about resource use (Hally 1981; Keepax 1977; Lopinot 1984; Minnis 1981; Schiffer 1987; Stein 1987; Wood and Johnson 1978).

The formation of the archaeobotanical record is still incompletely understood: distinguishing culturally from naturally deposited plant remains and delineating the transformational sequence from use to initial preservation, to deposition, to postdepositional modification, is difficult and often impossible. Nevertheless, we can obtain an understanding of the nature of the record, at least in part, by an examination of context (e.g., Dennell 1976; Pearsall 1988), by the use of information from a variety of sources (e.g., Rhode 1982; Whittaker 1985), and by eliminating suspect data, for example, noncharred plant materials from contexts other than those providing exceptional preservation conditions (Lopinot and Brussell 1982).

In addition, collection and analytical methods and techniques affect our ability to identify and interpret archaeological plant remains. For example, differential breakage among taxa (Brady 1989; Greenlee, Ch. 12, this volume; Rossen and Olson 1985) and appropriate sample size (e.g., Hastorf and Popper 1988; Pearsall 1989) must be considered. Incorrect conclusions may be drawn, particularly regarding taxonomic richness and diversity, if one does not control for sample size, or if it is inadequate to capture the variability within and between analytical units (Grayson 1984).

The British Camp site (45SJ24), San Juan Island, Washington, provides an opportunity to examine a variety of issues pertinent to the interpretation of the archaeobotanical record. The site is an extensive shell midden, occupied over approximately the past 2000 years, and has an extremely complex depositional history, typical of accretion-type middens (Waselkov 1987).

Each layer, lens, or feature in the site is considered generally equivalent to a sedimentary facies (Stein and Rapp 1985), and represents a single depositional event.

The event may have transpired over a long or short period of time, and the material may have several sources and transport agents, but as long as these depositional factors remain the same, they are considered to be a single deposit (Stein 1987). Each facies is distinguished from others in excavation by observed changes in sedimentary attributes, such as color, texture, composition, and shape of the deposit. Characteristics of the individual particles, as well as morphological attributes of the facies such as its size and physical configuration, should be indicative of depositional history. By treating each archaeological deposit as a sedimentary facies, one can study depositional and postdepositional processes as they relate to the archaeobotanical record for different kinds of deposits.

The hypothesis to be tested is that different deposits and different kinds of deposits, as lithostratigraphically defined (i.e., based on how they look and where they are located), have significantly different depositional and/or postdepositional histories. If they are the remains of particular kinds of events and are distinct, among other ways, in their archaeobotanical characteristics, then differences between classes in archaeobotanical content should be more pronounced than differences within individual deposits and within classes of deposits. If the kinds and number of taxa are highly variable between facies in the same descriptive class, no straightforward relationship between class and plant remains can be inferred.

The most obvious distinction in the deposits is between the upper and lower or "light" and "dark" layers of the site (Kenady 1972). Within each lithologically defined class (i.e., without reference to stratigraphy), facies from both layers have been examined. The light and dark strata are already known to be different from one another in terms of carbonate distributions (Nelson *et al.* 1986; Stein 1989) and lithic artifact content (Kornbacher 1989). This study will examine whether they are also different in terms of their archaeobotanical content, and if they are, whether those differences are more important between classes based strictly on lithology or between the two strata.

II. Methods

A. *Facies Classes*

Three kinds of facies that are broadly defined and commonly identified in the field have been chosen to test if observationally distinct facies are distinct in certain aspects of their cultural, depositional, and postdepositional histories. All are considered cultural deposits, as are virtually all shell midden deposits. None are, strictly speaking, features, although features do occur in the midden. Instead, the facies are representative of the kinds of deposits most common at the site: discrete or widespread lenses or layers of apparently similar content.

Plant remains from seven different facies from each of the facies classes have been examined. Additional samples from facies with more than one flotation sample have been included in order to determine the degree of variability within large facies. If there is as much variability within as between facies, then a variety of sources is indicated, possibly associated with mechanical mixing or a long depositional history.

Alternatively, the variability could be the result of postdepositional homogenization of originally distinct deposits through chemical alterations. In either case, materials initially laid down separately now could be, or appear to be, components of a single depositional event.

Class 1 facies are defined as deposits consisting of more than 50% matrix, "brown" when moist, "gray" when dry; less than 50% shell, the majority of which is fragmented; and less than 25% rock, the majority of which is angular. These kinds of facies occur in both of the major strata at the site, but are much more common in the lower dark layer. They belong to the uppermost portion of the dark layer, which comprises the lower level of many shell middens on the Northwest Coast. This layer is often referred to as black and greasy (e.g., Hester and Nelson 1978; Kenady 1972; Mitchell 1971; Stein 1985), and represents material that has been postdepositionally altered by prolonged submersion in brackish groundwater (Nelson *et al.* 1986; Stein 1989). Although most of the Class 1 facies are directly and transitionally above the deposits referred to as black and greasy, they appear to be continually or often moist, probably as the result of capillary action of the groundwater on the deposits. The moisture content of the Class 1 facies at this level is evidently somewhat more variable than those below them, because of tidally and seasonally influenced fluctuations in the groundwater level.

Field observations indicate that Class 1 facies, which are often both thick and widespread, are somewhat variable across space. If the Class 1 facies from the upper portion of the dark layer do represent the homogenization of once separate deposits, as hypothesized, then the plant remains from them should be quite variable in density or abundance and taxonomic richness, and should be relatively variable between samples from a single (large) facies as defined in the field. Class 1 facies from the portion of the site above the groundwater level may also reflect long periods of deposition, and therefore are also likely to be quite variable. However, the apparent absence of any agent of homogenization as active as groundwater suggests that the variability will be less pronounced.

Class 2 facies are defined as consisting of more than 50% matrix, "tan" in color; less than 50% shell, the majority of which is fragmented; and less than 25% rock, the majority of which is angular. They tend to occur as relatively small lenses with discrete boundaries and to look quite similar. If they represent the deposition of burned and oxidized material, as has been suggested (McCutcheon 1988), then the density and diversity of plant remains (probably as a function of sample size) should be very low, although the taxa represented could be quite variable between facies.

Class 3 facies are defined as consisting of less than 50% matrix, gray when dry, brown when moist; more than 50% shell, the majority of which is fragmented; and less than 25% rock, the majority of which is angular.

A greater degree of variability is expected from Class 3 than from Class 2 facies, both in kind and number of taxa represented, and in the density of plant remains across facies. Although these facies are initially described in the field as having the same major constituents in the same general proportions, it is obvious in excavation that there is variability across space in the shellfish taxa represented or in their relative proportions. There is reason to believe that the variability carries over to plant taxa as well. If they consist of materials from more than one source deposited over a period of time, then they could be fairly variable within single deposits.

Still, as a whole, facies of this class are not expected to have as many plant remains or to be as variable as Class 1 facies; they appear to represent generally similar kinds of activities, primarily shellfish processing. If this is so, then the density of plant remains in Class 3 facies should be relatively low. If percolation or down-washing through the porous matrix is actively occurring, then the distributions of seeds and small-sized charcoal should be indicative. This should be evident particularly in a comparison of samples from two levels of the same facies, the lower level having a greater density of remains than the upper level.

B. *Sample Collection*

The samples on which the present analysis is based were collected as part of the general procedure for the recovery of biological remains at the British Camp site. A sample is taken from each level of a facies identified in the field, and from each 2 × 2-m excavation unit in which the facies is present. This sample is usually taken as a randomly placed column and its size and location recorded. Sample size is somewhat variable, depending on the horizontal and vertical extent of the facies, but averages approximately 20 × 20 × 10 cm. In cases in which facies configuration will not allow removal of a column, an 8-liter bucket sample is removed. The samples discussed here were, with three exceptions (306/300 2E; 308/300 2B; 310/302 2H), column samples.

One-half of the sediment smaller than 1.5 mm is removed for chemical and faunal analyses. This is done so that faunal, floral, and chemical information can be compared and will consistently refer to the same excavated material. Although the one-fourth of the sediment designated for chemical analysis is lost to archaeobotanical study, the one-fourth designated for faunal analysis can be floated once the bone and shell have been removed (for a discussion of faunal sampling, see Ford, Ch. 13, this volume). Each sample is processed by water flotation, utilizing a modified SMAP-type (Shell Mound Archaeological Project) float tank (Watson 1976; Greenlee, Ch. 12, this volume).

For the present preliminary study, at least twenty pieces of charcoal were identified where sample size allowed. This is a somewhat standard sample size (Asch *et al.* 1972), although a preferable procedure is to create taxonomic redundancy curves for each sample or deposit type to assure that taxonomic richness has been adequately measured (Lyman 1982). Both counts and weights are recorded by taxon for each size fraction.

All seeds from all screens through which the samples are passed to facilitate sorting (4.0, 2.0, 1.0, and 0.5 mm) were separated and counted for the purposes of this analysis so that seed densities could be estimated (correcting for the fraction of the sediment removed). Counts, but not weights, are recorded by taxon for each sample.

III. Results

A. *Density*

Overall, Class 3 facies exhibit the highest densities of charred plant remains (Table 1, Figure 1), while Class 2 facies exhibit all the lowest values. The variability in density for charred plant remains between samples in Class 3 is fairly high: the average deviation from the mean is 24%. It is even higher in Class 1, where the average deviation is 47%, although the standard deviation itself is significantly higher for Class 3 than for Class 1.

Seed densities in Class 1 and Class 3 facies vary quite widely between samples, as do those of charred plant remains (Table 2, Figure 2). Although Class 1 values are slightly higher, there is neither a significant difference between mean seed densities, nor a consistent trend in the ranking between Classes 1 and 3. Class 2 facies are consistently lowest, however. The significant differences in seed density are between light and dark layers rather than between classes. Considering Classes 1 and 3 only, samples from the dark layer exhibit a significantly higher mean seed density than those from the light layer, both by class and when the two classes are combined and distinguished by layer.

B. *Taxonomic Richness*

Seeds, economically significant seed taxa, and charcoal (Tables 2, 3, and 4) have their highest individual and mean values for taxonomic richness in Class 1 facies, although again there is considerable variation between samples of Classes 1 and 3. And again, most of the lowest values are from Class 2 facies.

Sample size or volume can be plotted against the number of taxa to determine if differences in taxonomic richness are simply a function of sample size differences (Grayson 1984; Jones *et al.* 1983; Leonard 1985). A plot of number of seeds against number of taxa shows a generally linear relationship (Figure 3). So, although taxonomic richness values are highest for Class 1 facies, they are not significantly higher than average, given the number of specimens. The taxonomic richness of identified charcoal (Table 4) is somewhat more variable than that of seeds, although sample size is quite small in a number of cases.

C. *Taxa Represented*

Of the seed taxa represented in the samples, several are recorded in ethnobotanical studies as having been important in the historic period (Table 3) (e.g., Gill 1983; Gunther 1945; Norton 1985; Turner 1975). Most of these taxa [salal/huckleberry (*Gaultheria shallon*/*Vaccinium* spp.), Oregon grape (*Berberis* spp.), kinnikinnik or bearberry (*Arctostaphylos uva-ursi*), Nootka rose (*Rosa nutkana*), and serviceberry (*Amelanchier alnifolia*)], are inhabitants of open to dry forests or woodlands, although moist forest taxa are represented as well [thimbleberry (*Rubus parvifloris*), red elderberry (*Sambucus racemosa*)].

Another taxon is a disturbed woodland species, stinging nettle (*Urtica dioica*). Although it was utilized in the historic period, it is likely to have grown on the dis-

Table 1 Density of Charred Plant Remains

Unit	Facies/Level	Density (g/cm^3)	Rank
Class 1			
306/300	1L01	.0115	5
306/300	2E01	.0056	9.5
308/300	1E01	.0110	6
308/300	1J01	.0036	14
310/302	1N01	.0070	8
308/300	1O01	.0030	16
306/300	2F01[a]	.0140	4
306/300	2F02[a]	.0054	11
308/300	1W01[a]	.0052	12
$\bar{X}^b$ (all samples)		.0074	9.5
(light stratum)		.0077	
(dark stratum)		.0069	
s (all samples)		.0039	4.1
(light stratum)		.0034	
(dark stratum)		.0049	
Class 2			
306/300	1P01	.0012	18
306/300	1V01	.0002	21.5
308/300	1H01	.0003	20
310/302	2C01[a]	.00004	24
310/302	2I01[a]	.0001	23
310/302	2L01	.00001	25
310/302	2O01	.0004	19
310/302	2T01	.0003	21.5
$\bar{X}^b$ (all samples)		.00032	21.5
(light stratum)		.00057	
(dark stratum)		.00017	
s^c (all samples)		.00038	2.4
(light stratum)		.00055	
(dark stratum)		.00017	
Class 3			
310/302	1H01	.0077	7
310/302	1L01	.0056	13
310/302	1O01[a]	.0042	15
310/302	1O03[a]	.0151	3
308/300	2B01	.0022	17
308/300	2T01	.0161	2
310/304	2K01	.0056	9.5
310/304	2M01	.0248	1
$\bar{X}^b$ (all samples)		.0102	8.4
(light stratum)		.0082	
(dark stratum)		.0122	
s^c (all samples)		.0078	6.2
(light stratum)		.0049	
(dark stratum)		.0103	

[a]Within each class, denotes samples from the same facies.
[b]$\bar{X}$ = mean.
[c]s = standard deviation.

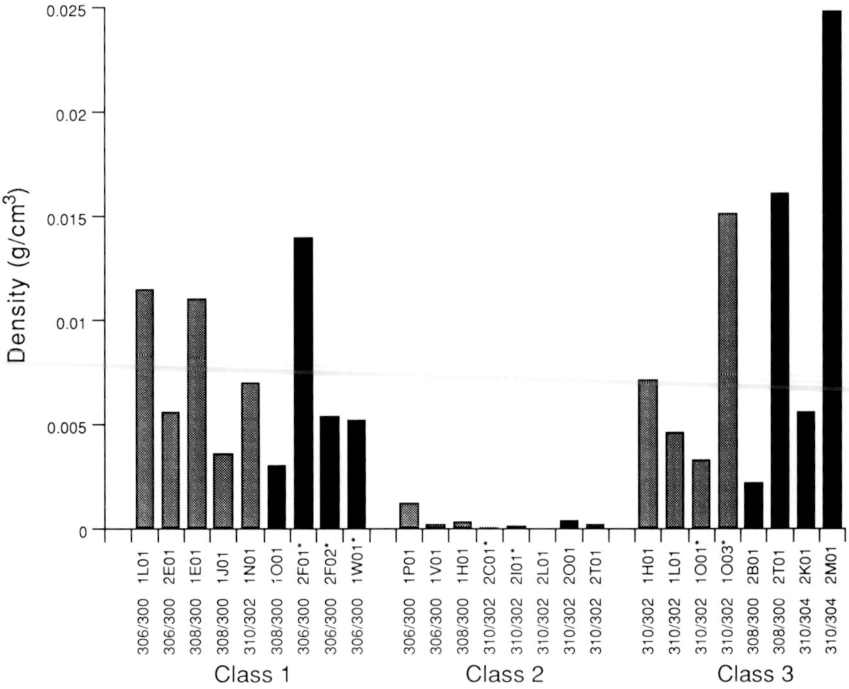

Figure 1 Density of charred plant remains. Density is calculated as g/cm^3 of flotation sample volume. Values are grouped by class, with unshaded bars representing samples from the "light" upper layer of the site, and shaded bars representing samples from the "dark" lower layer. Within each class, asterisks denote samples from the same facies.

turbed surface of the site itself or in the immediate environs, as its presence in all three kinds of facies suggests. Many of the other taxa, including various sedges, rushes, and seablite, are marsh inhabitants, some species of which were utilized historically (Gunther 1945; Turner 1979). These taxa also are found in varying amounts in all three classes of facies, with seablite (*Suaeda*) being particularly common. There was formerly a lake and marsh at the head of Garrison Bay, about half a mile from the site, first recorded in 1858 (Agee 1987). There was also, apparently, a marshy area directly adjacent to the site, from which the seeds could have come, either brought to the site for use, or blown there as seed rain.

The charcoal taxa represented (Table 4) come from several island habitats, principally moist forests dominated by western hemlock (*Tsuga heterophylla*), and including western red cedar (*Thuja plicata*) and red alder (*Alnus rubra*); dry to open forests/woodlands dominated by grand fir (*Abies grandis*), and including Douglas fir (*Pseudotsuga menziesii*) and Pacific madrona (*Arbutus menziesii*); and a savanna community, which includes Douglas fir, madrona, and Garry oak (*Quercus garryana*) (Agee 1987; Atkinson and Sharpe 1985; Fonda and Bernardi 1976). Sitka spruce (*Picea sitchensis*), also represented in several samples, is most commonly found in the Northwest in a narrow belt along the outer coast; however, it also grows in the rocky soils of the windswept southern end of San Juan Island.

Table 2 Seeds: Density and Taxonomic Richness

Unit	Facies/Level	Density (no./cm^3)	Rank	No. Seeds	No. Taxa
Class 1					
306/300	1L01	.0128	12	36	13
306/300	2E01	.0060	17	27	8
308/300	1E01	.0110	8	33	11
308/300	1J01	.0342	5	20	6
310/302	1N01	.0068	14	25	15
308/300	1O01	.0489	4	156	30
306/300	2F01[a]	.0700	1	110	15
306/300	2F02[a]	.0140	10	17	7
308/300	1W01[a]	.0134	11	62	11
$\overline{X}$ (all samples)		.0306	9.1	74	15.6
(light stratum)		.0142			
(dark stratum)		.0366			
s (all samples)		.0275	5.11	58.77	8.7
(light stratum)		.0116			
(dark stratum)		.0278			
Class 2					
306/300	1P01	.0026	6	7	4
306/300	1V01	0	24.5	0	0
308/300	1H01	.0017	23	2	1
310/302	2C01[a]	0	24.5	0	0
310/302	2I01[a]	.0019	21.5	11	4
310/302	2L01	.0019	21.5	5	4
310/302	2O01	.0065	16	17	7
310/302	2T01	.0021	20	8	6
$\overline{X}$[b] all samples)		.0025	19.6	8.2	4.2
(light stratum)		.0014			
(dark stratum)		.0025			
s[c] (all samples)		.0024	6.15	6.38	2.68
(light stratum)		.0013			
(dark stratum)		.0024			
Class 3					
310/302	1H01	.0097	13	18	8
310/302	1L01	.0055	18	7	4
310/302	1O01[a]	.0067	15	22	9
310/302	1O03[a]	.0259	7	56	14
308/300	2B01	.0035	19	22	6
308/300	2T01	.0663	2	36	14
310/302	2K01	.0184	9	83	20
310/304	2M01	.0649	3	43	13
$\overline{X}$[b] (all samples)		.0251	10.8	35.9	11.0
(light stratum)		.0120			
(dark stratum)		.0383			
s[c] (all samples)		.0261	6.52	24.54	5.21
(light stratum)		.0095			
(dark stratum)		.0321			

[a]Within each class, denotes samples from the same facies.
[b]$\overline{X}$ = mean.
[c]s = standard deviation.

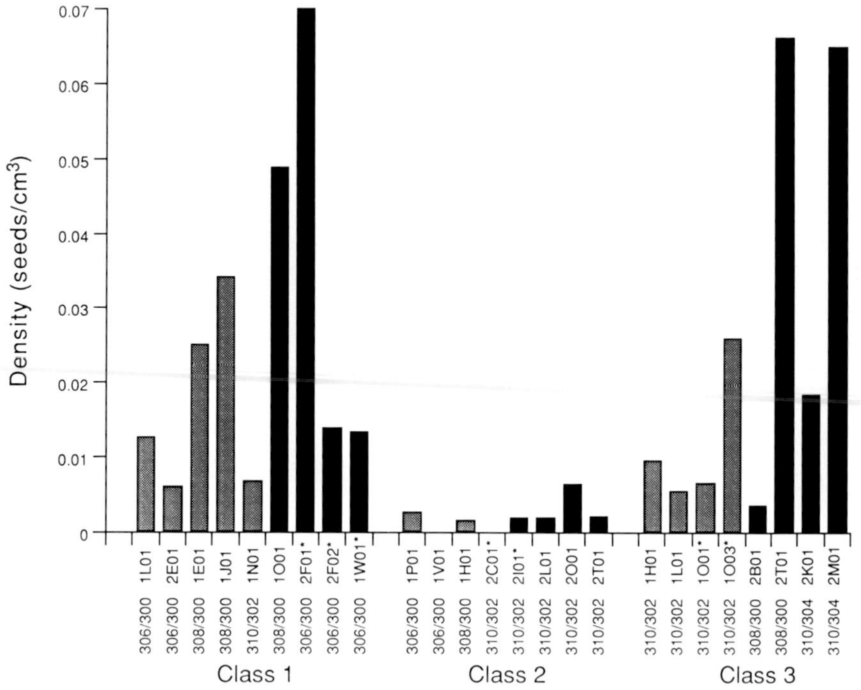

Figure 2 Seed density. Density is calculated as the number of seeds per cm^3 of flotation sample volume. Unshaded bars represent samples from the light layer of the site, and shaded bars represent samples from the dark layer.

Of the taxa identified, Douglas fir is most commonly represented, occurring in almost all samples from which charcoal was identifiable, and in all three kinds of facies. Hemlock and/or true fir (*Abies* spp.) are also represented in virtually all samples. Cedar is present in several samples, in both Class 1 and Class 3 facies, as is spruce. Western yew (*Taxus brevifolia*) is found in two to three of the Class 3 samples, in low quantities. Cedar, spruce, and yew were all economically significant taxa historically (Gunther 1945; Stewart 1984; Turner 1979), and are thought to have been important prehistorically as well (Gill 1983; Hebda and Mathewes 1984). However, all three taxa are minor components of woodlands and moist forests of the San Juan Islands. Hardwoods represented include alder, maple, madrona, salmonberry, and in one sample, Garry oak. The majority of the hardwood specimens—33 of the 61 pieces—come from Class 3 facies; 17 of the other 28 are from a single Class 1 sample.

IV. Discussion

A. *Class 1*

As noted in the results, Class 1 facies exhibit somewhat higher seed densities (Table 2, Figure 2), while density values for charred plant remains are slightly lower for

Class 1 than for Class 3 (Table 1, Figure 1). Density values can differ for a number of reasons, of course. They may reflect differences in the period of deposition, or preservational variability in different kinds of deposits or by taxon. Repeated wetting and drying of samples after their removal from the ground increases the probability of fragmentation (Jarman *et al.* 1972; Wagner 1988). Brady (1989) and Greenlee (Ch. 12, this volume) showed experimentally that the survival of charcoal processed by water flotation is, in part, a function of specific gravity. These studies and others (e.g., Rossen and Olson 1985) note, however, that specific gravity or bulk density is only in part taxon specific, and varies rather widely within an individual as well as within a particular taxon. Greenlee's measurement of specific gravity for these samples shows that there is no significant difference between facies from the light and dark layers (see Greenlee, Ch. 12, this volume). Variability in the density of archaeobotanical remains can serve as a means of judging the comparability of depositional conditions within and between facies having the same general characteristics as well as those that are observationally distinct (Miller 1988).

An examination of charred plant remains densities between Classes 1 and 3 shows that, as a whole, they do not differ significantly. Variability between samples is higher in Class 3, but is notable in both classes, so no clear distinction between them can be made on the basis of this evidence. When the two classes are subdivided into light and dark strata, average density values are higher for samples from the light stratum of Class 1, and from the dark stratum of Class 3, but again, the differences are not statistically significant.

Table 3 Economic Seed Taxa

Taxon	Class	Facies	Level	Stratum[a]
Amelanchier	1	2F	01	Dk
(serviceberry)	1	1W	01	Dk
	2	2T	01	Dk
	3	1H	01	Lt
Artostaphylos (kinnikinnick)	1	1O	01	Dk
Berberis	1	1N	01	Lt
Oregon grape	1	1O	01	Dk
	2	2I	01	Dk
Gaultheria/Vaccinium	1	2F	01	Dk
(Salal/huckleberry)	1	2F	02	Dk
	3	2T	01	Dk
Rosa (rose)	3	2K	01	Dk
Rubus	1	1N	01	Lt
(thimbleberry)	1	2F	01	Dk
Sambucus	1	1J	01	Lt
(elderberry)	2	2O	01	Dk
	3	2B	01	Dk

[a]Dk, Dark; Lt, light.

Table 4 Charcoal: Taxa Represented

	Class 1									Class 2								Class 3							
Facies	1L	2E	1E	1J	1N	1O	2F	2F	1W	1P	1V	1H	2C	2I	2L	2O	2T	1H	1L	1O	1O	2B	2T	2K	2M
Level	01	01	01	01	01	01	01	02	01	01	01	01	01	01	01	01	01	01	01	01	03	01	01	01	01
Stratum[a]	Lt	Lt	Lt	Lt	Lt	Dk	Dk	Dk	Dk	Lt	Lt	Lt	Dk	Dk	Dk	Dk	Dk	Lt	Lt	Lt	Lt	Dk	Dk	Dk	Dk
Taxon																									
Abies/Tsuga (fir/hemlock)	X[b]	X	X		X	X	X	X	X	X					X		X	X	X	X	X	X			X
Acer (maple)		X	X														X	X	X						
Alnus (alder)						X			X								X		X	X	X			X	X
Arbutus (madrona)							X	X																X	
Picea (spruce)			X		X	X		X										X			X			X	
Pseudotsuga (Douglas fir)					X	X	X	X	X						X		X	X	X	X	X	X			X
Quercus (Garry oak)																				X					
Rubus (salmonberry)					X															X	X				
Sambucus (elderberry)																									X
Taxus (yew)																		X	X	X					
Thuja (red cedar)	X		X							X									X	X	X				X
Unidentified conifer	X				X																	X			X
Unidentified hardwood									X											X	X	X			
No. taxa identified	1	2	4	0	6	5	3	5	5	2	0	0	0	0	2	0	3	4	6	9	8	3	4	4	5
No. specimens identified	20	2	8	0	20	15	13	11	19	2	0	0	0	0	2	0	9	9	14	20	21	8	20	20	12

[a]Lt, Light; Dk, dark.

[b]X, Present in sample.

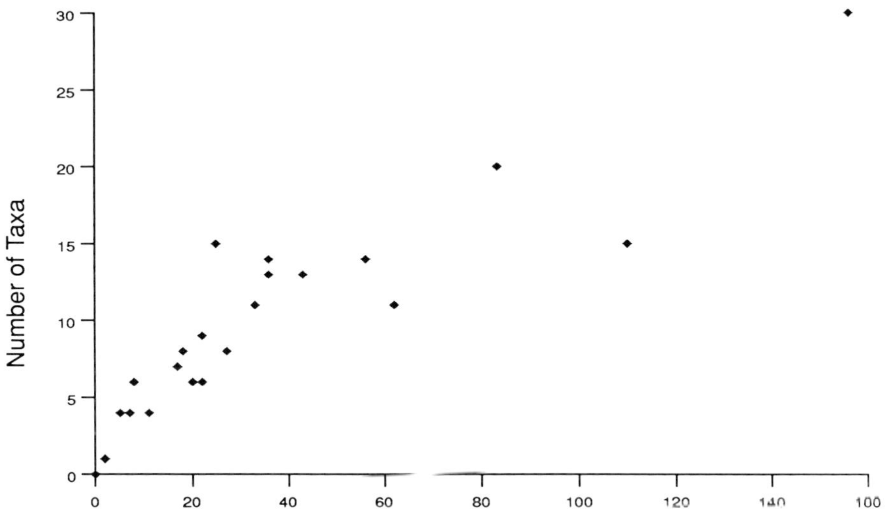

Figure 3 Plot of the number of seed taxa compared to sample size (number of seeds) as a measure of variability in taxonomic richness.

A comparison of seed densities between samples within the same facies indicates that, on the basis of this sample, large facies of Class 1 tend to be quite heterogeneous across space in the quantities of plant remains they contain. A comparison of levels from the same facies provides no evidence of a distribution such as might be expected if downward movement of particles was significant (Table 2, Figure 2). Dividing all samples by stratum, however, shows a significantly higher mean density and standard deviation for those in the dark stratum.

The variability between samples is consistent with the hypothesis that, in general, large facies were deposited accretionally (i.e., they have a long depositional history). The density values alone point to a variety of cultural/depositional events. The difference in standard deviation between light and dark strata suggests that post-depositional homogenization by chemical processes in facies in the dark stratum has masked variability in separate and possibly quite disparate depositional events.

The kinds and numbers of both seed and charcoal taxa should, in part, reflect the economic resource base and exploitation of particular habitat types. Determining which taxa were originally utilized is not possible, of course; only those that survived can be collected and identified. Nevertheless, in cases where preservation conditions appear to be generally equivalent (e.g., within a single facies or between facies of a single class), differences in taxonomic richness may be significant. Consistent differences between classes may reflect either preservational or depositional differences, and should be further examined.

Class 1 facies contain more seed taxa, representing a wider variety of habitats, than do Class 2 or Class 3 facies. Additionally, they include substantially more taxa deemed economically important than do facies from the other classes (Table 3). A plot of sample size (number of seeds) against number of seed taxa (Figure 3) shows that taxonomic richness is not higher than average given the number of seeds. When compared to sample size by volume collected, however, the numbers of both seeds and taxa are higher for Class 1 than for other classes. Hence, whether taxonomic richness is perceived to differ depends on how sample size is measured.

Some of the seed taxa, especially those represented in the smaller size classes in virtually all samples from all classes, are likely to be the result of noncultural factors. While there is no evidence for widespread burning in the part of the site from which the samples were taken, some taxa with no known economic value are ubiquitous and were probably blown into the site and charred incidentally.

The economic seed taxa identified generally are inhabitants of dry forests/woodlands, with some members of moist forest habitats represented. Most are berries, which in the ethnographic period were sometimes processed by drying over a fire (Turner 1975). A number of the Class 1 facies include several of these taxa (Table 3), in contrast to facies from Classes 2 and 3. The majority of the taxa are found in dark rather than light stratum facies; only one taxon (elderberry) is present exclusively in light stratum facies.

Class 1 charcoal includes representatives of moist forest and either dry woodland or savanna communities, or both. All of the taxa are found in the British Camp environs today, but the communities occur in somewhat distinctive settings (Agee 1987).

The ubiquity of Douglas fir charcoal in Class 1, and indeed in all classes is not surprising; it is a member of two of the three major vegetation communities represented, dry to open woodlands and savannas, and is more abundant in the San Juan Islands than in the rest of the Puget Trough (Franklin and Dyrness 1973). In addition, pollen cores from the San Juan Islands show that it has been extremely dominant over the past several thousand years (Hansen 1943; Nickmann and Leopold 1985).

Hemlock and/or true fir is also ubiquitous in occurrence in Class 1 as in the other classes from which charcoal was identified. Both are found on the island today, although hemlock is somewhat less common in the relatively dry San Juan Islands than in the rest of the Puget Trough, while grand fir is somewhat more common.

B. *Class 2*

Whereas Classes 1 and 3 cannot always be distinguished on the basis of their archaeobotanical content, Class 2 facies are consistently and significantly different, exhibiting all the lowest values for density and taxonomic richness. Relative size distributions of charred plant remains are also somewhat different from those of the other classes; several samples, from both light and dark layers, have no 4-mm fraction (Figure 4). However, the differences may be exclusively an artifact of preservation rather than a reflection of what was originally deposited. Plant parts are preserved by charring only in a reducing atmosphere; shell and bone from facies of this class have been heated to high temperatures and their outer surfaces oxidized

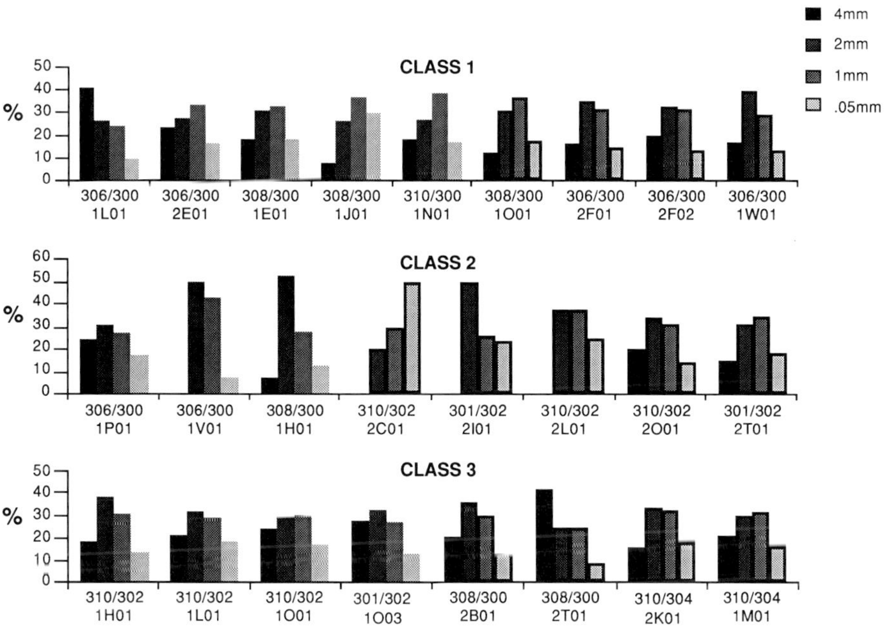

Figure 4 Size fraction distributions of charred plant remains. Distributions by size are measured by percent within each sample. Samples from the dark layer of the site are represented by bars outlined in black; samples from the light layer are not outlined.

(see McCutcheon, Ch. 15, this volume). In addition, organic matter values are significantly lower than for other classes (Stein 1989).

Whether or not they necessarily represent burned and oxidized materials, on the basis of their archaeobotanical content, they represent deposits that are distinctively different from those of other classes. While facies from Classes 1 and 3 appear to encompass a range of sources and in some cases to have had a rather long history of deposition, the Class 2 facies do not; they are much less variable across facies or between different levels of a single facies. They do not appear to have been significantly altered by groundwater. Facies from both sides of the light–dark stratigraphic boundary are generally similar in their archaeobotanical characteristics, although most of the identifiable charcoal and economic seed taxa recorded are from dark stratum facies.

C. *Class 3*

Class 3 facies, like those from Class 1, contain substantial quantities of charcoal and seeds. And like the Class 1 facies, those quantities vary quite widely across samples both within and between facies. In general, but particularly with reference to charred plant remains densities, they are more variable between samples than are Class 1 facies. Increased density of seeds and charcoal with depth is evident in a comparison of samples both from two levels of one facies and from light and dark stratum facies. Class 3 is the only class for which values for charred plant remains are higher in the dark stratum.

Mechanical processes commonly account for displacement of plant (and other) materials through midden matrix. Movement is dependent on both matrix pore size and on the size of the plant remains. Smaller particles, such as seeds and small fragments of charcoal, are most susceptible to downward movement by, among other agents, trampling, and percolation through porous matrix and cavities created by roots, earthworms, and burrowing animals (Keepax 1977; Spector 1970; Stein 1983).

There is no real difference between the two levels of 310/302 1O in relative charcoal size fraction distribution, however (Figure 4). In terms of both the density of charred plant remains and their size distributions, Classes 3 and 1 do not appear to represent consistently different kinds of cultural deposits. And, although charred plant remains densities are somewhat higher in the dark than in the light stratum facies of Class 3, the stratigraphic distinction does not extend to a consistent, significant difference in density or size distribution between facies in the two strata.

The average density of seeds is significantly higher in Class 3 facies from the dark than from the light stratum, however, as it is in Class 1. While an obvious explanation is the apparently lower pore space by volume in facies affected by groundwater, restricting movement by percolation, grain-size analysis suggests that is not the case. Although grain-size analyses have not yet been completed for many of the samples in this study, another series of samples from the same area show a virtually identical mean grain size above and below the boundary. These samples were not identified by facies class, but those in the upper stratum include a high concentration of shelly facies equivalent to those designated as Class 3. A grain-size distribution has been calculated for two levels of 310/302 1O (levels 1 and 2), and shows similar values for all grain sizes.

Class 3 facies both above and below the stratigraphic boundary also contain significantly fewer economically important seed taxa than do Class 1 facies (Table 3). The relative paucity of economic seed taxa may reflect, as has been assumed, that they tend to represent fish and shellfish processing activities and that consumption and processing of plants was not a significant activity.

Charcoal distributions by taxon for Class 3 facies also exhibit some differences from the Class 1 samples. Several taxa (alder, oak, and yew) are entirely or most consistently represented in Class 3. Overall, charcoal samples from Class 3 facies are slightly higher in taxonomic richness.

As is the case for Class 1, some taxa are present only in dark stratum facies, others only in light stratum facies. Considering both Classes 1 and 3, while there are no abrupt differences between strata in the character of the taxa represented, some changes can be seen. Taxa found only in light stratum facies (maple, salmonberry, yew) generally are associated with moist, cool conditions, while those found only in dark stratum facies (mainly madrona) are associated with drier conditions. This could reflect a shift in species availability resulting from the onset of slightly cooler weather, possibly the "Little Ice Age."

V. Summary and Conclusions

In summary, differences between Class 1 and Class 3 facies in the densities and size distributions of charred plant materials are not significant, although both have significantly higher mean densities than do Class 2 facies. Seed densities are significantly lower for Class 2 facies, while differences between Classes 1 and 3 are minor. However, there is a significant difference between light and dark strata samples in all three classes, the lower dark stratum having uniformly higher seed densities than the upper light stratum. Since the density of charred plant remains is higher in the dark layer only for Class 3 facies (and in no case is the difference statistically significant), and since grain-size distributions between strata do not differ, there appears to be a real difference in seed distribution.

Large facies of both Classes 1 and 3 are, in general, heterogeneous across space in terms of their archaeobotanical content, and may represent repeated dumpings of somewhat different kinds of cultural materials from somewhat different kinds of activities. These results are consistent with the conclusions reached by Campbell (1981) in an analysis of the taxonomic distributions of shellfish in a series of deposits from the Duwamish site (45KI01) in Seattle. Variability in seed density is significantly greater in facies from the dark stratum than from the light stratum of Classes 1 and 3, suggesting the postdepositional homogenization of originally separate deposits as a result of groundwater activity.

There are no strong temporal trends in the kinds of taxa represented, although there appears to be a shift toward greater use of wood species favoring cool, moist conditions above the stratigraphic boundary. Overall, considering both seed and charcoal data from all facies, I suggest that a variety of habitats was exploited: marsh, moist forest, dry woodland, and savanna. This is most evident in Class 1 samples, though Class 3 samples include a wider variety of wood taxa.

The forest and woodland communities are found directly adjacent to the site today, as is the site of a former marsh, but the savanna community is at some distance. Spruce is presently found in quantity only at a distance of approximately ten miles, in the vicinity of another large site, the Cattle Point site (45SJ01). A wide variety of plant-collecting activities was apparently pursued, which is somewhat at odds with the usual interpretation of limited and seasonal use of shell midden sites.

On the Northwest Coast, midden sites often are assumed to have been used mainly for purposes of shellfish collection and salmon processing, especially during the San Juan/Gulf of Georgia phase of the past 1500 years (e.g., Thompson 1978; Whitlam 1981). Facies located at or near the top of the dark layer do contain more seeds by volume and more economically significant taxa than those deposited more recently, but they have been dated at the British Camp site to about 1000 years ago, within what is considered to be the San Juan Phase. In any case, there is some evidence for plant utilization (aside from fuel collection) even in facies from the light upper layer.

Based on this preliminary analysis of a small number of samples, it is evident that plant remains occur in significant quantities in a variety of nonfeature shell midden deposits. In these cases, facies appearing to be shell rich contain fewer economically significant plants than those appearing to have smaller amounts of shell. However, plant use is reflected in deposits of all types, and samples should be collected from all depositional contexts and from several loci within each deposit when they are large. As the analyses described here have demonstrated, the archaeobotanical record is an important source of depositional and economic information for a variety of different kinds of midden deposits.

Acknowledgments

I would like to acknowledge the assistance of S. K. Duke, D. M. Greenlee, K. D. Kornbacher, J. Lemire, C. R. Lynn, P. J. Watson, and, in particular, J. K. Stein in the preparation of this manuscript.

References

Agee, J. K.
1987 The forests of San Juan Island National Historical Park. National Park Service Cooperative Park Studies Unit, Report CPSU/UW 88-1.

Asch, N. B., R. I. Ford, and D. L. Asch
1972 Paleoethnobotany of the Koster Site: The archaic horizons. Illinois State Museum Reports and Investigations No. 24.

Atkinson, S., and F. Sharpe
1985 *Wild plants of the San Juan Islands*. Seattle: The Mountaineers.

Bernick, K.
1983 Site catchment analysis of the Little Qualicum River site, DiSe1: A wet site on the east coast of Vancouver Island, British Columbia. National Museum of Man, Mercury Series, Archaeological Survey of Canada Paper No. 118, Ottawa.

Boas, F.
1925 Contributions to the ethnography of the Kwakiutl. Columbia University Contributions to Anthropology 3, edited by F. Boas. Columbia University Press, New York.

Boyd, R. T.
1985 The introduction of infectious diseases among the Indians of the Pacific Northwest 1774–1874. Ph.D. dissertation, Department of Anthropology, University of Washington, Seattle.

Brady, T. J.
1989 The influence of flotation on the rate of recovery of wood charcoal from archaeological sites. *Journal of Ethnobiology* **9**, 207–227.

Campbell, S. K.
1981 Horizontal variability in shell midden composition: Implications for the interpretation of stratigraphic changes. Paper presented at the annual meeting of the Canadian Archaeological Association.

Croes, D. R.
1976 The excavation of water-saturated sites (wet sites) on the Northwest Coast of North America (editor). National Museum of Man, Mercury Series, Archaeological Survey of Canada Paper No. 50, Ottawa.

Dennell, R. W.
1976 The economic importance of plant resources represented on archaeological sites. *Journal of Archaeological Science* **3**, 229–247.

Drucker, P.
1955 *Indians of the Northwest Coast*. New York: Natural History Press.

Ecklund-Johnson, D. J.
1984 Analysis of macroflora from the Hoko River rockshelter, Olympic Peninsula, Washington. Master's thesis, Department of Anthropology, Washington State University, Pullman.

Fonda, R. W., and J. A. Bernardi
1976 Vegetation of Sucia Island in Puget Sound, Washington. *Bulletin of the Torrey Botanical Club* **103**, 99–109.
Franklin, J. F., and C. T. Dyrness
1973 Natural vegetation of Oregon and Washington. U.S.D.A. Forest Service General Technical Report PNW-8.
Gill, S.
1983 *Ethnobotany of the Makah and Ozette People, Olympic Peninsula, Washington*. Ph.D. dissertation, Department of Botany, Washington State University, Pullman.
Grayson, D. K.
1984 *Quantitative zooarchaeology*. New York: Academic Press.
Gunther, E.
1945 Ethnobotany of Western Washington. University of Washington Publications in Anthropology, Vol. 10(1).
Haeberlin, H. E., and E. Gunther
1930 The Indians of Puget Sound. University of Washington Publications in Anthropology, Vol. 4(1).
Hally, D. J.
1981 Plant preservation and the content of paleobotanical samples: A case study. *American Antiquity* **46**, 723–742.
Hansen, H. P.
1943 A pollen study of two bogs on Orcas Island, of the San Juan Islands, Washington. *Bulletin of the Torrey Botanical Club* **70**, 236–243.
Hastorf, C. A., and V. S. Popper (editors)
1988 *Current paleoethnobotany*. Chicago: University of Chicago Press.
Hebda, R., and R. Mathewes
1984 Holocene history of cedar and native Indian cultures of the North American Pacific Coast. *Science* **225**, 711–714.
Hester, J. J., and S. M. Nelson
1978 Studies in Bella Bella prehistory. Department of Archaeology, Simon Fraser University Publication No. 5.
Hewes, G. W.
1973 Indian fisheries productivity in pre-contact times in Pacific Salmon area. *Northwest Anthropological Research Notes* **7**, 133–155.
Jarman, H. N., A. J. Legge, and J. A. Charles
1972 Retrieval of plant remains from archaeological sites by froth flotation. In *Papers in economic prehistory*, edited by E. S. Higgs. Cambridge: University of Cambridge Press. Pp.39–48.
Jones, G. T., D. K. Grayson, and C. Beck
1983 Artifact class richness and sample size in archaeological surface assemblages. In *Lulu linear punctated: Essays in honor of George Irving Quimby*, edited by R. C. Dunnell and D. K. Grayson. Museum of Anthropology, University of Michigan, Anthropological Papers Vol. 72. Pp.55–73.
Keely, P.
1980 Nutrient composition of selected important plant foods of the pre-contact diet of the Northwest Native American peoples. Master's thesis, Department of Nutritional Sciences, University of Washington, Seattle.
Kenady, S. M.
1972 Research design for the San Juan Islands, summer 1972. Manuscript on file, Department of Anthropology, University of Washington, Seattle.
Keepax, C.
1977 Contamination of archaeological deposits by seeds of modern origin with particular reference to the use of flotation machines. *Journal of Archaeological Science* **4**, 221–229.
Kornbacher, K. D.
1989 *Shell midden lithic technology: An investigation of change at British Camp (45SJ24), San Juan Island*. Master's thesis, Department of Anthropology and Sociology, University of British Columbia, Vancouver.

Leonard, R. D.
1985 Late Puebloan subsistence diversification: A product of sample size effects. Paper presented at the annual meeting of the Society for American Archaeology, Denver, Colorado.
Lopinot, N. H.
1984 Archaeobotanical formation processes and Late Middle archaic human-plant interrelationships in the mid-continental U.S.A. Ph.D. dissertation, Department of Anthropology, Southern Illinois University, Carbondale.
Lopinot, N. H., and D. E. Brussell
1982 Assessing uncarbonized seeds from open-air sites in mesic environments: An example from southern Illinois. *Journal of Archaeological Science* **9**, 95–108.
Lyman, R. L.
1982 Archaeofaunas and subsistence studies. In *Advances in archaeological method and theory, vol. 5*, edited by M. B. Schiffer. New York: Academic Press. Pp.331–393.
McCutcheon, P. T.
1988 A procedure for inferring the temperature at which archaeological bone has been burned. Paper presented at the annual meeting of the Society for American Archaeology, Phoenix.
Miller, N. F.
1988 Ratios in paleoethnobotanical analysis. In *Current paleoethnobotany*, edited by C. A. Hastorf and V. S. Popper. Chicago: University of Chicago Press. Pp.72–85.
Minnis, P. E.
1981 Seeds in archaeological sites: Sources and some interpretive problems. *American Antiquity* **46**, 143–152.
Mitchell, D. H.
1971 Archaeology of the Gulf of Georgia Area, a natural region and its culture types. *Syesis* **4**, suppl. 1.
Nelson, M. A, P. J. Ford, and J. K. Stein
1986 Turning a midden into mush: Evidence of acidic conditions in a shell midden. Paper presented at the annual meeting of the Society for American Archaeology, New Orleans.
Nickmann, R., and E. B. Leopold
1985 A pollen record for the San Juan Islands, Puget Sound, Washington. Manuscript on file, Department of Botany, University of Washington, Seattle.
Norton, H. H.
1979 The association between anthropogenic prairies and important food plants in western Washington. *Northwest Anthropological Research Notes* **13**, 434–449.
1985 Women and resources of the Northwest Coast: Documentation from the 18th and early 19th centuries. Ph.D. dissertation, Department of Anthropology, University of Washington, Seattle.
Pearsall, D. M.
1988 Interpreting the meaning of macroremain abundance: The impact of source and context. In *Current paleoethnbotany*, edited by C. A. Hastorf and V. S. Popper. Chicago: University of Chicago Press. Pp.97–118.
1989 *Paleoethnobotany: A handbook of procedures*. New York: Academic Press.
Rhode, D.
1982 An analysis of plant remains from three sites along the Black River, King County, Washington. Manuscript on file, Department of Anthropology, University of Washington, Seattle.
Rossen, T., and J. Olson
1985 The controlled carbonization and archaeological analysis of southeastern U.S. wood charcoal. *Journal of Field Archaeology* **12**, 445–452.
Schiffer, M. B.
1987 Formation processes of the archaeological record. Albuquerque: University of New Mexico Press.
Spector, J.
1970 Seed analysis in archaeology. *The Wisconsin Archeologist* **51**, 163–190.
Stein, J. K.
1983 Earthworm activity: A source of potential disturbance of archaeological sediments. *American Antiquity* **48**, 277–289.

1985 Facies in shell middens. Paper presented at the annual meeting of the Society for American Archaeology, Denver.

1987 Deposits for archaeologists. In *Advances in archaeological method and theory, vol. 11*, edited by M. B. Schiffer. New York: Academic Press. Pp.337–395.

1989 Formation processes of coastal sites: View from a Northwest Coast shell midden. Paper presented at the Circum-Pacific Prehistory Conference, Seattle.

Stein, J. K., and G. Rapp, Jr.

1985 Archaeological sediments: A largely untapped reservoir of information. In *Contributions to Aegean archaeology, publications in ancient studies, vol. 1*, edited by N. C. Wilkie and W. D. E. Coulson. Center for Ancient Studies, University of Minnesota, Minneapolis. Pp.143–159.

Stern, B. J.

1934 *The Lummi Indians of Northwest Washington*. New York: Columbia University Press.

Stewart, H.

1984 *Cedar*. Vancouver: Douglas and McIntyre.

Suttles, W. P.

1951a The early diffusion of the potato among the Coast Salish. *Southwestern Journal of Anthropology* **7**, 272–288.

1951b The economic life of the Coast Salish of Haro and Rosario Straits. Ph.D. dissertation, Department of Anthropology, University of Washington, Seattle.

Thompson, G.

1978 Prehistoric settlement changes in the southern Northwest Coast: A functional approach. University of Washington, Department of Anthropology, Reports in Archaeology No. 5.

Turner, N. J.

1975 Food plants of British Columbia Indians, part one: Coastal peoples. British Columbia Provincial Museum Handbook 34, British Columbia Provincial Museum, Victoria.

Turner, N. J.

1979 Plants in British Columbia Indian technology. British Columbia Provincial Museum Handbook 38. British Columbia Provincial Museum, Victoria.

Wagner, G. E.

1988 Comparability among recovery techniques. In *Current paleoethnobotany*, edited by C. A. Hastorf and V. S. Popper. Chicago: University of Chicago Press. Pp.17–35.

Waselkov, G. A.

1987 Shellfish gathering and shell midden archaeology. In *Advances in archaeological method and theory, vol. 10*, edited by M. B. Schiffer. New York: Academic Press. Pp.93–210.

Watson, P. J.

1976 In pursuit of prehistoric subsistence: A comparative account of some contemporary flotation techniques. *Mid-Continental Journal of Archaeology* **1**, 77–100.

White, R.

1980 *Land use, environment, and social change: The shaping of Island County, Washington*. Seattle: University of Washington Press.

Whitlam, R. G.

1981 Settlement-subsistence type, occurrence and change in coastal environments: A global perspective. Ph.D. dissertation, Department of Anthropology, University of Washington, Seattle.

Whittaker, F. H.

1985 Archaeobotany from an evolutionary perspective. Paper presented at the annual meeting of the Society for American Archaeology, Denver, Colorado.

Wood, W. R., and D. L. Johnson

1978 A survey of disturbance processes in archaeological site formation. In *Advances in archaeological method and theory, vol. 1*, edited by M. B. Schiffer. New York: Academic Press. Pp.315–381.

12

Effects of Recovery Techniques and Postdepositional Environment on Archaeological Wood Charcoal Assemblages

Diana M. Greenlee

I. Introduction

Most archaeobotanists recognize that some plant material is lost during the recovery process (e.g., Bohrer 1970; Struever 1968; Wagner 1982; Yarnell 1963, 1982). However, interpretations based on quantitative descriptions of plant remains (e.g., Asch and Asch 1975; Johannessen 1988; Miller 1985, 1988; Minnis 1978; Pearsall 1983) require that the botanical assemblage recovered be representative of the archaeological deposit. While processing techniques are suggested to influence the recovery of remains from sediment samples (e.g., Pearsall 1989; Wagner 1988), little attention has been focused toward directly assessing the kinds and amounts of recovery bias involved (cf Brady 1989). Using sediment samples from the shell midden at British Camp, San Juan Island, Washington, this paper assesses the significance of archaeological wood charcoal loss during water flotation by quantifying the charcoal content both before and after flotation. A statistically significant loss of charcoal in the recovery process should serve as a caution to researchers who quantify charcoal in archaeological studies.

The light and dark strata of the shell midden at British Camp are suggested by Stein (1984, Ch. 7, this volume; Nelson *et al.* 1986) to have been created by different moisture regimes. The appearance of the two strata is argued to be a postdepositional alteration associated with the groundwater table, influenced by tide levels in Garrison Bay. The upper light layer remains essentially dry due to its high porosity and its location above the water table, while the lower dark layer is frequently inundated by the fluctuating water table and is cyclically wet and dry. In the lower dark layer, the interaction of groundwater and organic materials produces carbonic acid, leading to the dissolution of carbonates. This is evidenced by a lower carbonate content in the fine-grained sediment and less identifiability of surface features of shell and bone in the dark layer.

The effects of these different postdepositional conditions on the charcoal in those strata should be evident in recovered charcoal assemblages. Because charcoal is very porous and has a large surface area, it is a potentially useful medium for reflecting postdepositional soil conditions. The amount and composition of postdepositional additions will reflect chemical conditions in the deposit. In the British Camp shell midden, alkaline compounds (e.g., calcium carbonate) in the soil solution may precipitate in the pores of charcoal fragments. Given the proposed differences in chemistry between the two strata, charcoal from the dark layer should have higher concentrations of such precipitates than the light layer. These acid soluble residues would not be dissolved and removed by the neutral pH water during flotation.

The postdepositional addition of material to charcoal is worthy of investigation not only because it provides information about chemical conditions in the deposit, but also because such additions affect the physical properties of the charcoal (e.g., fragment specific gravity, porosity, and strength), which in turn can influence recoverability and/or identifiability. The presence of precipitates, as opposed to air, in the pores of charcoal fragments will effectively decrease their porosity and increase their density. Separation from the heavy fraction by flotation may be hampered, depending upon how much material is added to the fragments. The relative volume of open pore space (V_{op}) is measured for several taxa of charcoal fragments from the two strata. The fragments are then soaked in a dilute acid solution and subsequent weight loss is monitored. Postdepositional conditions appear to provide an explanation for differences in the distribution of open-pore volume between the two strata.

II. Previous Work

While flotation has clearly increased our ability to retrieve archaeological materials, the technique is imperfect, especially in the case of fragile botanical remains. Knowledge of the factors influencing the content of archaeobotanical assemblages is critical to successful interpretation. The probability of charcoal recovery is governed by the strength of the charcoal fragments relative to forces associated with water flotation (Brady 1989). Those charcoal particles with strengths below the forces of flotation will break apart and the resultant smaller sized fragments will either be recovered or will pass completely through the retaining mesh. Several variables (e.g., structure, particle size and shape, specific gravity, and environment of deposition) may influence the strength of charcoal fragments, thus affecting their ability to withstand the forces of flotation.

Brady (1989) discussed two kinds of mechanical forces present in the flotation process: impact and internal static stresses. Impact stresses are those resulting from collisions between a fragment and the water surface, the flotation tank, or matrix particles. Internal static stress results from different rates of structural shrinkage and swelling due to changes in the fragment's moisture content, such as that which occurs as a result of wetting during flotation and subsequent drying. Jarman *et al.* (1972) suggested that internal static stresses were involved when they found that charred seeds displayed increasing levels of deterioration when exposed to cyclical wet/dry conditions.

In the case of archaeological wood charcoal, the stresses imposed by flotation may be quite destructive. In her research at Snaketown, Bohrer (1970) observed that mesquite charcoal essentially disintegrated upon immersion in water during flotation, most likely a result of internal static stress. In fact, it shattered to such an extent that no mesquite was identified in the analyzed archaeobotanical remains. Although the apparent loss of a taxon may not be a common occurrence, differential destruction of charcoal types in the flotation process remains a possibility.

In a study designed to assess charcoal loss as a result of flotation, Brady (1989) examined recovery rates for fragments of different species of modern charred wood from a pure sand matrix by the flotation process. His experiment failed to show significant losses of charcoal, despite variations in fragment density. One interpretation of his results is that charcoal recovery is not significantly affected by the flotation process; another interpretation offered by Brady is that modern charcoal is simply too strong relative to the stresses placed on it by the flotation conditions to experience measurable breakage.

Studies have shown wood to decrease in strength with time (Panshin and de Zeeuw 1980) depending upon conditions of temperature, pH, moisture content, and fungal attack. Similarly, charcoal that has been exposed to environmentally produced stresses (e.g., wet/dry and freeze/thaw cycles, postdepositional transport, fluctuating pH) over long periods of time is likely to be structurally weakened. Modern, unweathered, charred wood may not provide an appropriate analogue for archaeological charcoals in all cases; the initial identification of potential recovery bias is more easily demonstrated with archaeological materials. For this reason, I have chosen to use archaeological samples in this study. Explanations of recovery bias, however, should come from an understanding of the material and the stresses involved and do not require the use of archaeological charcoals.

III. Methods

A. *Determining Wood Charcoal Loss Rates*

1. Sample Selection and Treatment

Two 8-liter sediment samples collected from different deposits, 310/300 1B01 and 310/300 1C01 (hereafter B1 and C1, respectively), in the prehistoric shell midden at British Camp, San Juan Island, Washington, were used to examine wood charcoal loss. Both samples had reasonably large amounts of charcoal and were selected to contrast qualitatively different matrices (B1 had more rock and less shell than C1) from similar deposits (both were from light facies). Each sediment sample was divided into seven 1-liter and four 1/4-liter samples with a mechanical splitter. One-liter and 1/4-liter samples from deposit B1 are labeled 1-7 and 8-11, respectively; samples from

C1 are labeled 12-18 and 19-22. The splitter was used to ensure that the relative matrix composition of all eleven samples from each deposit was approximately equal. Each sample was gently sieved through nested screens to form five size fractions: $\geq$ 12.5 mm; 12.5 mm > X $\geq$ 6.0 mm; 6.0 mm > X $\geq$ 3.0 mm; 3.0 mm > X $\geq$ 1.5 mm; and the remaining < 1.5-mm matrix. For convenience, the size fractions will be referred to by the smaller retaining screen size (e.g., 3.0-mm fraction); however, it should be recognized that a range of particle sizes is actually involved.

To determine the amount of charcoal lost during flotation, the charcoal content of each sample was quantified both before and after flotation (Table 1). Charcoal weight was the parameter used to quantify abundance. The charcoal from each sample and size fraction was sorted from its matrix (the 1.5-mm size fraction was sorted only in the 1/4-liter samples), weighed (associated error ranged from 0.00003 to 0.00009 g), and gently remixed back into its matrix. This cycle was repeated at least three times until a standard error of less than 0.05 g could be attributed to the mean weight of each sample. This maximum error was determined in order to minimize the possibility of overlapping confidence intervals associated with the pre- and postflotation weights. The error terms reflect variability within the sorting/weighing process (e.g., inconsistent identification of wood charcoal fragments, low-level fragment breakage, weight variation due to daily changes in laboratory humidity and temperature). Prior to flotation, the size fractions were carefully recombined to form the original liter and 1/4-liter samples.

Table 1 An Example of Charcoal Loss Calculations for Sample 9

Sorting trials weight (g ± 1 SE)	Mean weight[a] (g ± 1 SE)	Weight difference[b] (g ± 1 SE)	% Loss[c]
Preflotation, 6.0 mm			
0.10984 ± 0.00005			
0.10803 ± 0.00007	0.10876 ± 0.00055		
0.10840 ± 0.00005			
Postflotation, 6.0 mm		0.04032 ± 0.00067	37.07
0.06800 ± 0.0			
0.06813 ± 0.00006	0.06844 ± 0.00038		
0.06920 ± 0.00006			
Preflotation, 3.0 mm			
0.44410 ± 0.00005			
0.37507 ± 0.00007	0.40796 ± 0.01999		
0.40470 ± 0.00005			
Postflotation, 3.0 mm		0.10254 ± 0.02040	25.13
0.30477 ± 0.00006			
0.30533 ± 0.00006	0.30542 ± 0.00040		
0.30616 ± 0.00006			

[a]Mean of sorting trials for pre- and postflotation weights ± 1 standard error.
[b]Difference between mean pre- and postflotation weights ± 1 standard error.
[c]Difference between mean pre- and postflotation weights expressed as a percentage of the preflotation weight.

Following flotation, the light and heavy fractions from each sample were allowed to dry at room temperature (moisture loss was monitored by repeatedly weighing samples until their weight stabilized) prior to further processing. The floated samples were then treated with the same sorting/weighing procedure as that used prior to flotation. Charcoal fragments were sorted from both the light and heavy fractions.

A processing schedule was designed to avoid the possibility of treatment bias between the two deposits. Samples were divided into two flotation batches, five samples from each deposit in flotation Batch 1, and six from each in flotation Batch 2. The sorting and floating procedures were performed in a random order within each batch to randomize any systematic error associated with the treatment of one deposit versus the other.

2. Water Flotation

The samples were processed with a flotation system similar to the Shell Mound Archaeological Project (SMAP) machine described by Watson (1976). The flotation tank (Figure 1) is equipped with a submerged showerhead that provides turbulance once the tank is filled with water. Samples are poured into an insert tub inside the filled tank. The light fraction floats out a tube onto chiffon fabric draped over geological screens (Figure 2). The heavy fraction sinks to the bottom of the insert and is retained by a 1.5-mm mesh screen; when separation is complete, the insert is removed (Figure 3) and the heavy fraction is washed out. Sediments < 1.5 mm fall through the mesh to the bottom of the tank and are periodically removed through drains located there.

Intersample contamination is a potential problem. The possibility of contamination during flotation was evaluated by periodically draining and rinsing the tank. The water was filtered through fine mesh chiffon fabric as it drained and the captured materials were dried and examined. No charcoal fragments ⩾ 1.5 mm were recovered from this source. Therefore, intersample contamination is not likely to distort the results of this project.

Loss during flotation can occur by two mechanisms: (1) charcoal fragments are accidentally removed from the samples by randomly occurring events such as air currents or splashing water; and (2) charcoal fragments experience fracturing and reduction in size due to the stresses of flotation, appearing as slightly smaller fragments in the same size fraction, in smaller size fractions, or passing through the screens completely. Although care was taken to minimize the probability of loss through the former (fragment removal), it could not be completely eliminated; accidental fragment loss is always a potential problem in flotation.

3. Other Considerations

Before the difference between pre- and postflotation charcoal weights can be attributed to charcoal loss during flotation, other potential sources of loss must be considered. In this study, the sorting/weighing procedure provided one opportunity for charcoal loss. The pattern of variation in charcoal weights through the

Figure 1 The flotation system used by the San Juan Island Archaeological Project. Sediment samples are carefully poured into an insert tub inside the filled tank. A showerhead, located below the insert, provides turbulence to assist in separation. The degree of agitation and water level can be controlled by adjusting water inflow through the showerhead and water outflow through spigots at the bottom of the tank.

sorting/weighing process was used to evaluate the significance of this potential source of loss. A nonrandom pattern of decreasing charcoal weight through time should characterize the series of sorting/weighing trials if this step resulted in charcoal loss. If the pattern of variation in charcoal weights through the sorting/weighing process is random, then it is unlikely that this procedure is accompanied by significant loss.

Limitations inherent in the sieving process provide an additional source of error. DallaValle (1943) and others (e.g., Cadle 1955; Herdan 1960; Whitby 1958) note that the probability of a particle passing through a sieve is a function of particle shape, sieve opening properties, and shaking time. Given the irregular, often elongated shape of many charcoal fragments and the gentleness required in sieving them, not all particles small enough to pass a given screen may have done so. Sieving procedures were standardized to maintain consistency between samples.

B. *Open-Pore Volume Measurement*

1. Sample Selection and Treatment

Based on the chemical regimes of the light and dark strata, charcoal from the dark layer should have higher concentrations of postdepositionally added alkaline precipitates than in the light layer. This material, precipitated in the pore spaces of the charcoal fragments, will decrease their porosity and increase their density. Differences in the volume of open-pore spaces will reflect different amounts of precipitated deposits; some variation will also arise as a result of differences in anatomical features (e.g., cell wall thickness, and cell size and shape).

The open-pore volumes of 103 charcoal fragments from four light (310/302 1O01, 310/302 1L01, 310/302 1O03, and 310/302 1N01) and four dark (310/302 2H01, 308/300 1O01, 308/300 1W01, and 306/300 2F01) facies were measured and compared. These samples had already undergone flotation by SJIAP personnel. Wood charcoal fragments (> 3.0 mm) had been previously identified to taxon (see Nelson, Ch. 11, this volume), thus enabling comparisons of open-pore volume between taxa.

Open-pore volume was measured for individual charcoal fragments in the following manner: all particles were (1) placed in a drying oven for 5 days to remove atmospheric moisture; (2) weighed three times for mean and standard error; (3) allowed to equilibrate with atmospheric moisture; (4) placed in deionized H_2O in a desiccator flask under vacuum for 5 days; (5) daubed with a moist sponge to remove excess surficial water; and (6) weighed three times for mean and standard

Figure 2 Light fraction particles float out a tube and land in fine chiffon fabric draped over geological screens. PVC tubing is used for the water and light fraction outlet, eliminating possible contamination and overflow problems associated with conventional sluices.

Figure 3 When separation is complete, the insert is removed and the > 1.5-mm heavy fraction is rinsed out.

error. The wet mass of a fragment is equal to its dry mass plus the mass of water filling its pores; thus, the volume of H_2O in the pores is obtained by subtracting the dry mass from the wet mass of the charcoal. The volume of H_2O is divided by the dry mass of the fragment to express it relative to fragment size (Table 2). This measurement

Table 2 A Sample of Open-Pore Volume (V_{op}) Calculations for Charcoal Fragments

Facies	Specimen	Dry mass (g ± 1 SE)	Wet mass (g ± 1 SE)	V_{op}[a] (± 1 SE)
310/302 1001	5	0.07330 ± 0.00010	0.19040 ± 0.00012	1.59754 ± 0.00144
310/302 1L01	8	0.08170 ± 0.00012	0.28080 ± 0.00029	2.43696 ± 0.00351
310/302 2H01	20	0.01120 ± 0.00006	0.04383 ± 0.00009	2.91339 ± 0.00033
308/300 1W01	3	0.03360 ± 0.00010	0.11023 ± 0.00003	2.28065 ± 0.00078

[a] $V_{op} = \frac{\text{wet mass} - \text{dry mass}}{\text{dry mass}}$.

of open-pore volume does not reflect the true open-pore volume of the particle because more time is required for water to completely replace air in the pores. However, since the primary concern involves *patterns* of pore volume and all samples were treated equally, relative measurements are adequate.

Similar to the procedures discussed previously, a processing schedule was organized to decrease the probability of treatment bias between light and dark strata. The open-pore volume measurements were obtained in two processing batches, each comprised of two light and two dark facies.

C. *Residue Measurement*

To measure the relative influence of acid-soluble precipitants on open-pore volume, 50 charcoal fragments for which open-pore volume measurements are available were randomly selected from the dark and light strata (25 fragments each). Each fragment was soaked in deionized H_2O for 24 hours, dried for 48 hours, and weighed. After being allowed to equilibrate with atmospheric moisture for 24 hours, each fragment was placed in 30 ml of a 3% HCl solution to soak for 24 hours. Each fragment was then rinsed with deionized H_2O, dried for 48 hours, and weighed. As an additional precaution, each beaker was weighed prior to and following the procedures to monitor weight gain related to the deposition of acid-soluble precipitants in the bottom of each beaker.

IV. Results

A. *Charcoal Loss Data*

Figures 4, 5, and 6 plot the amount of charcoal within each size fraction of each sample before and after flotation. The 12.5-mm size fraction is excluded from analysis because only one sample (Sample 1) contained charcoal in this size range prior to flotation; none of the charcoal in this fraction remained after flotation. The line shows the theoretical relationship which would be expected if there were no weight differences between pre- and postflotation charcoal assemblages. A weight gain (where the postflotation weight is higher than the preflotation weight) will appear

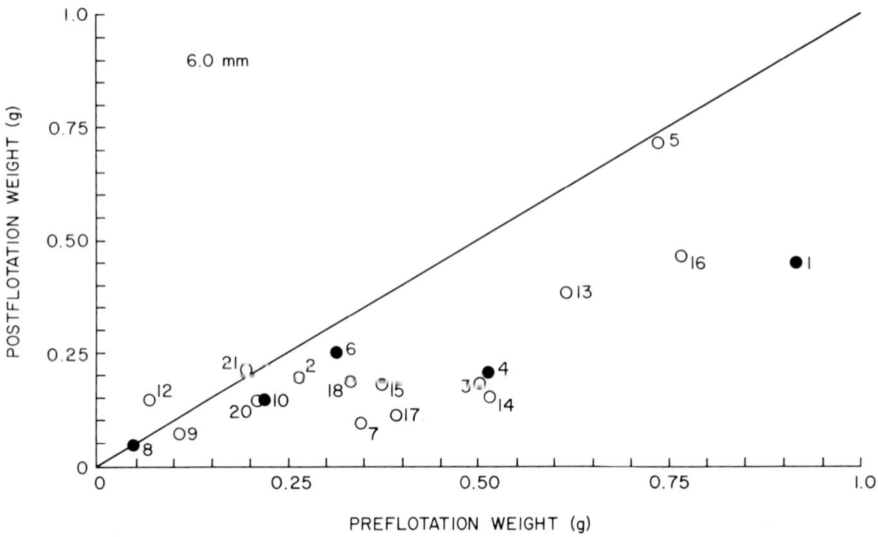

Figure 4 Pre- versus postflotation weights for the 6.0-mm size fraction of each sample. The line has a slope of 1.0; the slope was theoretically obtained if there were no differences between the pre- and postflotation weights. The filled circles represent samples removed from the statistical analysis (see discussion in Effect of Treatment Bias).

as a point above the line, and a weight loss (where the postflotation weight is less than the preflotation weight) will appear as a point below the line.

The confidence intervals for the preflotation measurements are frequently larger than their postflotation counterparts. This is due to a difference in precision related to the greater difficulty involved in sorting "dirty" unfloated samples as opposed to "clean" floated samples. Also, a reduction in precision is associated with decreased particle size, presumably reflecting greater difficulty in identifying the more numerous small fragments of wood charcoal.

There are seven instances of overlapping 95% confidence intervals out of 52 cases (3.0 mm: Samples 3, 5, 7, 13, and 14; 1.5 mm: Samples 9 and 20). This is higher than statistically expected, suggesting that the calculated maximum measurement error necessary for separating pre- and postflotation weights was overestimated. In four cases (3.0 mm: Samples 7 and 14; 1.5 mm: Samples 9 and 20) the confidence interval for one of the measurements overlaps the mean of the other; the means are not significantly different at $\alpha = 0.05$. In the majority of cases, however, the differences between pre- and postflotation weights appear to be significant and are not attributable to measurement error.

Three observations (6.0 mm: Samples 12 and 21; 3.0 mm: Sample 3) display a net gain rather than loss. This probably reflects the addition of one or two fragments to the postflotation fraction as a result of error associated with the sieving process. Errors from the same source probably occur as frequently in the opposite direction, but cannot be distinguished from fragment breakage.

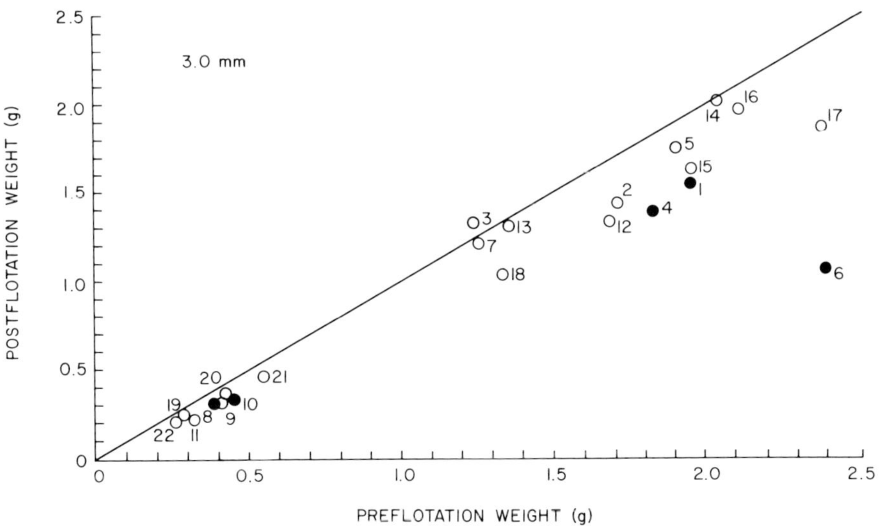

Figure 5 Pre- versus postflotation weights for the 3.0-mm size fraction of each sample.

B. *Open-Pore Volume Data*

The distribution of open-pore volume measurements for all charcoal fragments is skewed toward particles with low open-pore volumes. Figure 7 shows the open-pore volume measurements of charcoal specimens from the two strata. Particles of higher open-pore volumes are better represented in the light stratum than in the dark.

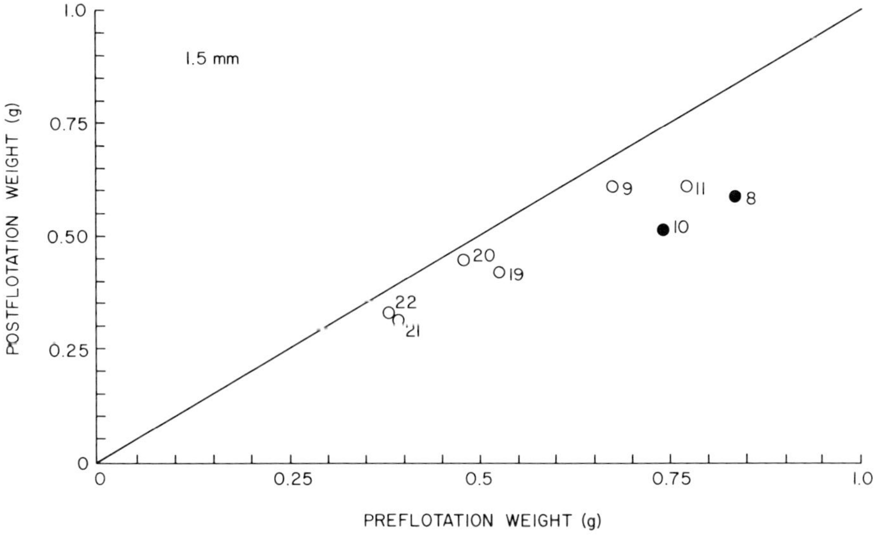

Figure 6 Pre- versus postflotation weights for the 1.5-mm size fraction of each 1/4-liter sample.

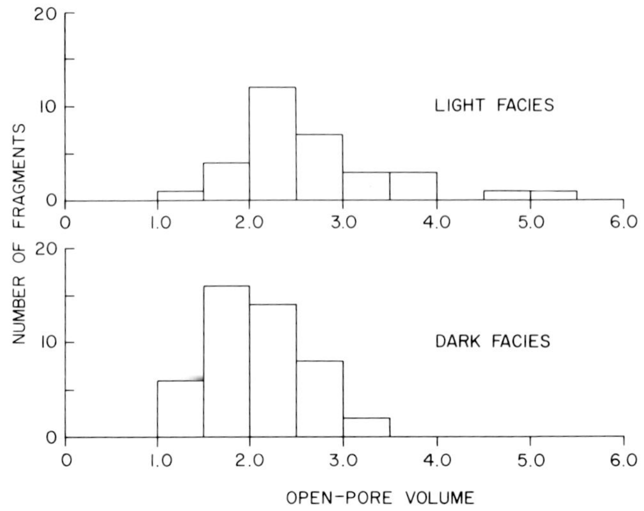

Figure 7 The distribution of open-pore volumes of wood charcoal fragments from light and dark facies. Note the asymmetrical distributions, biased toward particles with lower volumes.

C. *Residue Data*

Following the acid treatment, a white precipitate was observed in the bottom of the beakers. The precipitate was not crystalline and reacted strongly with HCl, suggesting it was a carbonate deposit. Figure 8 shows the relationship between the percent weight loss shown by fragments after the acid treatment and the open-pore volume of those fragments. While there is considerable variance, the fragments from the dark stratum appear to have lost a greater proportion of their weight during the treatment than those from the light stratum.

V. Discussion

A. *Documenting Charcoal Loss*

1. Effect of Sorting/Weighing Procedures

To evaluate the possibility of charcoal loss associated with the sorting/weighing procedure, the occurrence of increased or decreased charcoal weight between the three sorting trials of each sample is classified into one of four classes: steady increase (+,+); steady decrease (−,−); increase, then decrease (+,−); and decrease, then increase (−,+). For example, the preflotation sorting/weighing trials of the 3.0-mm fraction of Sample 6 produced three measurements: 2.40176 g, 2.3988 g, and 2.40266 g. The weight initially decreases between the first and second trials, and it increases between the second and third trials. This sample belongs to the (−,+) class.

A chi-square test of the distribution of cases across the classes was performed (Table 3). Only those samples sorted/weighed three times are included in this test;

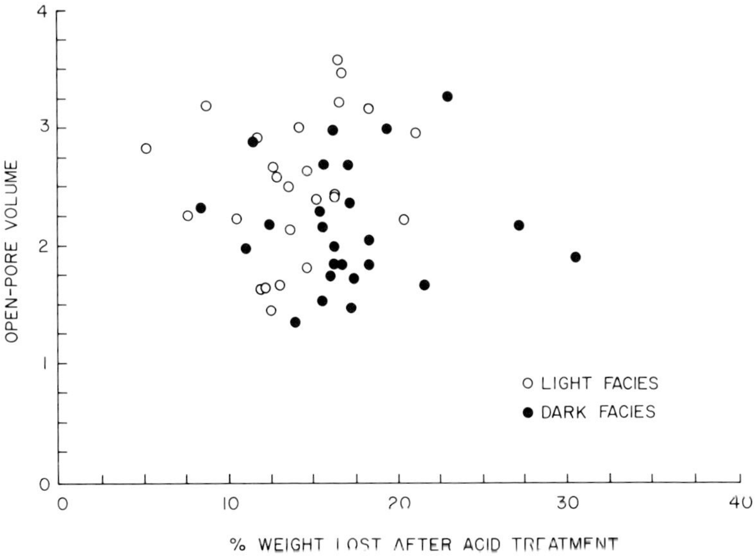

Figure 8 A plot of percent weight loss experienced by fragments during the dilute acid treatment versus their original open-pore volume.

six samples that required a fourth sorting/weighing trial to lower the standard error of their means to an acceptable level were excluded because their inclusion would significantly bias the chi-square calculations with empty and low frequency classes (Zar 1984). The results indicate that the sorting/weighing procedure is not characterized by significant charcoal loss or, if it occurs, loss is not of a sufficient magnitude to be identifiable beyond the limits of measurement error.

2. Effect of Treatment Bias

In spite of attempts to prevent the occurrence of systematic error in sample processing, the results display a suspicious pattern. Figure 9 shows the percent of charcoal loss in samples of both deposits in the two treatment batches. The percent of charcoal loss in samples from deposit B1 (Figure 9A) appears higher for flotation Batch 1 than Batch 2, while there is essentially no difference in loss rates between the two processing batches for deposit C1 (Figure 9B). A two-factor analysis of

Table 3 Results of a Chi-Square Test of the Effect of Sorting/Weighing Procedures on Charcoal Loss

	(+,+)	(−,−)	(+,−)	(−,+)	*n*
f_1	12	25	26	29	98

Notes: $H_\varnothing$: The cases are distributed randomly across the classes. H_A: The cases are distributed nonrandomly across the classes. $\chi^2 = 7.3913$; d.f. $= 3$; $0.05 < p < 0.10$; therefore, do not reject $H_\varnothing$.

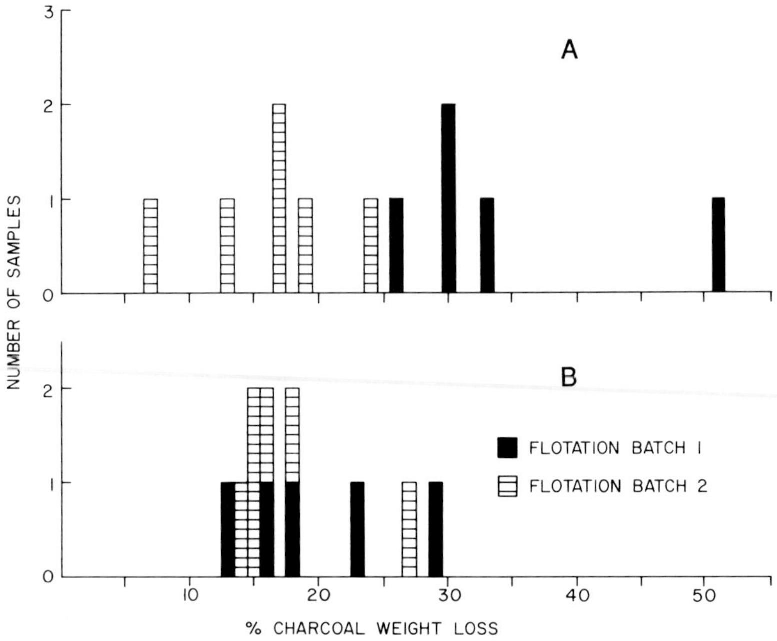

Figure 9 A, a histogram of the percent of charcoal loss for samples from deposit B1; B, a histogram of the percent of charcoal loss for samples from deposit C1.

variance with unequal, but proportional, replication (Zar 1984) was performed to assess the apparent difference in mean percent of charcoal loss between the two processing batches and the two deposits (Table 4). There is a significant effect on percent charcoal loss between the two batches, there is no difference in charcoal loss between the two deposits, and there is a significant interaction effect between the batches and deposits (i.e., the effect of batch treatment is not independent of the deposit).

In order to identify where the differences are located, the two-factor analysis of variance was followed by a Tukey test for multiple comparisons, modified for unequal sample sizes (Zar 1984). Comparison of the four classes (Deposit B1, Batch 1; Deposit B1, Batch 2; Deposit C1, Batch 1; and Deposit C1, Batch 2) indicates that the mean percent charcoal loss in the first (Deposit B1, Batch 1) is significantly dif-

Table 4 Results of a Two-Factor ANOVA to Assess the Effects of Deposit and Treatment Batch on Charcoal Loss

Effect	F	*p*	Conclusion
Batch	12.2494	$0.005 > p > 0.0025$	Reject $H_\emptyset$
Deposit	4.1917	$0.10 > p > 0.05$	Do not reject $H_\emptyset$
Interaction	7.11	$0.025 > p > 0.01$	Reject $H_\emptyset$

Table 5 Tukey Test for Multiple Comparisons to Identify Differences in Means between Batches 1 and 2 of Deposits B1 and C1

Comparison	Difference	SE	q	Conclusion
1 vs 2[a]	17.9168	2.9055	6.1665	Reject $H_\varnothing$
1 vs 4	16.7974	2.9055	5.7612	Reject $H_\varnothing$
1 vs 3	14.3761	3.0347	4.7372	Reject $H_\varnothing$
3 vs 2	3.5407	2.9055	1.2186	Do not reject $H_\varnothing$
3 vs 4	2.4213	2.9055	0.8334	Do not reject $H_\varnothing$
4 vs 2	1.1194	2.7703	0.4041	Do not reject $H_\varnothing$
	$1 \neq 2 = 3 = 4$			

[a] 1, Batch 1, B1; 2, batch 2, B1; 3, batch 1, C1; 4, batch 2, C1.

ferent from the mean percent charcoal losses of the other three, between which there are no significant differences (Table 5).

There is no reason to believe that the B1 samples in Batch 1 are *a priori* different from those in Batch 2, given that they originated from the same sediment sample and were divided with a mechanical splitter. The only difference in the treatment between the Deposit B1, Batch 1 samples and the remainder of samples processed was the set of sieves used in the preflotation sieving. Prior to flotation, B1 samples of Batch 1 were sieved with a set of nested screens different from the set used for the remainder of the sample processing. The apparent difference in charcoal loss rates most likely represents an error in charcoal measurement, rather than an actual difference in the rate of charcoal recovery. To avoid any biasing effects, those samples from Deposit B1, Batch 1 are excluded from further analysis.

3. Statistical Analysis of Loss Data

The significance of the difference in charcoal weight before and after flotation is tested using variations of the *t*-test. Applications of this statistical test require that samples come from a normally distributed population (Zar 1984). However, the *t*-test is known to be quite robust with respect to nonnormal distributions (Ryan *et al.* 1985; Thomas 1986; Zar 1984), the effect being minor deviations in reported versus actual *p*-values.

A preliminary examination of the data reveals a significant positive correlation at $\alpha = 0.05$ between the preflotation weight and the amount of loss for the 6.0-mm and 3.0-mm size fractions (6.0 mm: $r = .602$, $n = 14$, $p = 0.023$; 3.0 mm: $r = .559$, $n = 17$, $p = 0.020$). The more charcoal there was, the more there was to lose. Because I am working with different sample sizes (liter and 1/4-liter), the preflotation charcoal weights are bimodally distributed (e.g., Figure 5), resulting in a distinctly bimodal distribution of loss weights. Thus, the assumption of normality of differences is clearly not met. One solution to this problem involves dividing the samples into two groups (eleven 1-liter samples and six 1/4-liter samples) and treating them separately; another involves expressing the relative amount of charcoal weight loss for each sample as a percentage of preflotation charcoal weight.

Table 6 Statistical Analyses of Charcoal Loss Data

Sample size	Size fraction	*t*	*n*	*p*	Conclusion
Absolute weights:					
1 liter	6.0 mm	4.67	11	0.0000	Reject $H_{\varnothing}$
	3.0 mm	3.70	11	0.0021	Reject $H_{\varnothing}$
.25 liter	6.0 mm	1.28	3	0.16	Do not reject $H_{\varnothing}$
	3.0 mm	7.43	6	0.0003	Reject $H_{\varnothing}$
	1.5 mm	4.41	6	0.0035	Reject $H_{\varnothing}$

Size fraction	*t*	*n*	*p*	Conclusion
Percentages:				
6.0 mm	2.41	14	0.016	Reject $H_{\varnothing}$
3.0 mm	5.93	17	0.0000	Reject $H_{\varnothing}$
1.5 mm	6.06	6	0.0009	Reject $H_{\varnothing}$
All	13.69	17	0.0000	Reject $H_{\varnothing}$

Notes: H_ϕ: The mean difference in charcoal weight is significantly less than or equal to 0. H_A: The mean difference in charcoal weight is significantly greater than 0.

When the samples are divided into 1-liter and 1/4-liter groups, the degrees of freedom are substantially reduced, but the bimodality of loss weights is eliminated. Paired-sample *t*-tests (Table 6) are used to evaluate the significance of the difference between pre- and postflotation charcoal weights. For the 1-liter samples, the paired-sample *t*-test allows rejection of the null hypothesis for both 6.0-mm and 3.0-mm size fractions. As previously noted, the 1.5-mm fraction was not quantified in the 1-liter samples. In the 1/4-liter samples, the null hypothesis is not rejected in the 6.0-mm fraction, while it is in the 3.0-mm and 1.5-mm fractions. Failure to reject the null hypothesis for the 6-mm fraction of 1/4-liter samples probably stems from the fact that half of the samples (Samples 11, 19, and 22) contained no charcoal in this size fraction and one (Sample 21) showed a net gain in weight after flotation. Thus, significant weight loss was demonstrated in four of the five size fractions examined.

The loss weights are also converted to percentages of the preflotation charcoal weights. Because standard transformations are not appropriate for percentage data (they tend to be irregular with respect to extreme values) and the presence of negative values prevents the use of the accepted arcsine transformation or its modifications (Zar 1984), the distributions remain slightly nonnormal. A *t*-test for each size fraction allows rejection of the null hypothesis in each case (Table 6). A *t*-test for percent total charcoal loss (all size fractions combined for each sample) allows rejection of the null hypothesis as well. Again, significant losses were identified between pre- and postflotation charcoal weights.

4. Examination of Loss Rates

Charcoal recovery is determined by the physical characteristics of the charcoal, the composition and depositional history of the sediment matrix, and the flotation conditions. Comparison of loss rates between seven 1-liter and four 1/4-liter subsamples from one sediment sample (C1) shows a slightly skewed distribution of loss rates, ranging from 13.9 to 29.1% (Figure 9B). This range is surprisingly large, given that flotation conditions were standardized and the subsamples were compositionally equivalent (having been obtained from a single sample with a mechanical splitter) and had identical postdepositional histories.

Loss rates from four 1-liter and two 1/4-liter subsamples from B1 (Figure 9A) appear to be more normally distributed, ranging from 7.3 to 24.8%. When the loss rates from C1 are compared with those of B1, they are not significantly different ($t = .3728$; $n = 17$; $p > 0.50$). These two deposits had very similar postdepositional histories and varied only slightly in sediment composition [preliminary analyses failed to indicate a significant correlation between charcoal loss and sediment composition (e.g., amount and size distribution of rock and shell) for the samples].

5. Changes in Size Distribution of Fragments

Flotation has a visible impact on charcoal fragment size distribution. Figure 10 compares the percent by weight of the different size fractions of charcoal fragments in all samples before and after flotation. A general trend is apparent in which the relative frequency of fragments in the small size fractions (3.0 mm and/or 1.5 mm) increases following flotation, presumably at the expense of larger fragments. While at first glance, increases in the 1.5-mm fraction appear negligible, these calculations reflect only the 1/4-liter samples, thus representing only 1/8 of the total 1.5-mm charcoal. Differences would have been more dramatic had the 1.5-mm size fraction been measured for all samples.

B. *Patterns in Open-Pore Volume*

1. Effect of Treatment Bias

Charcoal fragments were processed in two batches, each composed of two light and two dark facies, in an effort to eliminate treatment bias in open-pore volume determinations. A two-sample t-test of the mean open-pore volumes of the two

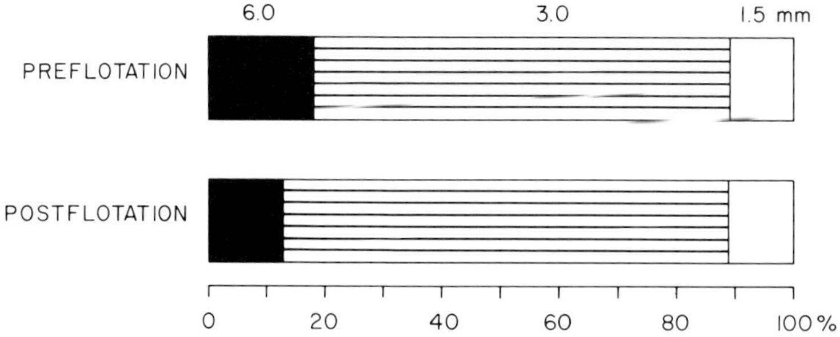

Figure 10 Charcoal fragment size distribution (percent by weight) for all samples before and after flotation.

processing batches indicates that there is no significant difference between the two ($t = 1.156$; $n = 103$; $0.50 > p > 0.20$). Thus, the distributions of open-pore volume do not reflect error introduced by differential treatment of samples from the two strata.

2. Statistical Analysis of Open-Pore Volume Data

A Kolmogorov–Smirnov goodness of fit test is used to assess the normality of the distribution of open pore volumes from all facies. The results (D = .119; $n = 103$; $p < 0.01$) indicate that the distribution is significantly different from normal. The distribution is heavily weighted toward low open-pore volumes. Two sample *t*-tests are used to determine the significance of differences in the mean open-pore volume of charcoal fragments between the light and dark strata (Table 7). The null hypothesis is rejected, suggesting that the mean open-pore volume of fragments from the dark stratum is significantly less than that from the light. Although the sample means are influenced by the skewed distributions, other measures of central tendency less susceptible to asymmetry (e.g., sample median and mode) also suggest the two strata represent different populations (light median = 2.43; light mode = 2.4; dark median = 2.03; dark modes = 1.8, 2.2).

Because the open-pore volume is also influenced by anatomical features of the charcoal, this data must be examined relative to the taxa comprising the assemblages of the two strata. With the exception of *Pseudotsuga*, which has a slightly bimodal tendency [perhaps a function of early wood/late wood differences (Panshin and de Zeeuw 1980)], the distribution curves of open-pore volume data for individual classes are skewed toward lower open-pore volumes. The mean open-pore volumes of different taxonomic classes are compared between the two strata (Table 7). The mean open-pore volumes of the three conifer classes are significantly different between the two strata, while the distributions of the hardwoods do not differ significantly. The hardwoods in the light layer display a large range of open-pore volumes (1.98–5.20).

The differences in open-pore volume between the two strata are suggested to reflect the presence of alkaline residues precipitated from the soil solution. This condition was expected to be more pronounced in the lower deposits of the midden.

Table 7 Statistical Analyses of Open-Pore Volume Measurements from Light and Dark Strata

Taxon	*t*	*n*	*p*	Conclusion
All taxa	4.606	103	$p < 0.00005$	Reject $H_{\emptyset}$
Pseudotsuga	1.886	34	$0.05 > p > 0.025$	Reject $H_{\emptyset}$
Abies/Tsuga	3.916	41	$p > 0.00005$	Reject $H_{\emptyset}$
Thuja	3.126	13	$0.005 > p > 0.0025$	Reject $H_{\emptyset}$
Hardwoods[a]	.803	15	$0.25 > p > 0.10$	Do not reject $H_{\emptyset}$

Notes: $H_{\emptyset}$: The mean open-pore volume of charcoal fragments from facies of the light stratum is significantly less than or equal to the mean from the dark stratum. H_A: The mean open-pore volume of charcoal fragments from facies of the light stratum is significantly greater than the mean from the dark stratum.

[a] Includes *Alnus, Acer,* and *Rubus*.

C. *Differences in Residue Precipitation*

There is a strong positive correlation ($r = .95$) between fragment weight loss and beaker weight gain as a result of the acid treatment. Some variability was present, possibly as a result of slight differences in moisture content when weighed. Several charcoal fragments suffered minor breakage during the treatment; in a few cases, a negligible amount of charcoal powder remained in the beaker following fragment removal.

The two strata differ significantly in the percent of weight loss shown by fragments during the acid treatment. A two-sample t-test is used to compare the mean loss between the strata ($t = 2.621$; $n = 48$; $0.01 > p > 0.005$). The results indicate that fragments from the dark stratum lost a greater percentage of their weight as acid-soluble residues than those from the light stratum. No patterns are visible in comparisons of the percent of weight loss between the three conifer taxa and the hardwoods; there is a great deal of variability and samples sizes are too small.

The effects of the presence of calcium carbonate precipitates on the density of charcoal fragments were evaluated. Published values of specific gravity for wood charcoal (1.43; Baileys and Blankenhorn 1982) and calcium carbonate (2.93; Weast 1989) were used, along with dry and wet masses and the percent of weight lost as acid-soluble residues for charcoal fragments, to calculate the density of several particles. The increase in particle density resulting from the addition of carbonate precipitates was less than 0.001 g/cm^3 in all cases examined. Differences in density on this scale are not likely to severely affect recoverability.

VI. Conclusions

A significant difference was demonstrated in the weight of pre- and postflotation wood charcoal assemblages. This difference, identifiable in statistical analyses of absolute and relative weights for the 6.0-mm, 3.0-mm, and 1.5-mm size fractions, is interpreted to reflect charcoal loss during the recovery process. That this charcoal loss is a result of fragment breakage and not simply random accidental fragment removal is indirectly supported by evidence for alteration of the size distribution of charcoal particles. Postflotation assemblages are composed of relatively greater percentages of fragments in smaller size fractions. Thus, flotation is accompanied by charcoal fragmentation.

Charcoal loss rates can vary considerably, even among samples of similar composition from a single deposit. While the number of samples was too small to draw any general conclusions, these results clearly suggest that further research is critical to understanding the influence of different factors on charcoal recovery. Even under "ideal" conditions, multiple samples from a single deposit experienced variable loss rates.

Patterns of recovery will influence interpretations of archaeological wood charcoal assemblages. If charcoal loss is a completely random event (i.e., all fragments have the same probability of recovery from the flotation process), then quantitative measurements are simply less precise and interpretations are unbiased. However, if the flotation procedure is characterized by systematic breakage of particular taxa of charcoal fragments, then accuracy of the measurements is affected and subsequent interpretations will be biased. This phenomenon can be easily monitored by comparing the frequency of taxonomic classes in small samples from a deposit prior to and following flotation. If the sampling scheme is adequate, differing taxonomic frequencies will reflect bias in recovery.

Variation in charcoal loss rates has important implications for the use of absolute charcoal weight as a standardizing measure of occupational intensity or preservation conditions between deposits (e.g., Asch and Asch 1975; Bohrer 1970; Johannessen 1984, 1988; Miller 1988; Pearsall 1983). The general assumption that charcoal assemblages from similar deposits are comparable has been challenged. Charcoal experiences variable rates of recovery as a result of differences in structural properties, depositional and postdepositional histories, and recovery conditions. Successful comparisons between deposits require these factors to be identical. As demonstrated, even under the best of conditions there is significant variability in results.

The characteristics of archaeological wood charcoal assemblages and physical properties of charcoal fragments are influenced by postdepositional conditions within a site (see also Nelson, Ch. 11, this volume). Compounds in the soil solution of the midden were precipitated as residues in the pores of charcoal fragments. The carbonates are not dissolved during the neutral pH water flotation process. However, later treatment with a dilute acid solution removed the precipitants and allowed their measurement. The results of the open-pore volume and acid treatment portions of this research support arguments by Stein and others (Stein 1984, Ch. 7, this volume; Nelson *et al.* 1986) about the postdepositional chemical conditions in the light and dark strata of the shell midden at British Camp.

The potential of archaeological charcoal analysis for obtaining information beyond taxonomic frequencies has gone largely unexploited (cf Hesse and Rosen 1988; Nelson, Ch. 11, this volume; Rosen 1989; F. Whittaker, personal communication). When viewed as a sedimentary particle, charcoal provides unique information about depositional and postdepositional environments. Not only can analyses of charcoal properties be used to answer questions of depositional history, but such analyses can only help to increase our knowledge of the material. This knowledge is necessary to understand biases in the assemblages recovered.

Acknowledgments

I would like to thank several people who assisted me during the various stages of this project: J. K. Stein for providing the sediment and flotation equipment; M. A. Nelson for providing unpublished data and charcoal samples; C. T. Greenlee for technical assistance; A. Cagle, E. Lester, J. L. Libby, and M. A. Nelson for assistance with floating; and T. J. Brady, R. C. Dunnell, J. L. Libby, M. A. Nelson, D. M. Pearsall, and J. K. Stein for discussing the project and/or commenting on earlier drafts of the manuscript.

This material is based upon work supported under a National Science Foundation Graduate Fellowship.

References

Asch, N. B., and D. L. Asch

1975 Plant remains from the Zimmerman site - Grid A: A quantitative perspective. In *The Zimmerman site: Further excavations at the grand village of Kaskaskia*, edited by M. K. Brown. Illinois State Museum, Reports of Investigations 32. Springfield, Illinois. Pp.116–120.

Baileys, R. T., and P. R. Blankenhorn
1982 Calorific and porosity development in carbonized wood. *Wood Science* **15**, 19–28.

Bohrer, V. L.
1970 Ethnobotanical aspects of Snaketown, a Hohokam village in southern Arizona. *American Antiquity* **35**, 413–430.

Brady, T. J.
1989 The influence of flotation on the rate of recovery of wood charcoal from archaeological sites. *Journal of Ethnobiology* **9**, 207–227.

Cadle, R. D.
1955 *Particle size determination.* New York: Interscience Publishers.

DallaValle, J. M.
1943 *Micromeritics.* New York: Pitman Publishing.

Herdan, G.
1960 *Small particle statistics, 2d ed.* London: Butterworths.

Hesse, B., and A. Rosen
1988 The detection of chronological mixing in samples from stratified archaeological sites. In *Recent developments in environmental analysis in Old and New World archaeology*, edited by R. E. Webb. BAR International Series 416, Oxford. Pp.117–129.

Jarman, H. N., A. J. Legge, and J. A. Charles
1972 Retrieval of plant remains from archaeological sites by froth flotation. In *Papers in economic prehistory, vol. 1*, edited by E. S. Higgs. Cambridge: Cambridge University Press. Pp.39–48.

Johannessen, S
1984 Paleoethnobotany. In *American Bottom archaeology, a summary of the FAI-270 project contribution to the culture history of the Mississippi River Valley*, edited by C. J. Bareis and J. W. Porter. Urbana: University of Illinois Press. Pp.197–214.
1988 Plant remains and culture change: are paleoethnobotanical data better than we think? In *Current paleoethnobotany*, edited by C. A. Hastorf and V. S. Popper. Chicago: University of Chicago Press. Pp.145–166.

Miller, N. F.
1985 Paleoethnobotanical evidence for deforestation in ancient Iran: a case study of urban Malyan. *Journal of Ethnobiology* **5**, 1–19.
1988 Ratios in paleoethnobotanical analysis. In *Current paleoethnobotany*, edited by C. A. Hastorf and V. S. Popper. Chicago: University of Chicago Press. Pp.72–85.

Minnis, P. E.
1978 Paleoethnobotanical indicators of prehistoric environmental disturbance: a case study. In *The nature and status of ethnobotany*, edited by R. I. Ford. Anthropological Papers No. 67, Museum of Anthropology, University of Michigan, Ann Arbor. Pp.347–366.

Nelson, M. A., P. J. Ford, and J. K. Stein
1986 Turning a midden into mush: evidence of acidic conditions in a shell midden. Paper presented at the 51st Annual Meeting of the Society for American Archaeology, New Orleans, Louisiana.

Panshin, A. J., and C. de Zeeuw
1980 *Textbook of wood technology.* New York: McGraw-Hill.

Pearsall, D. M.
1983 Evaluating the stability of subsistence strategies by use of paleoethnobotanical data. *Journal of Ethnobiology* **3**, 121–137.
1989 *Paleoethnobotany: A handbook of procedures.* San Diego: Academic Press.

Rosen, A. M.
1989 Ancient town and city sites: a view from the microscope. *American Antiquity* **54**, 564–578.

Ryan, B. F., B. L. Joiner, and T. A. Ryan, Jr.
1985 *Minitab handbook, 2d ed.* Boston: Duxbury Press.

Stein, J. K.
1984 Interpreting the stratigraphy of Northwest shell middens. *Tebiwa* **21**, 26–34.

Struever, S.
1968 Flotation techniques for the recovery of small-scale archaeological remains. *American Antiquity* **33**, 353–362.

Thomas, D. H.
1986 *Refiguring anthropology*. Prospect Heights, Illinois: Waveland Press.
Wagner, G. E.
1982 Testing flotation recovery rates. *American Antiquity* **47**, 127–132.
1988 Comparability among recovery techniques. In *Current paleoethnobotany*, edited by C. A. Hastorf and V. S. Popper. Chicago: University of Chicago Press. Pp.17–35.
Watson, P. J.
1976 In pursuit of prehistoric subsistence: a comparative account of some contemporary flotation techniques. *Midcontinental Journal of Archaeology* **1**, 77–100.
Weast, R. C.
1989 *CRC handbook of chemistry and physics* (editor). Boca Raton, Florida.: CRC Press.
Whitby, K. T.
1958 The mechanics of fine sieving. American Society for Testing Materials Special Technical Publication 234, 3–25.
Yarnell, R. A.
1963 Comments on Struever's discussion of an early "Eastern Agricultural Complex". *American Antiquity* **28**, 547–548.
1982 Problems of interpretation of archaeological plant remains of the Eastern Woodlands. *Southeastern Archaeology* ***1***(1), 1–7.
Zar, J. H.
1984 *Biostatistical analysis, 2d ed.* Englewood Cliffs, New Jersey: Prentice-Hall.

13

Interpreting the Grain Size Distributions of Archaeological Shell

Pamela J. Ford

I. Introduction

Archaeologists working on the southern and central Northwest Coast of North America have described shell-bearing sites in which two distinct stratigraphic layers are visible (e.g., Bernick 1983; C. Carlson 1979; R. Carlson 1979; Conover 1978; Luebbers 1978; Schwartz and Grabert 1973; Stein 1984, 1989). A lower dark brown to black layer is often described as containing little shell and a matrix that appears to be fine grained and compact. The overlying layer is described as shell rich and lighter in color. The two layers do not occur in all sites, but where they do appear together, the boundary is dramatic and the stratigraphic division has been used by some to define cultural components (C. Carlson 1979; Luebbers 1978).

The distinct dark and light strata appear at British Camp and questions about their origin have yet to be answered. Nelson *et al.* (1986) propose that the difference between the dark and light layers in the shell midden at British Camp may be explained by the presence of groundwater. Moisture retained in the midden at and below the water table may support a microbial community essential to the breakdown of organic matter. The organic acids produced by this breakdown would attack the surfaces of shells (calcium carbonate) in the midden. As the shells are etched, they lose the distinctive surface features that allow taxonomic identification. Additionally, the etching process dissolves $CaCO_3$ with byproducts of Ca, CO_2, and H_2O, which are leached from the midden by groundwater. Because they are dissolved and removed in the groundwater, a concentration of translocated $CaCO_3$ is not found in lower facies (see also Ham 1976). Above the water table, organic matter remains essentially dry and therefore fails to etch the shell. The conditions under which organic acids are produced and translocated are not present in the upper dry part of the site.

Deciphering a Shell Midden

The working hypothesis described above has yet to be falsified but other explanations for the dark and light distinction have not been ruled out. One explanation involves the shell that both layers contain. Several invertebrates in the midden have different environmental requirements during life and different breakage patterns after death. Since environmental requirements are well known for these shellfish, knowledge of the kinds of shellfish in the deposits provides information about the kinds of places from which they were harvested. Observations of shell and its conditions in these deposits suggests that little of it came directly from an intertidal zone to its final resting place in the midden. Some shell is crushed, some appears to be burnt, some is in large fragments and some is whole (Figure 1). The various conditions of shell suggest that the shells seen in the facies have been broken differentially after collection from an intertidal habitat. The amount of breakage may be related either to the type of taxa, the type of cultural activity, or the number of times the shell was deposited, moved, and deposited again.

This analysis attempts to recognize differences in source for the shell in the facies at British Camp. Here, the term "source" refers not only to the natural habitat of the shellfish population, but also to the number of depositional cycles (deposition-removal-deposition) the shell has experienced before coming to rest as part of the facies recovered by archaeological excavation. Differences between the dark and light layer facies may be explained by differences in the source or sources for the shells the facies contain.

II. Theoretical Expectations

When archaeological specialists in the analysis of faunal remains and in the analysis of sediment ask about the source for shell in midden deposits, they refer to two different definitions for the term "source." The specialist in faunal analysis considers the source to be the population and environment from which the materials originally came, and asks questions about the nature of the population in which the fauna lived. (The faunal analyst's interest in a turkey bone from the dump focuses on whether the turkey it represents was wild or domesticated, local or nonlocal, one of many or an isolated occurrence.) The concern of the geoarchaeologist is the location and situation from which shell was removed immediately prior to final deposition. (The geoarchaeologist's interest in a turkey bone from the dump focuses on whether it last sat in the trash can of an individual dwelling or in the back of a trash truck.) Both analyses are concerned with source but define source differently.

The biological source is the environment or population from which the animals, in this case shellfish, were harvested. Different shellfish inhabit different intertidal and/or subtidal environments according to wave action and substrate. The identification of the taxa present in archaeological facies informs about the environment they occupied during life. This paper treats the habitat from which shellfish were harvested as the "biological source."

The source investigated by geoarchaeologists is the situation from which these shells were retrieved just prior to deposition in the most recent depositional episode.

Figure 1 Profile view of the variable condition of shell in archaeological facies, 45SJ24.

Shells harvested from the biological source may be collected and combined with other shells, moved to a new location, and treated in any of a variety of ways (e.g., shucking, heating, moving, trampling) before they are incorporated in the most recent depositional episode. The source from which they are retrieved before the last deposition observed in the archaeological facies is labeled here as the "sedimentological source." As the biological source is inferred from the population and/or environment from which living shellfish were harvested, the sedimentological source is inferred from the sedimentary attributes of the shell. The sedimentary attribute of shell measured in this study is grain size. In the analysis of deposits as the remains of subsistence and habitat reconstruction, the biological source is the source of interest. In the analysis of deposits as sediment, the sedimentological source is the source of interest. This paper will show that recognition of sedimentological source depends upon an understanding of biological source.

The durability of skeletal material varies, and appears to be related to skeletal microstructure (Chave 1964; Discroll and Weltin 1973). Because of this, the hard parts of various invertebrate taxa carry individual characteristics that are reflected in differential breakage patterns, variable susceptibility to loss of surface features, variable robustness, and different mean size and shape. Because overall shell morphology influences shell fragmentation, grain size distributions of shell and taxonomic composition of archaeological deposits may be correlated in such a way that the composition affects the grain size distribution of the shell in the deposit. Perhaps the difference between the dark and light layers, as perceived qualitatively by the archaeologist, is caused by a difference in shellfish taxa and a biologically determined difference in shell traits.

This analysis tests a hypothesis of source for archaeological shell from the facies at British Camp. Two kinds of source, biological and sedimentological, are distinguished and will be examined in different ways. The analysis of biological source is not unusual for archaeological analyses and is a major focus of most reports on faunal materials (e.g., Ham 1982; Hamblin 1984; Wessen 1988). This analysis involves the identification of the kinds of invertebrate taxa represented by shell and an understanding of the particular habitat requirements of each. The biological composition of each facies reflects the kind or kinds of local and/or nonlocal resources that were utilized by site occupants. The analysis of sedimentological source as defined here is unusual for archaeological analyses of shell, although there have been other attempts to distinguish various breakage patterns or cultural patterns that influence the observable archaeological record of shellfish (Muckle 1985).

Sedimentologists use grain size distributions of sediment to interpret depositional histories of geological deposits (Blatt *et al.* 1980; Stein 1987; Visher 1969). Although archaeologists are concerned with reconstructing the depositional history of archaeological sites, they rarely examine grain size distributions of artifactual material (see Madsen, Ch. 9, this volume; Stein and Teltser 1989), but archaeologists have analyzed size-frequency distributions of archaeological shell where the size of invertebrates at death is the variable measured (see Ford 1989). To interpret depositional histories, the size of shells as they are removed from the archaeological context is the variable measured. Grain size distributions reflect the manner in which the shell entered the archaeological record (biological and sedimentological source) and the manner in which the shell has been altered since deposition occurred (see Dunnell and Stein 1989). Most artifact types enter the archaeological record as large objects and become smaller while in archaeological context. Smaller sizes may be the result of chemical and physical weathering that alter artifacts. The materials that became smaller after they were deposited provide information about changes in the archaeological deposit since deposition, but the materials that entered the record already small provide information about source, transport agents, and environments of deposition (Dunnell and Stein 1989). Because people carry materials in suspension, the grain size distributions produced when people are the transport agent frequently reflect the grain size distributions of the source (Stein and Teltser 1989). When transported in suspension, sediments are subject to relatively little damage from transport itself and the grain size distributions of the archaeological facies should reflect the sources of the materials rather than sorting effects of transport.

To determine the biological source for the shell, the shells are identified and classified according to biological taxon and the various taxa present are described in terms of the environments that they inhabit, then variations in biological source are compared between facies and between dark and light layers. To determine the sedimentological source for the shell, the grain size distributions of shells are examined and these are compared between facies and between the dark and light layers to test for differences between the two layers. This analysis will test a hypothesis of no difference in source, biological or sedimentological, for the dark layer facies and light layer facies at British Camp. If there is no difference in either source, then I expect to see (1) the same set of invertebrate taxa represented in both dark layer and light layer facies; (2) the same intertidal communities represented in both the dark layer and light layer facies; and (3) the same grain size distributions represented in both dark layer and light layer facies.

III. Procedure

All material utilized for this analysis was collected from the Biological Samples taken from within individual facies in adjacent 2 × 2-m horizontal excavation Units 310/300 and 310/302 (Figure 2). During excavation, a portion of each facies was pedestaled for separate excavation as a biological sample. Each biological sample was collected in such a way that the volume of the sample (width, depth, and height) and its three-point provenance were recorded. Most biological samples are approximately 20 × 20 × 10 cm or 25 × 25 × 5 cm in size, but each sample varies in size and shape according to the size and configuration of the facies from which it was taken. Depending upon facies size (weight and volume), biological samples represent from 2 to 100% of the individual facies from which they were taken. These biological samples provide data on facies structure (bulk density, grain size distribution, relative abundance of material types), content (plants, vertebrates, and invertebrates), and chemical composition of the matrix (see Greenlee, Ch. 12, this volume; Nelson, Ch. 11, this volume; Stein, Ch. 7, this volume). Since facies do not necessarily

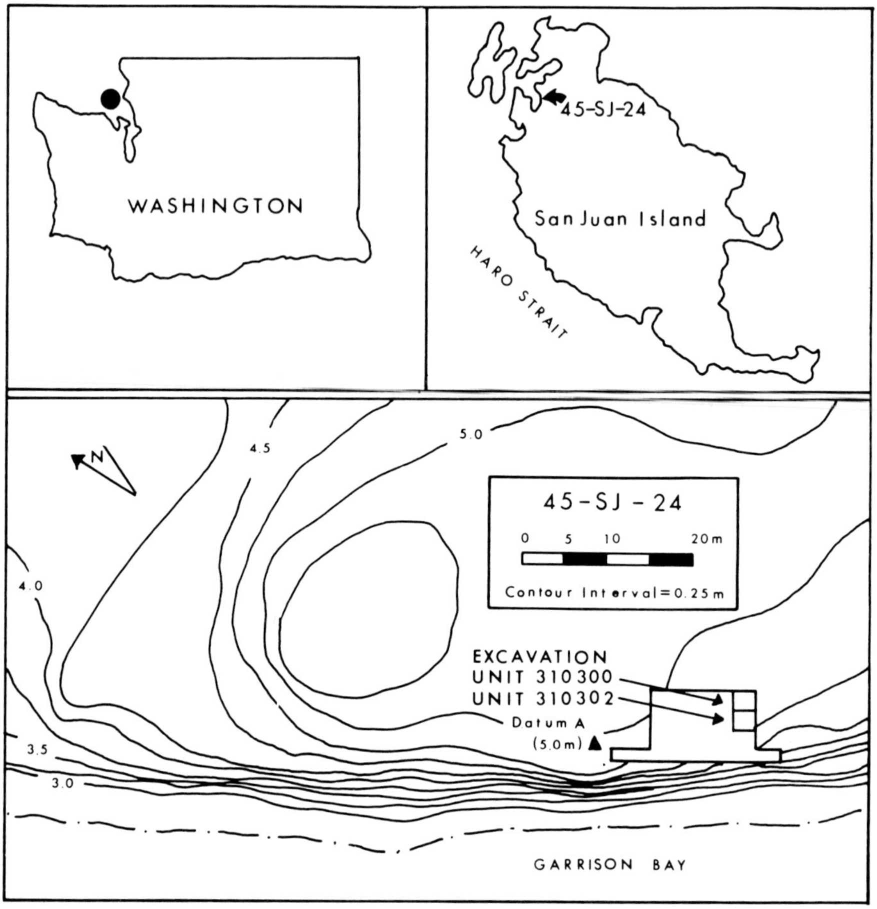

Figure 2 Excavation unit locator map.

intersect profiles (excavation unit boundaries), column sampling alone would be inadequate to sample facies variability and midden structure.

The biological sample provides invertebrate faunal remains as well as materials critical to other analyses. To obtain all of these materials from one sample, the sample is processed in a standardized manner (Figure 3). Matrix is defined as the material which passes through the 1.5-mm mesh: a portion of it was retained for chemical analyses and the rest of the biological sample was processed through a flotation system. After flotation, shell greater than the 1.5-mm mesh was dried and retained for faunal analysis. For the analysis described herein, the shell was poured into a series of nested screens with mesh sizes of 25 mm, 12.5 mm, 6 mm, and 3 mm. For taxonomic identification, the shell from each size class (determined by the mesh size of the screens) was compared to modern comparative specimens of local invertebrates. The weight in grams was recorded for all material; the condition of valves (broken or whole) was noted; and relative abundances of taxa were calculated. Grain size distributions of shell from each facies are based upon the four size classes defined by the size of the mesh in the nested screens. Grain size distributions were calculated for all shell within each facies and for particular individual invertebrate taxa within each facies. Grain size distributions are based solely upon the size of a particular piece of shell and are not a biological measurement of animal size.

Before grain size distributions are interpreted as descriptors of sedimentological source, the relationship between kinds of shellfish and their grain size distributions is assessed. To determine whether or not the taxonomic content of each facies affects the grain size distribution of the shell in the facies, the relative abundances of individual shellfish taxa in a facies were compared with the relative abundances of all the shell recovered in each size class for individual facies. The comparison of relative abundances (percentages) rather than absolute abundances requires a nonparametric test of correlation: Spearman's Rank correlation coefficient was utilized (Zar 1974). The relative abundances of taxa within each facies were compared separately with the relative abundances of all shell within each of the four size classes. Many taxa are present in facies but only a few are present in very large amounts. The shellfish taxa examined were: mussel (Family Mytilidae, *Mytilus* spp. and *M. edulis*); cockle (*Clinocardium* sp. and *C. nuttalli*); horse clam (*Tresus* spp. and *T. gapperi*); bent-nose and sand clams (*Macoma* spp., *M. nasuta*, and *M. secta*); venus clams (Family Veneridae, *Saxidomus* spp., *S. giganteus*, *S. nuttalli*, and *Protothaca staminea*); barnacles (*Balanus* spp.); and sea urchins (*Strongylocentrotus* spp.). Taxonomy and nomenclature follow Abbott (1974). Data from 79 facies in excavation Units 310/300, 310/302, 310/304, and 310/306 were utilized for this comparison.

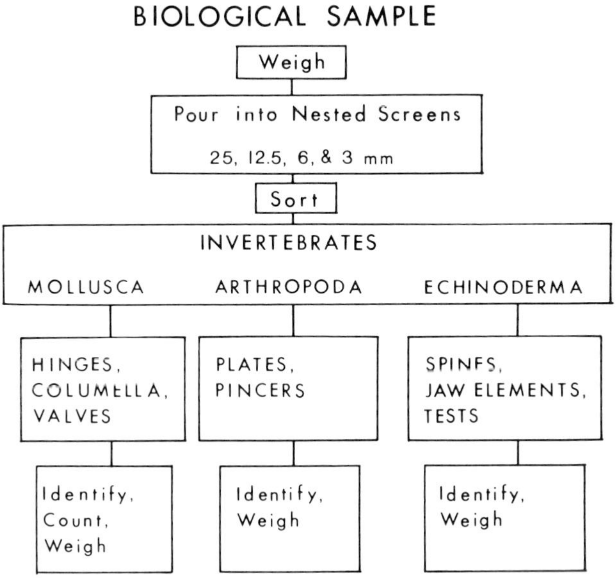

Figure 3 Recovery of material: biological sample.

Because each taxon is not necessarily present in each archaeological facies, the number of facies involved in each comparison varied from 72 to 79.

Facies are labeled alphabetically following the order of their excavation within each 2 × 2-m excavation unit. Since excavation occurs independently within each unit, the letter designation 1C in Unit 310/300 does not necessarily refer to the same deposit designated as 1C in Unit 310/302 (see Stein *et al.*, Ch. 6, this volume). Facies are labeled with Unit, Facies, and (for those facies thicker than 10 cm) Level designations. For instance, the label 310/3001R03 refers to excavation Unit 310/300, Facies 1R, and Level 03 (the top of which is 20 cm below the surface of Facies 1R). Several excavated facies in Unit 310/302 were not considered in this analysis for a variety of reasons. Facies 1A in all units is considered a plowzone, containing prehistoric aboriginal material as well as historic period Euro-American materials, and is not included.

IV. Data

A. *Relative Abundance of Taxa*

The relative abundances of taxa from the facies in the two excavation units are indicated in Figures 4 and 5. The percentages of shell for each facies as shown do not sum to one hundred because additional taxa (not considered in this study) are present in each facies in very small amounts. Their relative abundance has been determined but left out of this report. These additional taxa include unidentifiable mollusk shell (Phylum Mollusca), snails and limpets (Class Gastropoda), tusk shell (Class Scaphopoda), chitons (Class Polyplacophora), and crab (Phylum Arthropoda, Class Crustacea). Additional bivalves are also present in the facies but excluded from this summary. With the exception of the unidentifiable mollusks, the additional shells are present in very low relative abundances. Mussel, cockle, venus clams, and barnacles are present in every facies, and sea urchin is present in all but one of the facies.

The relative abundances appear upon first glance to vary equally among the light layer facies and among the dark layer facies (Figures 4 and 5). The range of relative abundance of each taxon for dark layer and for light layer facies is summarized in Table 1. To test the null hypothesis that relative abundances of particular taxa do not differ between the light and dark layer facies, the Mann–Whitney test (a nonparametric analogue to the two-sample t-test) was utilized (Zar 1974). Values for U and U′ were calculated based upon the ranks of relative abundances of each taxon within the dark layer facies and within the light layer facies. The values of U and U′ calculated for mussel, horse clam, venus clams, and barnacles indicate that the relative abundance of these taxa do not differ significantly between facies of the two layers (see Table 2 for various p values). However, the relative abundance of cockle shell is significantly higher in the light layer facies ($p < 0.02$). Similarly, the relative abundance of bent-nose/sand clams ($p < 0.05$) and of sea urchins is greater in the dark layer facies ($p < 0.05$). See Table 2 for a summary of the calculated statistical values.

A previous analysis reported in Nelson *et al.* (1986) indicated that shell surface condition varies between the dark and light layer facies. This was recognized by com-

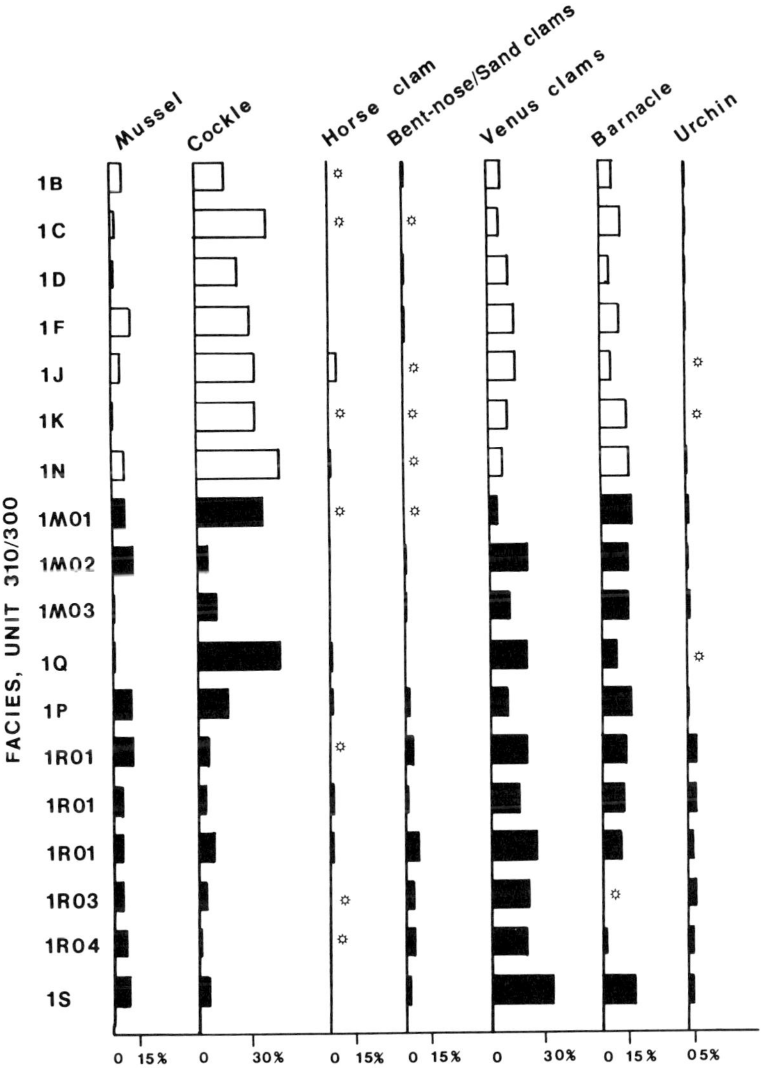

Figure 4 Relative abundance of taxa: Unit 310/300. Facies within the dark layer are indicated by shading. Asterisks indicate relative abundance of less than 1%.

paring the identifiability of shell from both layers in an analysis of shell from a single excavation unit (Unit 310/300). Because the taxonomic classification for cockle includes both genus-level and species-level identifications (*Clinocardium* sp. and *C. nuttalli*), a lower abundance of cockles in the dark layer facies could be the result of the factor of "reduced identifiability" as observed in the earlier analysis. If this is so, then the relative abundance of genus-level identifications as opposed to species-level identifications would be greater in the dark layer facies. The hypothesis that

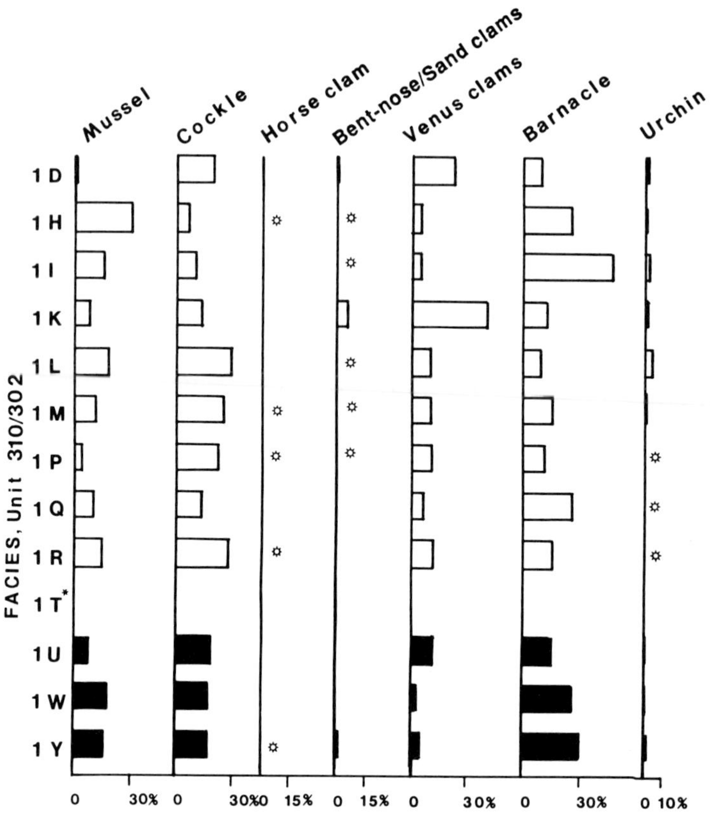

Figure 5 Relative abundance of taxa: Unit 310/302. Facies within the dark layer are indicated by shading. Asterisks indicate relative abundance of less than 1%.

cockle identifiability is greater in the light layer was tested by examining the relative abundance of genus-level (*Clinocardium* sp.) identifications to species-level (*C. nuttalli*) identifications in the light layer and dark layer facies. A two-tailed Mann–Whitney test was utilized to test the null hypothesis of no difference between the relative abundance of genus-level identifications of cockle in the light and dark layer facies. The relative abundance of genus-level identifications to species-level identifications for cockle shell was calculated for each facies and ranked for the dark and light layers. The results of the Mann–Whitney test indicate that there is no significant difference regarding cockle shell identifiability between dark and light layer facies ($p > 0.10$). In other words, all cockle shell appears in greater relative abundances in the light layer facies, which are more recent in time, and that change apparently bears no relationship to the etching of shell surface features observed in a small sample in earlier studies (Nelson *et al.* 1986).

The relative abundance of sea urchins is greater in the dark layer facies. Sea urchin spines are very tiny, tapered cylinders. Urchin jaw elements and test fragments are

Table 1 Summary of Relative Abundance of Taxa in Dark and Light Layers (%)

	Dark layer Minimum	Maximum	Light layer Minimum	Maximum
Mussel	1.6	17.6	0.6	29.1
Cockle	1.5	43.9	6.8	44.1
Horse clam	0	2.5	0	3.6
Bent-nose/Sand clams	0	7.5	0	5.6
Venus clams	2.6	33.2	4.0	40.2
Barnacles	33.0	77.0	30.0	79.0
Sea urchin	0.6	5.1	0	3.0

flattened and irregularly shaped. Because of their size and shape, questions arise about the possible downward movement of sea urchin spines in archaeological facies. If urchin spines move downward, crossing facies boundaries, the relative abundance of spines compared to other sea urchin body parts would be greater in the lower (dark) layer. I calculated the abundance of sea urchin spines relative to sea urchin jaw elements and test fragments for each facies and ranked the percentages for facies from the dark and light layers. A Mann–Whitney test indicates no significant difference in the relative abundance of sea urchin spines compared with other skeletal elements in the dark and light layer facies ($p > 0.20$). Consequently, the smaller amount of urchins in the light layer facies is not a function of postdepositional downward movement of spines. This conclusion is consistent with Madsen's analysis indicating that sand-sized grains are present in all facies (Madsen, Ch. 9, this volume).

B. *Grain Size Distributions*

The second data set constructed for this analysis is the grain size distribution calculated separately for all shell from each facies. These grain size distributions are displayed as histograms based upon the percent, by weight in grams, of all shell within each of four size classes (Figures 6 and 7). The largest size class includes all shell greater than that able to pass through the 25-mm mesh of the top screen; the second

Table 2 Comparison of Relative Abundance of Taxa within Dark Layer Facies and Light Layer Facies

	U	U′	p
Mussel	101.5	122.3	$p > 0.20$
Cockle	46.5	177.5	$p < 0.01$
Horse clam	82	142	$p > 0.20$
Bent-nose/Sand clams	55.5	168.5	$p < 0.02$
Venus clams	107.5	116.5	$p > 0.20$
Barnacles	101	123	$p > 0.20$
Sea urchin	50	174	$p < 0.01$

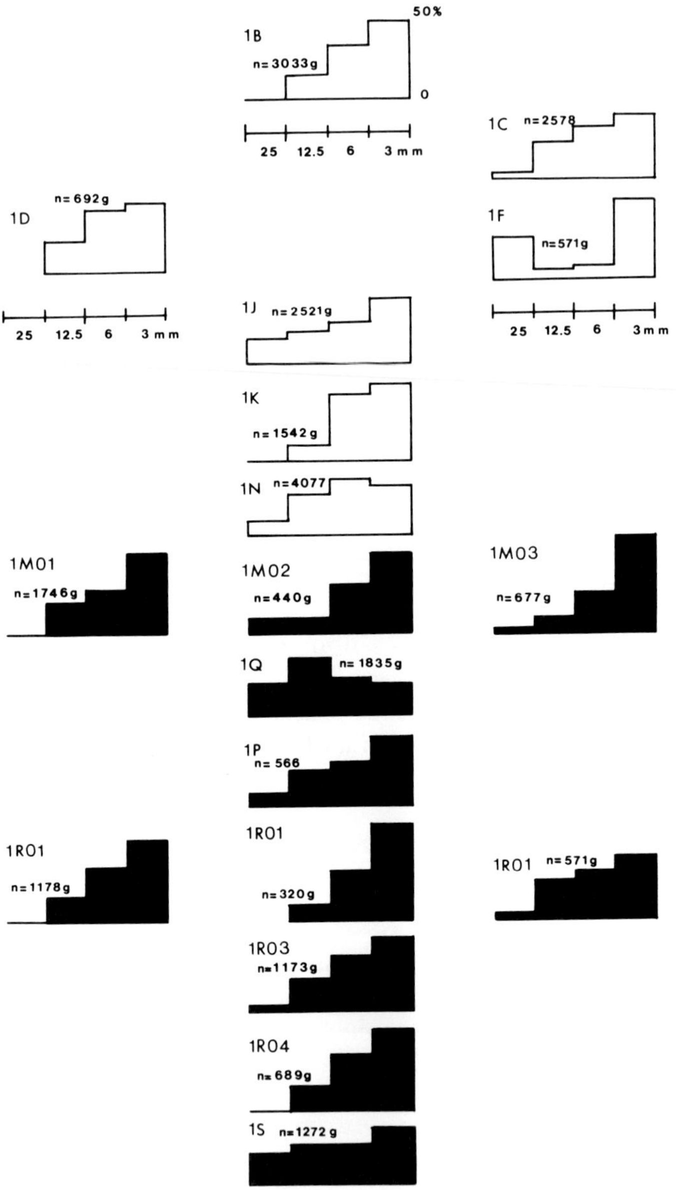

Figure 6 Grain size distributions, all shell, Unit 310/300. The histograms are displayed in the order of facies deposition following the Harris Matrix. They are presented from top to bottom according to increasing depth below surface.

size class includes all shell that passed through the 25-mm mesh but which could not pass through the 12.5-mm mesh; the third size class includes all shell that passed through the 12.5-mm mesh but which could not pass through the 6-mm mesh; and the fourth size class includes all that passed through the 6-mm mesh but which could

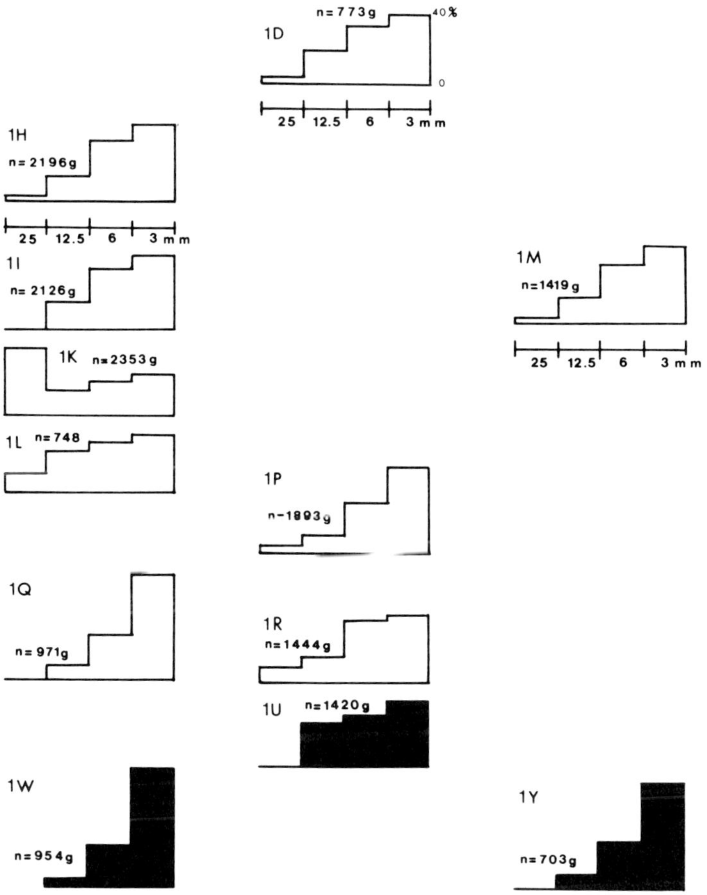

Figure 7 Grain size distributions, all shell, Unit 310/302.

not pass through the 3-mm mesh. The histograms have been displayed in the order of facies deposition following the Harris Matrix (see Figures 8 and 9) (Kornbacher 1989; Stein *et al.*, Ch. 6, this volume).

To assess the relationship between the taxonomic composition of the facies and the grain size distribution for all shell in each facies, grain size distributions were constructed for individual taxa from the facies used in this analysis. These distributions are illustrated as histograms based upon the percent of shell for individual taxa in each facies (by weight) and are displayed in Figures 10–23. The figures are structured in the same order as the Harris Matrix. The taxa described are mussel, cockle, horse clam, bent-nose/sand clams, venus clams, barnacles, and sea urchins. The taxon venus clams (Family Veneridae) includes native little-neck clam (*Protothaca staminea*) and butter clams (*Saxidomus* spp., *S. giganteus*, and *S. nuttalli*). Separate grain size distributions for native little-neck and butter clams were calculated for an additional clarification of the grain size distribution of venus clams (Figures 24–27). Histograms for the taxon Mollusca are also presented in Figures 28 and 29. This

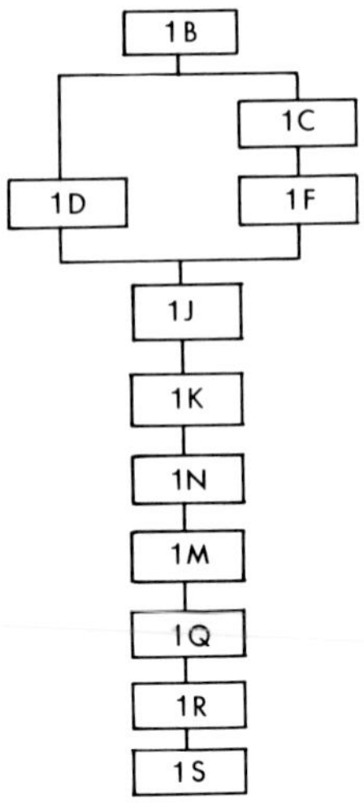

Figure 8 Harris Matrix, Unit 310/300.

taxon consists of all shell identifiable only as mollusk and contains all unidentifiable fragments from gastropods, bivalves, scaphapods, and chitons.

Grain size distribution histograms show little systematic difference between dark and light layer facies for each taxon. However, grain size distribution histograms (Figures 20 and 21) for barnacles within each facies indicate that there is a reduction in the relative abundance of barnacle shell in the 3-mm size class in the light layer facies. The relative abundances for this size class from all facies were ranked and compared across the dark/light boundary using the Mann–Whitney test. The test confirms that there is a significant difference between layers in the relative abundance of barnacle shell falling in the smallest size class ($p < 0.005$).

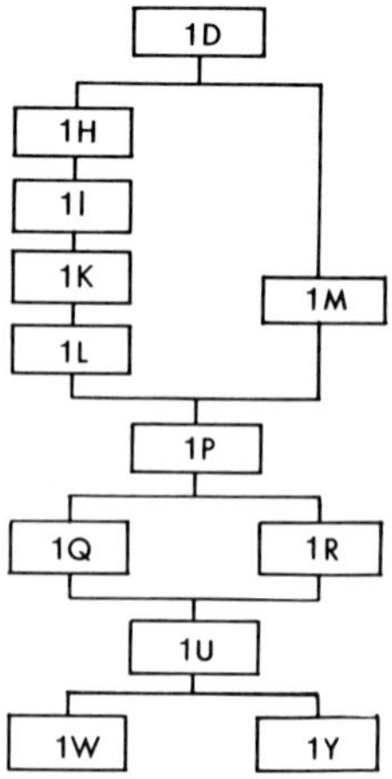

Figure 9 Harris Matrix, Unit 310/302.

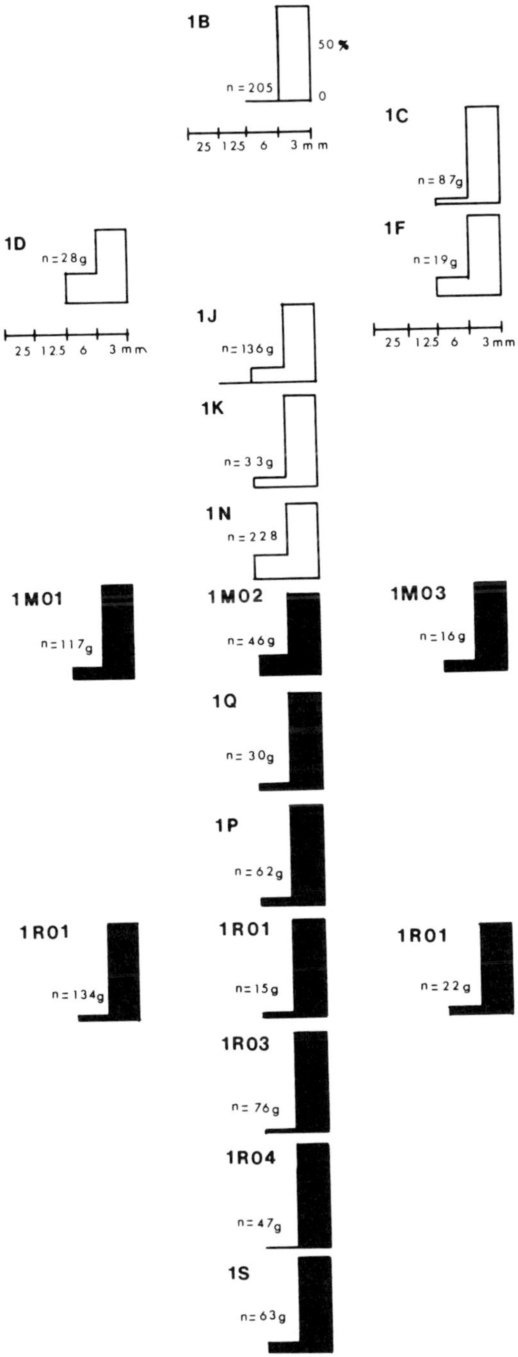

Figure 10 Grain size distributions, mussel shell (Mytilidae), Unit 310/300.

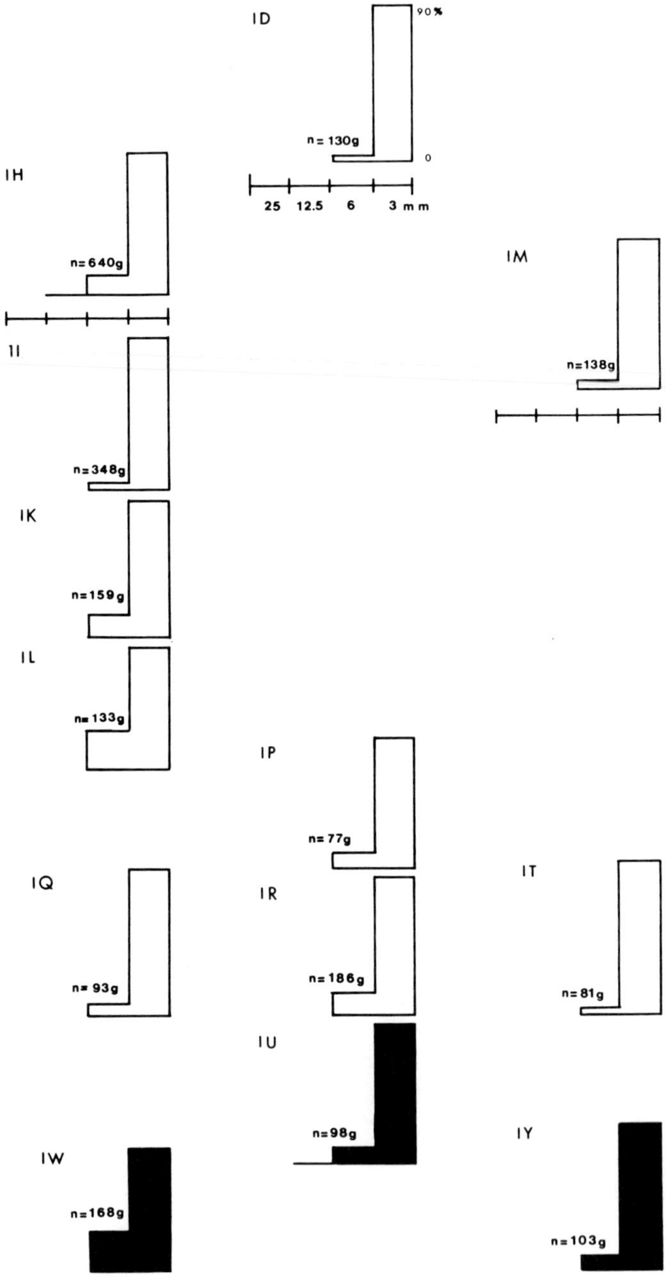

Figure 11 Grain size distributions, mussel shell (Mytilidae), Unit 310/302.

The grain size distributions for all shell within each facies are predominately skewed toward the large size class and exhibit a mode in the smallest size class (see Figures 6 and 7). At 45SJ24, the mode in the 3-mm category for all shell appears in 26 of 30 cases. Analysis of grain size distributions by taxon may allow us to interpret this situation.

When grain size distributions of sediments are separated by their compositional types (in this case, biological taxon), objects that enter the record at various sizes can be identified (Stein 1987; Stein and Teltser 1989). The various shellfish taxa present in these deposits enter the record at different sizes, as seen in the histograms for various taxa (Figures 10–27). Cockle, bent-nose/sand clams, and venus clams enter the record in the largest size class and continue to contribute material to all four size classes. Mussel enters at the 12.5-mm size class and continues to contribute material to smaller size classes. Barnacles enter at the 12.5-mm class and continue to either break or to disarticulate and contribute to the smallest size classes. Sea urchin elements enter the record at the 6-mm class and continue to break and/or disarticulate to contribute to the smallest class.

C. *Relative Abundance and Grain Size Distributions*

I have compared the amount of individual shellfish taxa in a facies separately with the grain size distribution of all shell in the facies in an attempt to assess the relationship, if any, between taxonomic content (composition) and grain size distribution (texture) of the facies. The result of comparisons of the relative abundance of a taxon in the facies (x) and the relative abundance of material in each grain size class for all shell (y) provides a third set of data for this assessment. The results of nonparametric measures of correlation (Spearman's Rank Correlation Coefficient) indicate that significant correlations exist between the relative abundances of certain shellfish taxa and the relative abundances of all shell material within certain grain size classes. The results of comparisons for each taxon and grain size class are described below and summarized in Table 3. To illustrate these comparisons more clearly, consider the variable relative abundance of cockle shell in all facies (Figures 4 and 5). Consider as well variations in the grain size distribution shape for all shell in each facies (Figures 6 and 7). The results of this analysis indicate that there is a relationship between the percentage of cockle shell in a facies and the percentage of all shell that falls in the largest size class in the grain size distribution. The following correlations are shown for specific size classes in the grain size distributions of all shell in each facies and the relative abundance of individual shellfish taxa in each facies: mussel shell and the 3-mm size class ($p < 0.001$); cockle and the 25-mm and 12.5-mm size classes ($p < 0.001$); horse clam and the 12.5-mm size class ($p < 0.002$); bent-nose and sand clams and the 25-mm and 12.5-mm size classes ($p < 0.02$); venus clams and the 25-mm and 12.5-mm size classes ($p < 0.001$); barnacles and the 3-mm size class ($p < 0.001$); and sea urchins and the 3-mm size class ($p < 0.01$) (see Table 3). These correlations will be examined below, following a discussion of the characteristics of shell for the various shellfish taxa involved in this analysis.

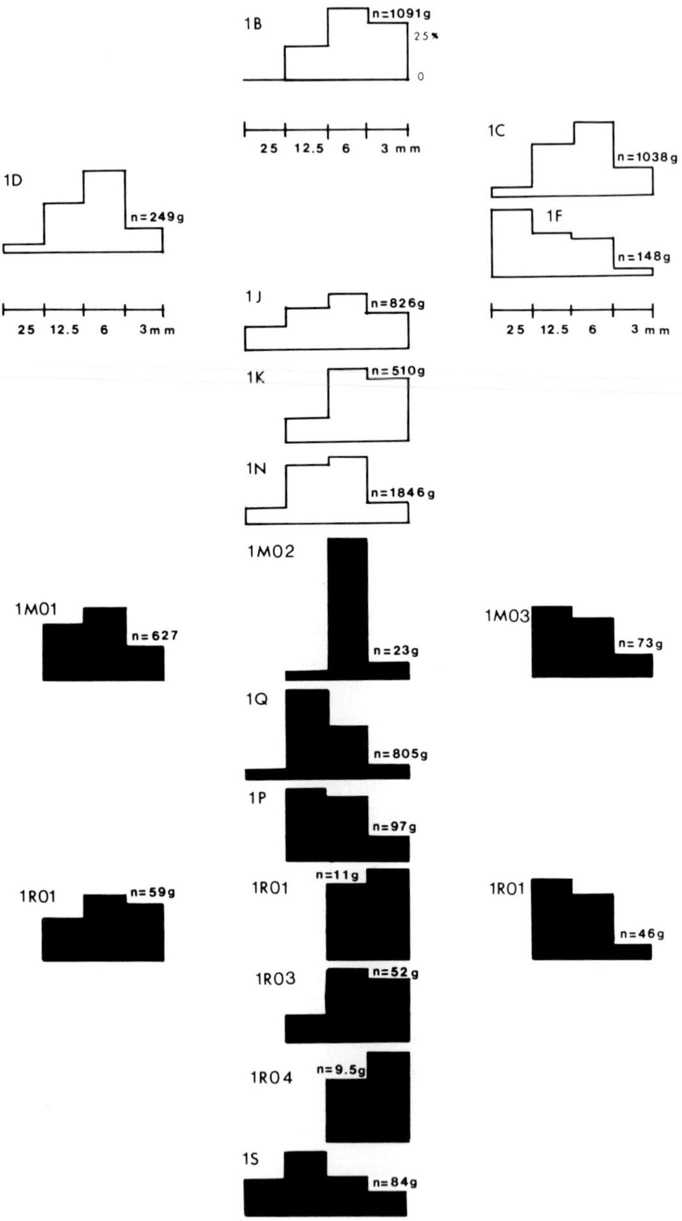

Figure 12 Grain size distributions, cockle shell (*Clinocardium*), Unit 310/300.

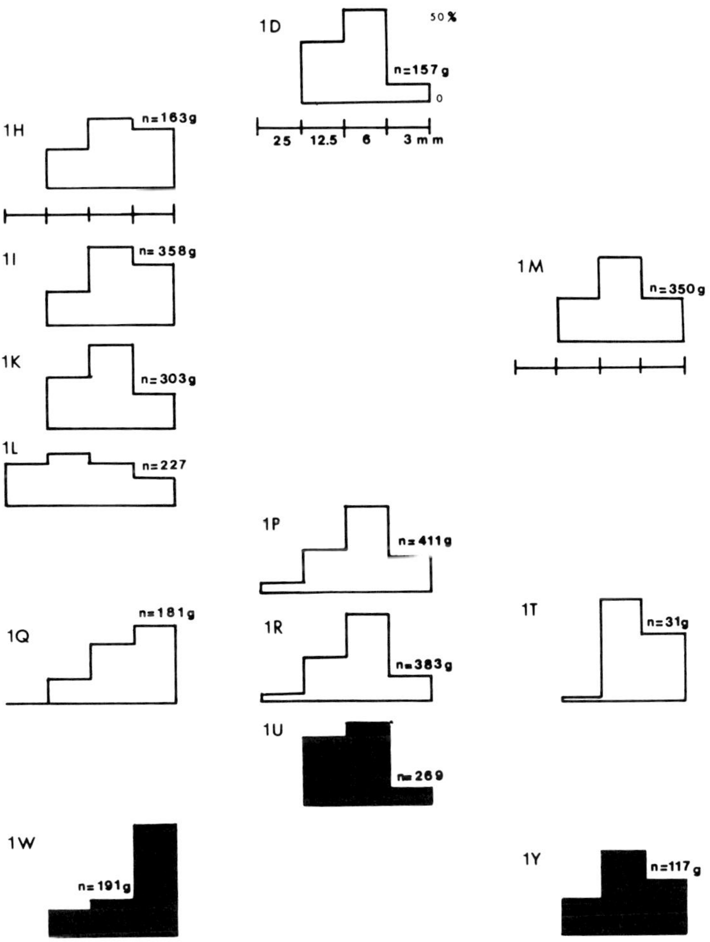

Figure 13 Grain size distributions, cockle shell (*Clinocardium*), Unit 310/302.

V. Summary of Individual Invertebrate Taxa

A. *Mussel (Family Mytilidae,* Mytilus *spp.,* M. edulis*)*

The only species of mussel identified in these deposits is *Mytilus edulis*, but the possibility that certain valve fragments represent *M. californianus* or *Modiolus rectus* cannot be ruled out. All three species are available in the San Juan Archipelago and *M. edulis* is available today in Garrison Bay, immediately adjacent to the archaeological site. Mussel shells have been recovered in every facies but not in the 25-mm mesh. Very few are recovered in the 12.5-mm mesh (see Figures 10 and 11). Mean width for a sample of 85 modern *M. edulis* valves is 35 mm. The minimum width of valves in the modern sample is 16 mm and the maximum width is 70 mm. The mussel valves from these archaeological deposits exhibit the nacreous structure for

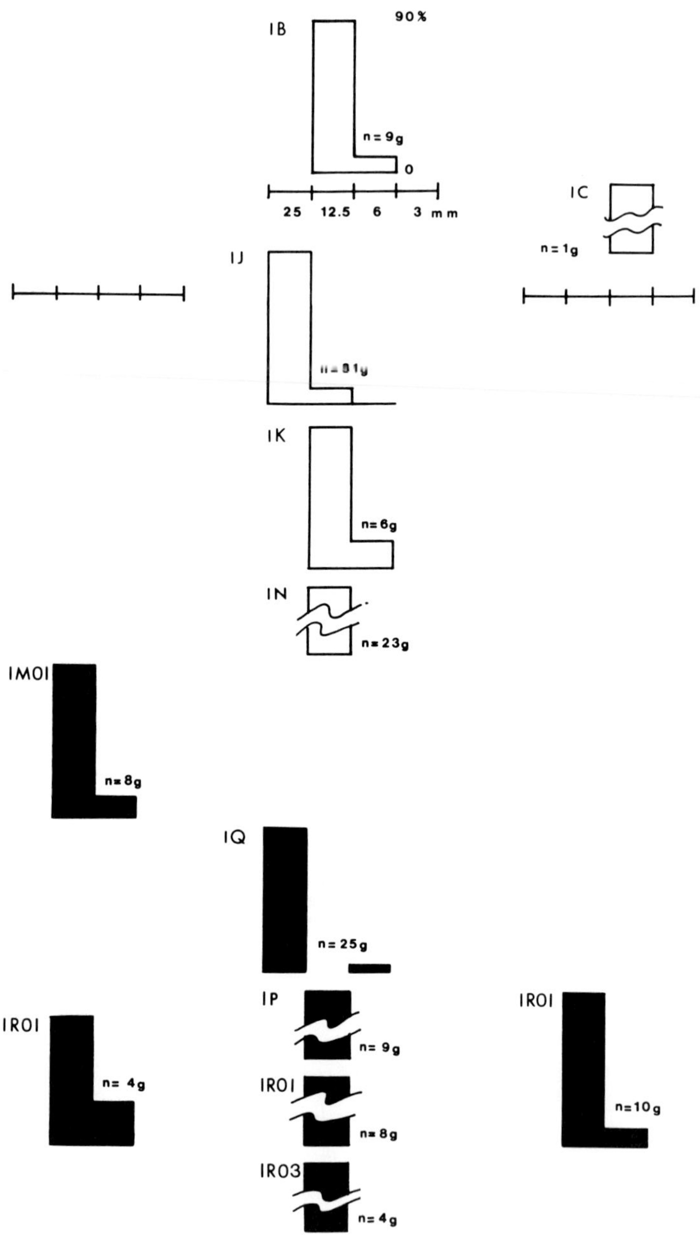

Figure 14 Grain size distributions, horse clam (*Tresus*), Unit 310/300.

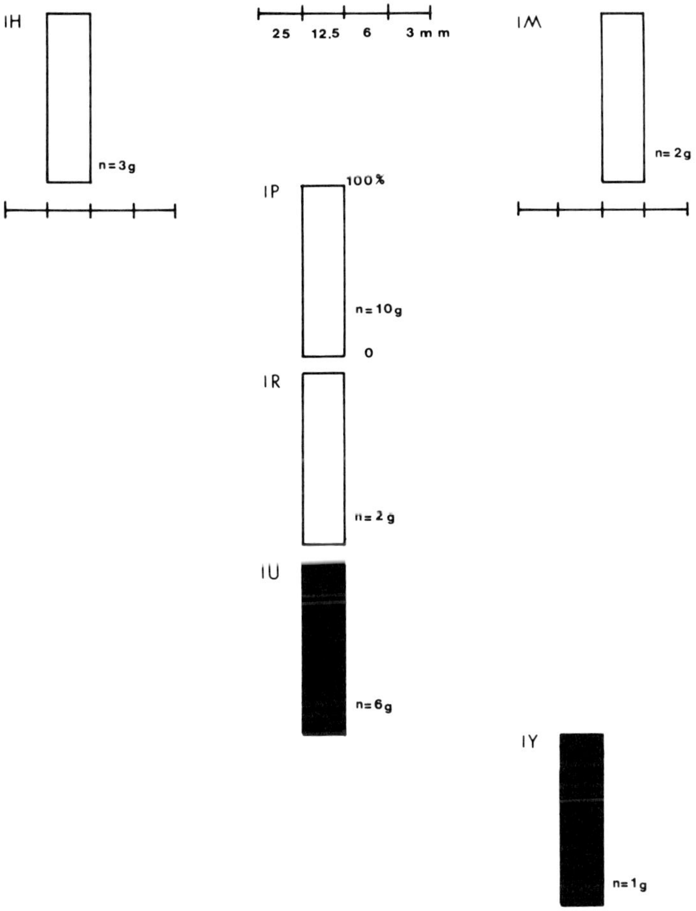

Figure 15 Grain size distributions, horse clam (*Tresus*), Unit 310/302.

which mussel is well known. The layers of nacre forming the valves are thin and brittle. Although losing luster to fire and/or heat treatment, the small thin shell fragments remain identifiable to the family level. Valves are fragile and none has been recovered whole. The abundance of mussel shell in a deposit shows a correlation with the amount of all shell in the 3-mm size class of the grain size distributions.

B. *Cockles* (Clinocardium *sp.*, C. nuttalli)

The basket cockle of the Northwest Coast (*Clinocardium* sp., *C. nuttalli*) is commonly found in the intertidal zone of the protected embayment adjacent to the site. Cockle valves are available in a wide size range. Mean width for a sample of 84 modern cockle valves is 60 mm with a minimum width of 23 mm and a maximum width of 91 mm. The valves are frequently large and contribute to the largest grain size

Table 3 Correlations between Relative Abundance of Individual Taxa and Relative Abundance of Grain Size Classes

Taxon	Grain size (mm)	Spearman's RHO	*p*
Mussel	12.5	−.44	$p > 0.50$
	6	−.006	$p > 0.50$
	3	.507	$p < 0.001$
Cockle	25	.456	$p < 0.001$
	12.5	.466	$p < 0.001$
	6	.014	$p > 0.50$
	3	−.52	$p > 0.50$
Horse clam	25	.219	$p > 0.05$
	12.5	.348	$p < 0.002$
	6	.075	$p > 0.50$
Bent-nose/Sand clam	25	.282	$p < 0.02$
	12.5	.274	$p < 0.02$
	6	.059	$p > 0.50$
	3	−.308	$p > 0.50$
Venus clams	25	.563	$p < 0.001$
	12.5	.555	$p < 0.001$
	6	.04	$p > 0.50$
	3	−.647	$p > 0.50$
Barnacles	12.5	−.44	$p > 0.50$
	6	−.006	$p > 0.50$
	3	.507	$p < 0.001$
Sea urchins	6	−.217	$p > 0.50$
	3	.317	$p < 0.01$

class for many facies but they tend to break easily and are very rarely recovered whole. The radial ridges on the outside of the valve enable genus-level identification of small valve fragments and almost any portion of the shell is identifiable. The abundance of cockle shell in a facies shows a correlation with the abundance of all shell in the two largest size classes, 25 mm and 12.5 mm. The observed increase in cockle shell abundance in the light layer does not appear to be a result of postdepositional processes acting differentially on cockle shell in the dark layer.

C. *Horse Clam* (Tresus *sp.*, T. gapperi)

The horse clam (*Tresus* sp., *Tresus gapperi*) is a local resident of the site area and the San Juan Archipelago, and, although not as abundant as other invertebrates, it is available in the bay adjacent to the site. Horse clam has a large, robust, and thick valve, but, with the exception of the hinge area, characteristic landmarks are absent from many areas of the valve. Consequently, large and small valve fragments are not distinguishable from the valve of the butter clam or the gooeyduck (*Panope generosa*). The horse clam hinge is frequently recovered broken from the rest of the valve. Whole or fragmentary hinges can pass through the larger screens to be recovered in the 6-mm class. All horse clam is identified in the form of either whole valve, hinge, or hinge fragment. No valve fragments without hinge can be identified.

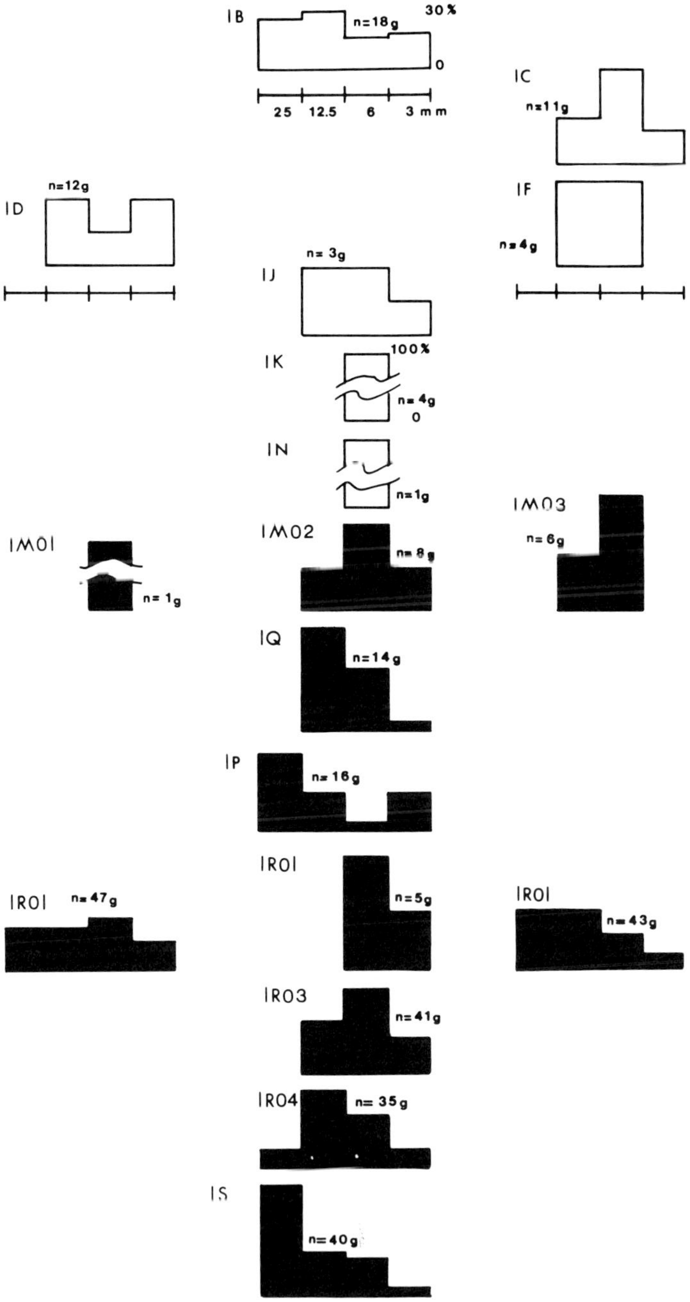

Figure 16 Grain size distributions, bent-nose/sand clams (*Macoma*), Unit 310/300.

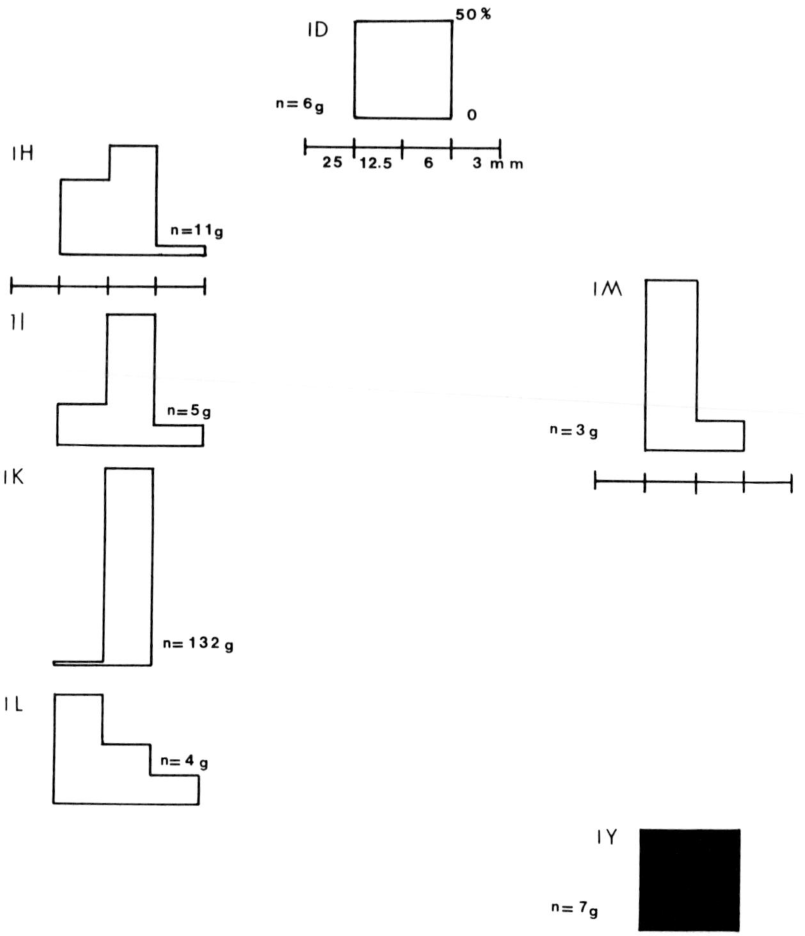

Figure 17 Grain size distributions, bent-nose/sand clams (*Macoma*), Unit 310/302.

The shell is large and heavy, and when present, fragments increase the relative abundance of unidentifiable mollusks in the facies. The relative abundance of horse clam shows a correlation with the amount of all shell falling in the 12.5-mm size class of the grain size distribution.

D. *Bent-nose/Sand Clams* (Macoma *spp.*, Macoma nasuta, M. secta)

The sand clam (*Macoma secta*) and the bent-nose clam (*M. nasuta*) and all other unidentifiable fragments of *Macoma* are included under this taxonomic label. This clam is available in the intertidal zone that abuts the archaeological deposits of British Camp. Although the bent-nose clam has a thin and fairly friable valve, the characteristic tweak of the right valve (the "bent-nose") allows certain valve fragments to be identified to species whether or not the hinge is present. None of the other

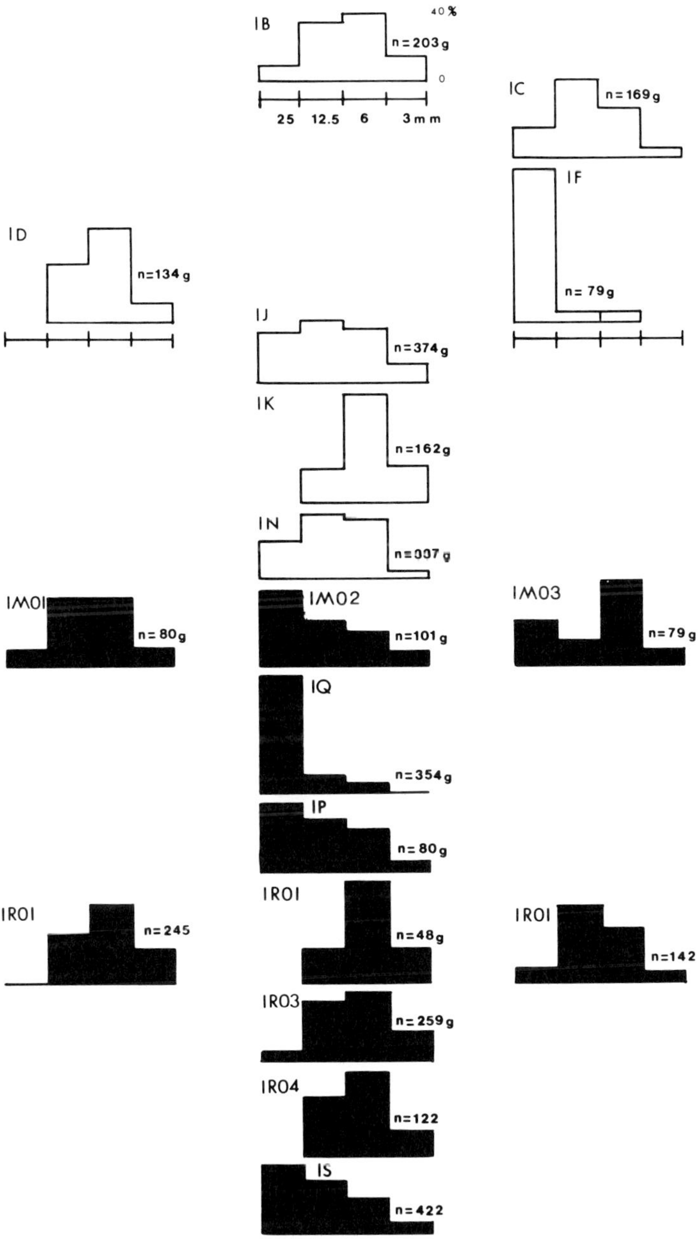

Figure 18 Grain size distributions, venus clams (Veneridae), Unit 310/300.

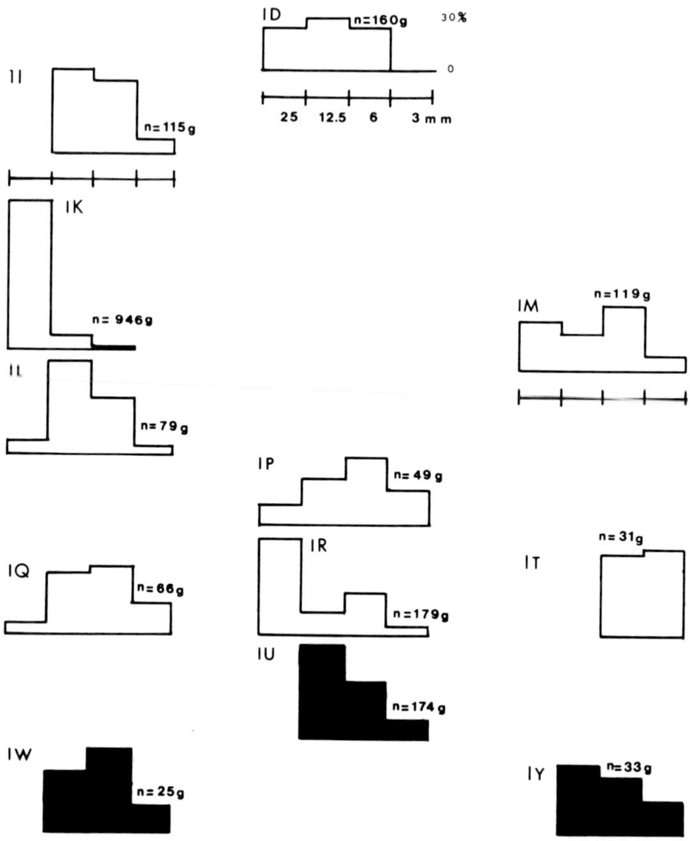

Figure 19 Grain size distributions, venus clams (Veneridae), Unit 310/302.

bivalves (except the cockle) have hingeless valve fragments that are identifiable to species. The hinge itself is fairly small and can be identified to genus in the 6-mm and 3-mm size classes. The relative abundance of this genus shows a correlation with the amount of all shell within the two largest size classes of the grain size distribution. The shell of *Macoma* is very light. The decrease in shell of this taxon in the light layer may be a simple reflection of the problem that haunts all closed arrays: since percentages must sum to 100, it is possible that slight increases in the relative abundances of other taxa (including unidentifiable shell) in the light layer are reflected in a decrease in the relative abundance of this taxon. A larger sample of facies should provide clarification. This taxon is present in 25 of 31 facies.

E. *Venus Clams (Family Veneridae,* Protothaca staminea, Saxidomus *spp.,* S. giganteus, S. nuttalli*)*

There are four species of venus clams identified to date from the British Camp deposits, all of which are locally available in Garrison Bay. These include the little

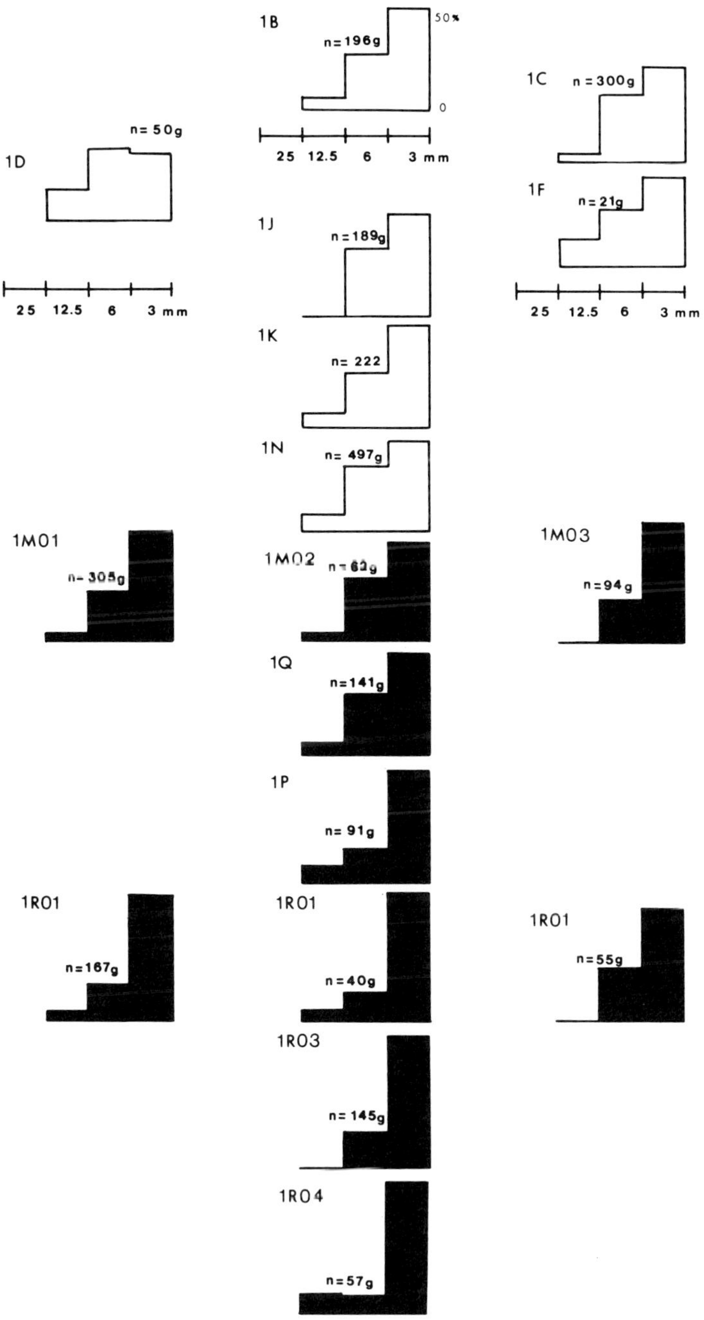

Figure 20 Grain size distributions, barnacles (*Balanus*), Unit 310/300.

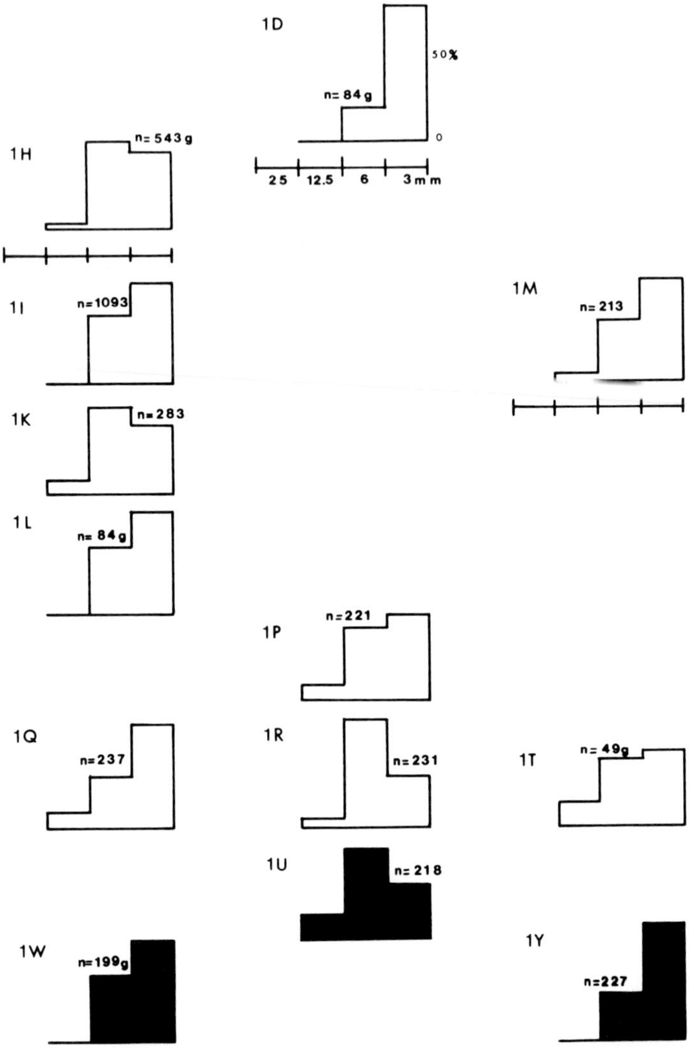

Figure 21 Grain size distributions, barnacles (*Balanus*), Unit 310/302.

transenella (*Transenella tantilla*), native little-neck clam (*Protothaca staminea*), and butter clams (*Saxidomus giganteus* and *S. nuttalli*). Little transenella is so tiny and so rarely recovered that it has not been differentiated for this discussion. The native little-neck clam is robust and has a wide size range, although it does not get as large as the butter clam (see below). The mean width for a sample of 72 modern native little-neck clam valves is 45 mm. The minimum width for the sample is 22 mm, the maximum width 74 mm. This taxon is abundant within the midden. The butter clam valves are very robust and attain great width. The mean width for 68 modern *Saxidomus giganteus* is 80 mm with a minimum width of 54 mm and a maximum width of

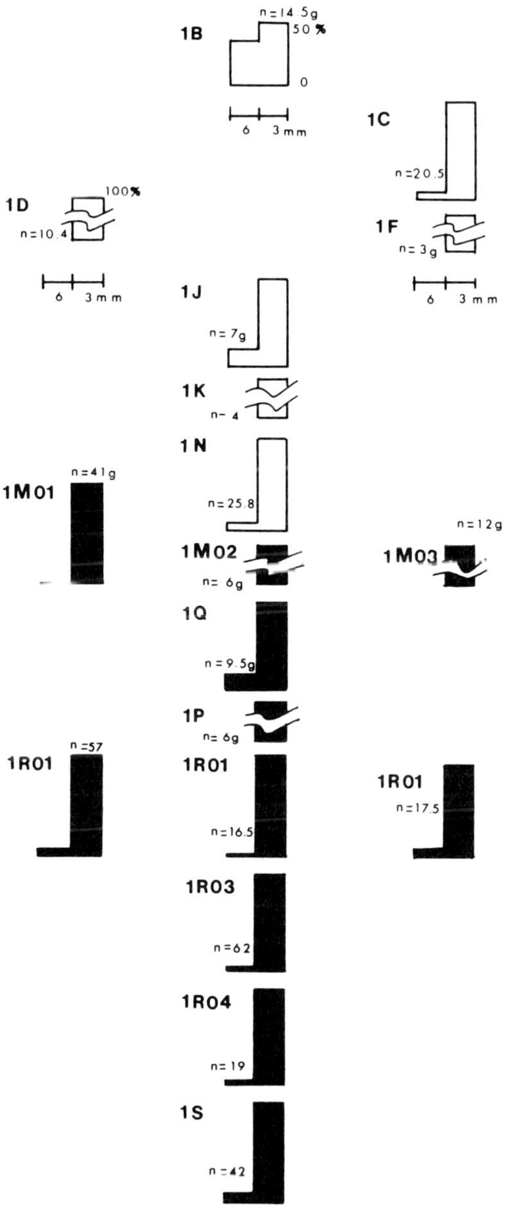

Figure 22 Grain size distributions, sea urchins (*Strongylocentrotus*), Unit 310/300.

117 mm. *S. nuttalli* is generally smaller and less robust than *S. giganteus*. Both whole valves and valve fragments containing most of the hinge are identifiable, at least to the genus level. If the butter clam is fragmented enough to pass through the 12.5-mm mesh, the fragments usually lack identifying landmarks. The valve does not

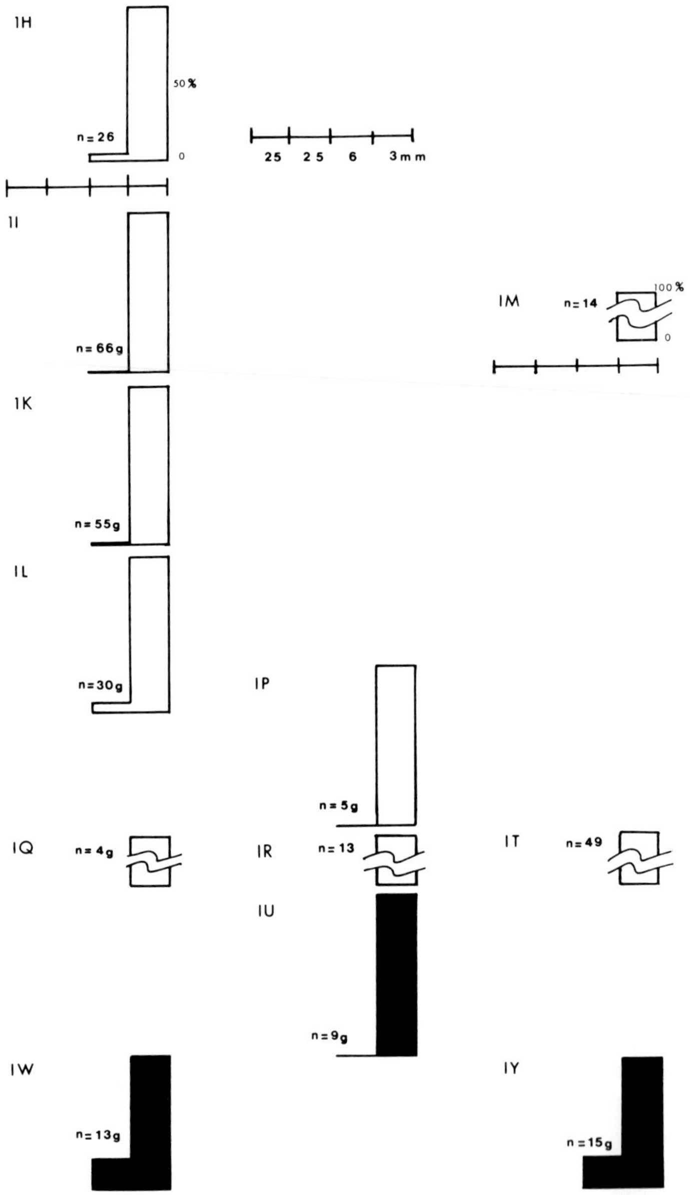

Figure 23 Grain size distributions, sea urchins (*Strongylocentrotus*), Unit 310/302.

have radial markings so when the hinge is not present, identification beyond the class level is precluded. The relative abundance of all venus clams shows a correlation with the amount of all shell in the two largest grain size classes. To understand grain size distributions further, the individual contributions of native little-neck and/or butter clam may be examined. Grain size distribution histograms are presented for all venus

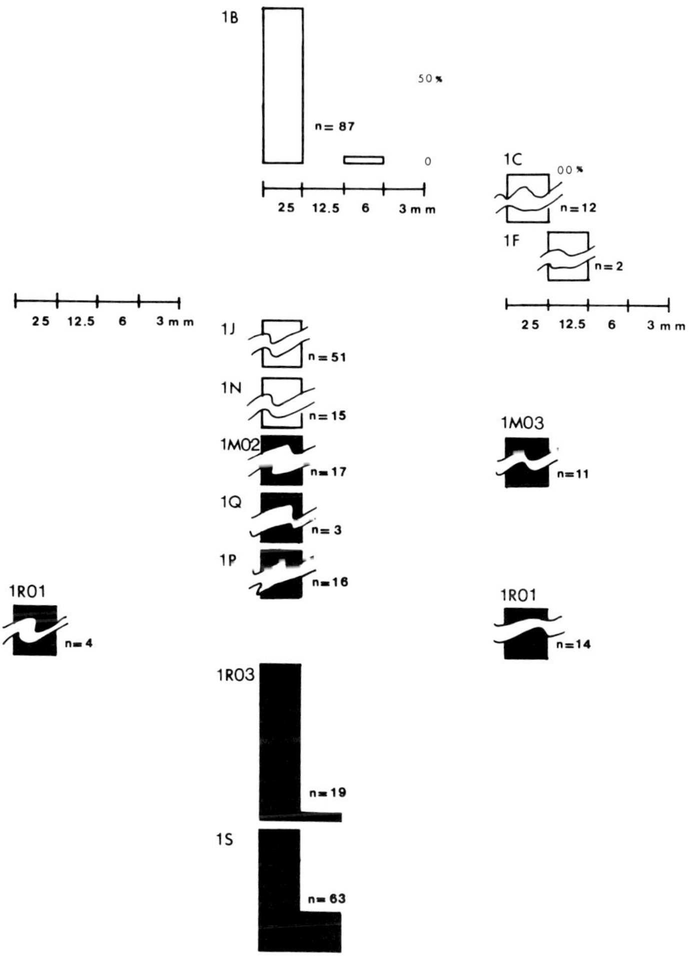

Figure 24 Grain size distributions, native little-neck clam (*Protothaca staminea*), Unit 310/300.

clams (Figures 18 and 19) and separately for those specimens identified as native little-neck clams and as butter clams (Figures 24–27).

F. Barnacles (Balanus *spp.*)

The barnacles (*Balanus* spp.) recovered at this site are available from the Garrison Bay intertidal zone and are represented by basal, body, and opercular plates. These plates are separated from each other once soft tissue breaks or decays. The shape of basal plates and body plates for each individual barnacle depends upon the conditions in which a barnacle lived (substrate, amount of crowding, and wave action). Consequently, body plates may be long and thin, short and stubby, or very irregular in shape. Basal plates are fragile and rarely recovered in entirety. Opercular plates are very small and usually recovered within the 3-mm size class. Body plates may

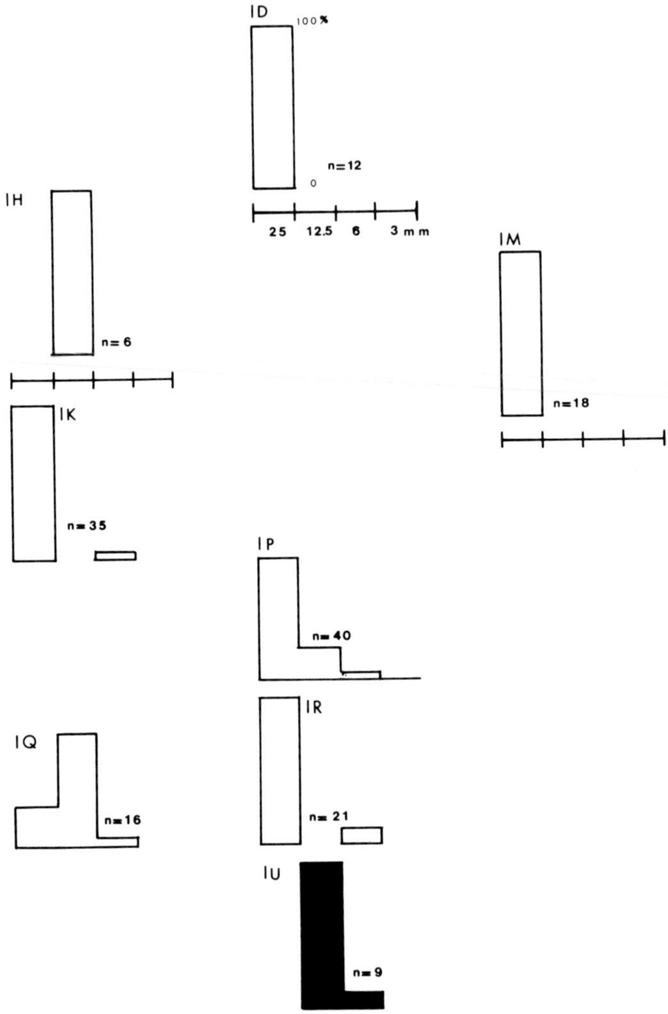

Figure 25 Grain size distributions, native little-neck clam (*Protothaca staminea*), Unit 310/302.

appear in any size class. Barnacle fragments are abundant and the taxon is present within all facies. The percentage of barnacle shows a correlation with the amount of all shell falling in the 3-mm size class of the grain size distribution. The observed decrease in amount of barnacle in the 3-mm class in the light layer may be the result of differential postdepositional processes in the dark and light layers. The small size of opercular plates and the wrinkled surface of body plates provide a relatively large amount of surface area for fragments of this taxon. This allows for increased exposure to organic acids in solution and etching may occur more rapidly on these than on other shell fragments. This would cause barnacle elements in the dark layer to lose surface material, thereby becoming smaller in size and increasing the portion of barnacles recovered in the smallest size class.

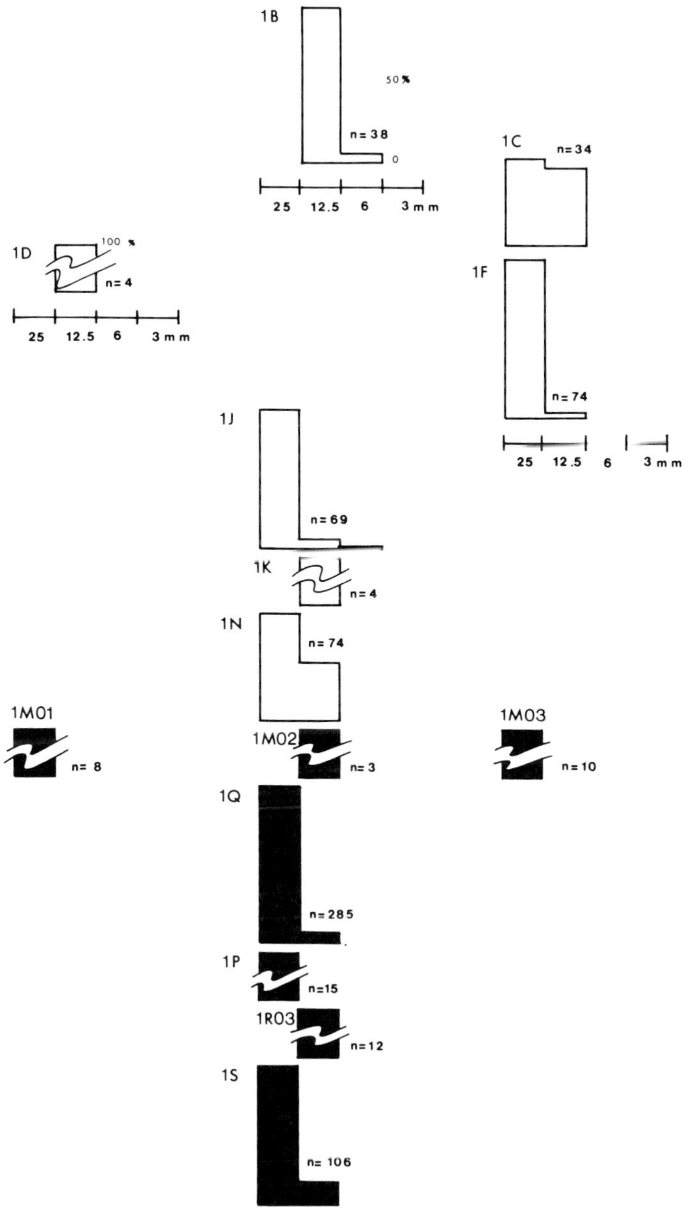

Figure 26 Grain size distributions, butter clam (*Saxidomus*), Unit 310/300.

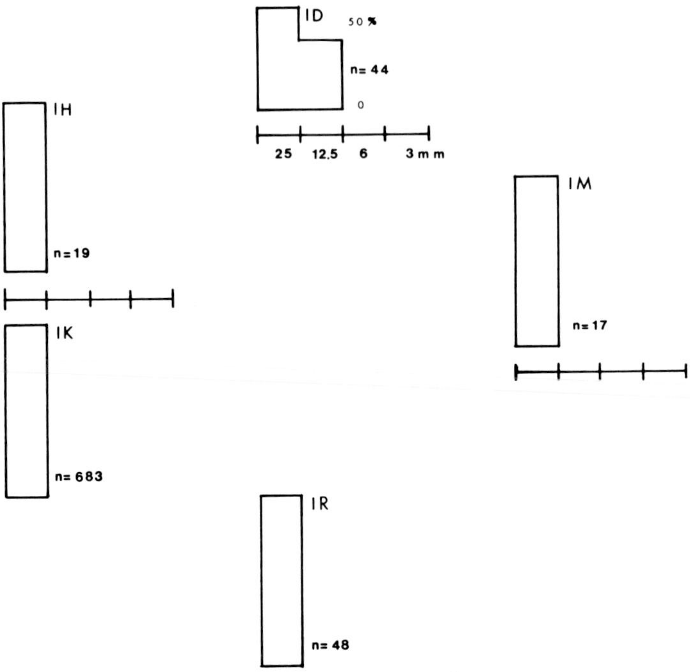

Figure 27 Grain size distributions, butter clam (*Saxidomus*), Unit 310/302.

G. *Sea Urchin* (Strongylocentrotus *spp.*)

There are four species of sea urchins (*Strongylocentrotus droebachiensis*, *S. franciscanus*, *S. pallidus*, and *S. pupuratus*) in the San Juan Archipelago, but Garrison Bay and Westcott Bay do not provide appropriate habitat for this animal. The tests of sea urchins are fragile, have a characteristic breakage pattern (Smith 1984), and are always recovered from these archaeological facies in fragmented condition. Even if the urchin test entered the record whole (and thus greater than 25 mm in size), it would not remain intact in archaeological deposits (Smith 1984). The spines and jaw elements are separated from the test once soft tissue has decayed. Fragments of sea urchins in archaeological deposits are rarely identifiable to the species level (Norman 1985). The comparison of sea urchin abundance with the grain size distribution for all shell indicated a significant correlation exists for the 3-mm size class. As I have discussed, relative abundance of sea urchin remains in the upper light layer does not appear to be a function of downward movement of these cylindrical elements. It may instead be related to a change in species. *S. franciscanus* is notably larger than *S. droebachiensis* or *S. pallidus*. The addition of this species would be reflected by an increase in materials in the 6-mm size class. Since the data is presented as relative abundances (percentages), a closed array exists and any increase in abundance in the larger size class will cooccur with a decrease in abundance in the smaller size class.

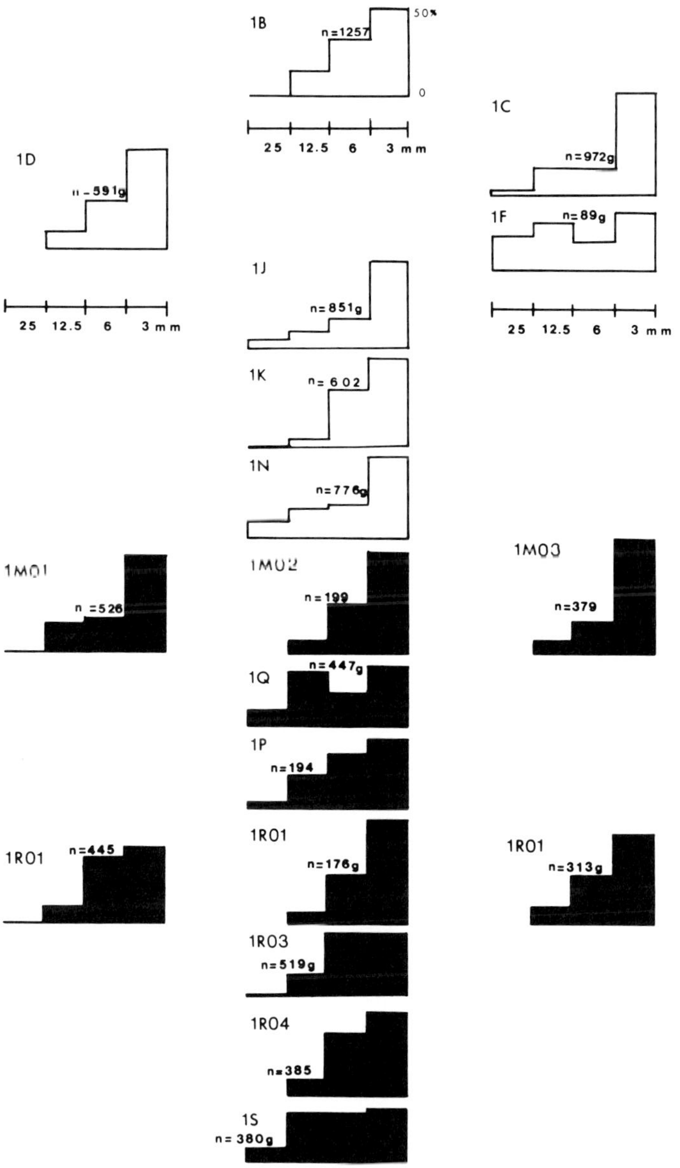

Figure 28 Grain size distributions, unidentifiable mollusks (Mollusca), Unit 310/300.

VI. Discussion

A. *Biological Source*

There are two biological sources for the shell in the excavation units and facies examined. Most of the taxa identified in the archaeological facies are taxa that

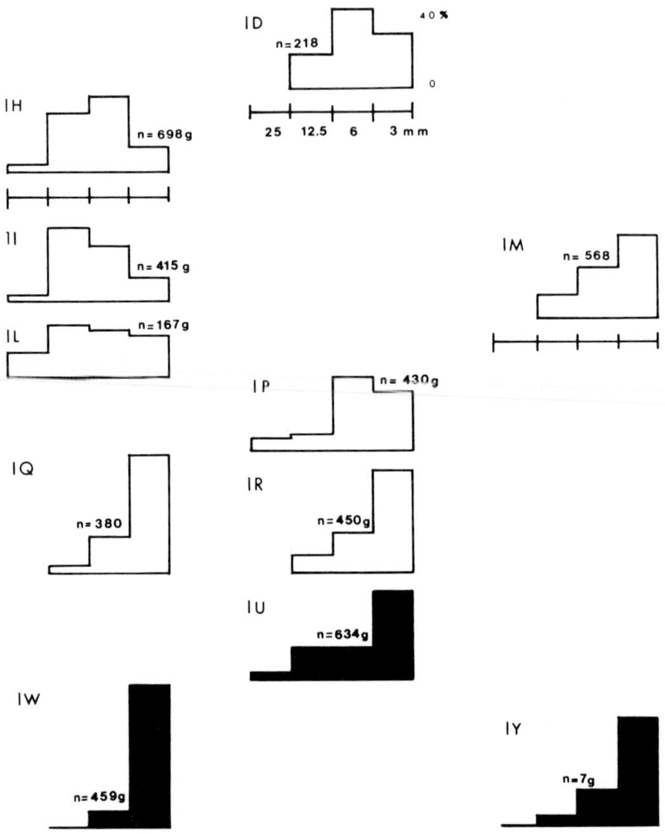

Figure 29 Grain size distributions, unidentifiable mollusks (Mollusca), Unit 310/302.

occupy protected embayments in which wave action is low or minimal. Some of these taxa occupy a sandy substrate and some require rocky substrate but both sand and rock are found in most of the protected embayments in the region including those on San Juan Island. All but one of the taxa discussed could have been harvested at adjacent Garrison Bay and/or nearby Westcott Bay and/or any of several similar settings on this or any of the other islands of the San Juan Archipelago.

Sea urchins do not occupy the same quiet-water habitat. Sea urchins require a rocky substrate and exposure to wave action. Barnacles and mussels can share this habitat but the archaeological mussels identified to date belong to a species that does not require it to the same extent as the urchins. On San Juan Island today, sea urchins are seen at a few open coast locations such as Cattle Point. The archaeological sea urchins from British Camp always appear in facies along with cockles, mussels, venus clams and barnacles. Urchins, in fact, are present in all but one of the facies examined and represent a second biological source. This combination of shellfish indicates a minimum of two biological sources for the materials in 29 of 30 facies: the intertidal zones of the rocky coast and protected embayments.

B. *Influence of Biological Characteristics on Grain Size Distributions*

It is necessary to identify natural characteristics of shell that may influence the shape of grain size distributions. Several significant correlations were produced when the percentages of each taxon in a facies were compared with the percentages of all shell in each grain size class. The kind of shellfish present in a facies has an observable affect on the structure of grain size distributions. For instance, mussel, which is brittle and easily broken, is present in higher abundances where the amount of all shell in the smallest size class is high. The assumption on which this study is based is that variable characteristics of robusticity, shell size, friability, and content influence shell breakage patterns and ultimately influence the shape of grain size distribution histograms. If only one shellfish taxon were present at this site, then any variation in grain size distribution would reflect variation at the sedimentological source rather than variation at the biological source.

C. *Sedimentological Source*

If the influence of taxon-specific characteristics on the texture of facies can be held constant, then sedimentological sources for shells in archaeological deposits can be recognized. The biological source is consistent and most of the grain size differences between facies can be explained by variations in abundance of taxa or in condition of shell at the sedimentological source. The biological source of shell plays a major role in determining which materials are available for harvest. Cultural practices determine the condition of the shell (as sediments) when they are finally deposited. The range of possible cultural practices operating on shell is infinite and the particular practices may not be identifiable in archaeological shell, but there are differences between the sedimentological attributes of shell in facies that are recognizable (see Figure 1).

D. *Individual Facies and Their Grain Size Distributions*

Of the grain size distributions for all shell in 30 facies, 26 histograms are predominately skewed to the largest size class (Figures 6 and 7). This can be explained in part by the presence of large amounts of small and/or friable taxa (mussel, barnacle, and sea urchin, but especially barnacle) whose presence is correlated significantly with the amount of shell in the smallest size class. The relative abundance of barnacle shell in the analyzed facies varies from 30 to 79%. The next most abundant taxon is cockle, which is also very friable. In most facies, the shells are predominantly broken, thus contributing higher amounts to the smaller size classes than they would if they remained whole.

Four facies exhibit grain size distributions that differ from the majority. The shape of the histogram for all shell in 310/3001F is controlled by the relative abundances of individual taxa within the facies (Figure 4). The histograms for both cockle and venus clams exhibit modes in the largest size class (Figures 12 and 18). Cockle makes up 54% of all shell in the 25-mm class and venus clams make up 46% of the

25-mm class. Although none of the valves recovered from this sample were recovered whole, fragments of these two taxa were large enough and robust enough to remain in the 25-mm size class in spite of processes affecting other shell. The shells identified as venus clams in facies 310/3001F include butter clam in the 25-mm class (Figure 26) and native little-neck clam in the 12.5-mm class (Figure 24). The histogram for unidentifiable mollusks shows a mode in the 3-mm class and a second mode in the 12.5-mm class (Figure 28). Since butter clam valves can break in such a way that large fragments lacking identifying landmarks are created, it is likely that the relatively large amount of unidentifiable material in the 12.5-mm class is from butter clam. None of these valves are whole, but even when broken the large size of venus clam and cockle valves can dominate large size classes. Additionally, fragments of butter clam that would pass through the 25-mm mesh to contribute to the 12.5-mm size class have not been identified because valve fragments of this taxon are rarely identifiable without specific landmarks. Based on taxonomic composition, the biological source for this facies is similar to that of other facies. It is an unusually large amount of shell consisting of cockle and butter clam in the 25-mm size class that makes the grain size distribution differ from that of other facies. This amount of cockle and butter clam could reflect conditions of the biological source (e.g., an abundance of these taxa or accessibility) or conditions that appear at the sedimentological source (e.g., separation of taxa during treatment).

Facies 310/3001N is the only facies to exhibit a mode in the 6-mm size class for all shell (Figure 6). Cockle and the bent-nose/sand clams are the individual taxa represented in histograms exhibiting modes in the 6-mm class in this facies (Figures 12 and 16). There are 4077 g of shell in the biological sample of which 1290 g (32%) fall in the 6-mm class (Figure 6). Of the 6-mm shell, 1 g is identified as bent-nose/sand clams and 759 g (59%) are cockle shell. Although the relative abundance of cockle shell did not show a correlation with the relative abundance of shell in the 6-mm class in earlier comparisons, it is apparent that for this facies, the presence of cockle and its condition (broken to this size) have a strong effect on the grain size distribution of shell. The biological source for this facies is also similar to that for all other facies but in other facies cockle appears in the larger size classes. Because most of it occurs in small-sized pieces, the cockle shell in this facies appears to have been treated differently than that of other facies between the time of harvest at the biological source and incorporation in the depositional episode. Treatment variations could include, for example, different food processing techniques or different uses for shell in which it is broken into smaller pieces. The grain size distribution for cockle differs from those of other facies and indicates that the sedimentological source for cockle is different in this facies than it is in other facies.

Facies 310/3021K resembles other facies with a skew towards the larger size classes, but it has a mode in the 25-mm size class (Figure 7). Histograms for individual taxa from this facies indicate that venus clam is the only taxon present exhibiting a mode in the 25-mm class (Figure 19). There are 909 g of shell in the 25-mm class. Of these, 829 (91% of the size class) are venus clams. These include both native little-neck clam and butter clam (Figures 25 and 27). Whole valves make up 61% of the material in the 25-mm class. Since whole valves account for most of the material in the 25-mm class for this facies, there are few fragments of large valves to contribute material to the 12.5-mm size class. Native little-neck clam was identified in the 25-mm class, but none appeared in the 12.5-mm class (Figure 25). A small amount appears in the 6-mm class (Figure 25). Butter clam was identified in the largest size class but not in smaller size classes (Figure 27). It is interesting that 24% of the shell in this biological sample consists of whole valves from bivalve mollusks. For most facies at this site, the relative abundance of whole valves in a sample is 1% or less. Additionally, the mode for barnacles, usually in the 3-mm size class, lies in the 6-mm class for facies 310/3021K (Figure 21). The large amount of whole valves and the unusual 6-mm class mode for barnacles suggest that the material in this facies comes from a different sedimentological source than does the material in all other facies. The taxa are not unusual but the shell in this facies has not been subjected to the same set of processes between harvest and deposition that have operated upon shell observed in the other facies. The fact that so many mollusks are whole and so many barnacle fragments are relatively large suggests that little has happened to these materials between the time they were harvested at the biological source and deposited in what is now an archaeological facies. The differences between this and other facies suggest that postharvest factors and repeated sediment transport and deposition (e.g., cooking, construction, dumping) contributed to the grain size distributions observed in all other facies. Because the kinds of shellfish remain constant throughout the excavation units, the kinds of intertidal and/or subtidal zones utilized by island inhabitants for shellfish gathering have remained the same for the facies included in this analysis.

The grain size distribution for all shell from Facies 310/3001Q varies from those of other facies in that its mode is in the 12.5-mm size class for all shell and it is skewed toward the smallest size class (Figure 6). It contains small amounts of the taxa whose abundances correlate with the abundance of material in the 3-mm class: 2% of the shell from mussel, 8% from barnacles, and less than 1% from sea urchins. There are 652 g of shell in the 12.5-mm size class. Of that, 438 g (67%) are cockle shell. Predictably, the high amount of cockle shell and the low amount of small-sized materials influence the structure of the grain size distribution. Facies 310/3001Q appears to be an intrusive pit and less than 1% of the bivalve shell is made up of whole valves. Most shell in this facies falls in the large size classes. The biological source for this facies is similar to that of most other facies except for the low relative abundance of those taxa that correlate with abundance in the smallest size classes (mussel, barnacles, and sea urchins). The sedimentological source provided a large amount of large size materials for this deposit.

VII. Summary and Conclusions

Each of the British Camp facies examined in this study contains shellfish taxa that are common throughout the San Juan Archipelago. More importantly, a consistent set of locally available taxa (mussel, cockle, horse clam, venus clam, and barnacle) is present in each facies. In 29 of 30 facies studied, sea urchin is also present and indicates that non-locally available resources are included in the facies. The presence of locally available taxa and the non-locally available urchin indicates that a minimum of two biological sources are shared by 29 of 30 facies. Grain size distributions also indicate a similarity in sedimentological source, because 26 of 30 distributions for all shell are of the same general shape and modality. The fact that most of these shells entered the archaeological record broken and small reflects the kinds of treatments to which shellfish were subjected prior to incorporation in the most recent depositional episode: heating and shucking (for food), dumping (for use in pathways, drainage, building stabilization, building insulation, or as garbage), and/or trampling (in garbage piles or on pathways). Examination of the grain size distributions of the other four facies indicates that three of these facies contain a similar set of taxa but that large numbers of large-size shell have skewed the grain size distributions. The biological source for those facies is the same as that of the other 26, but relative abundances of the taxa differ. Facies 310/3021K is unusual in comparison to all of the other analyzed facies because 26% of the shells in the facies are whole valves.

Most facies (26 of 30) appear to share the same sources, both biological and sedimentological, because they contain the same set of taxa and the shells exhibit similar grain size distributions. Three of the 30 facies contain the same set of taxa as the other 26 but in differing distributions: they share the same biological source but not the same sedimentological source. The last facies (310/3021K) is different from the others in grain size distribution and amount of whole valves (sedimentological source differs) but is similar to the others in the kinds of taxa it contains (biological source is similar).

There are significant differences in the relative abundances of urchin, cockle, and bent-nose/sand clams between the dark and light layers. These differences at first suggest cultural differences between the dark and light layers (e.g., cultural preference) but variable relative abundances of taxa may also represent vagaries in biological source at the time of harvest. All three taxa remain present in both dark and light layer facies, and the grain size distributions of cockle and bent-nose/sand clams do not vary between the layers. There is a change in abundance of material in the 3-mm size class between dark and light layers for urchin. This could be due to a minor change at the biological source. Before concluding that changes in relative abundance are related to the change in stratigraphy between the dark and light layers, additional areas of the site need to be examined. A larger sample of facies and wider areal coverage will help to illustrate the relationship between taxa present and sedimentological and biological sources.

For three taxa (cockle, barnacle and urchin), the possibility of postdepositional alteration has been examined. Barnacle data suggest the loss of surface features and size reduction in dark layer facies. Greater percentages of barnacle shell are found in the 3-mm size class for light layer facies than for dark layer facies. Not only do barnacle shells enter the record in small pieces (either broken or disarticulated) and continue to disarticulate as soon as soft tissue decays, but the body and basal plates are irregularly shaped and more susceptible to rapid loss of surface area. Loss of surface area leads to reduced size which could result in the observed difference in this taxon between the dark and light layer facies. The presence of groundwater and organic acids may be responsible for etching of barnacle shell surfaces leading to size reduction in barnacle shells in the dark layer.

More cockles appear in the light layer facies than in the dark layer facies. Examination of relative abundance of species-level versus genus-level identifications indicated that identifiability does not vary between layers, so postdepositional factors are not the cause of this difference. More sea urchins appear in the dark layer facies than in the light layer facies. Analysis indicates that this is not the result of postdepositional downward movement of cylindrical skeletal elements.

Examination of postdepostional processes on three taxa ruled out such processes as an explanation for differential abundances of the taxa in dark and light layer facies for two of the three taxa. This observation cannot be taken as an indicator that postdepositional processes do not act on this midden, however. In fact, it is possible that the grain size distributions for particular taxa or particular facies are greatly influenced by some postdepositional processes. Until additional taxon-specific tests are devised and executed for the purposes of recognizing the existence and extent of postdepositional processes, it will not be possible to separate all of the potential effects of postdepositional processes from the conditions of shell at the sedimentological source.

The research reported here illustrates that the taxonomic composition of shellfish remains for each of these archaeological deposits correlates with the shape of the grain size distributions of individual shellfish taxa for those deposits. The biological and sedimentological sources for the shellfish have produced facies in both the light and dark layers of the site that look alike. With little exception, the biological source is consistent for the dark and light layer facies. The sedimentological source shows some variation (one facies) and there are three facies in which extant variation could be due to variation in sedimentological or biological source. The information presented here is consistent with the explanation that the stratigraphic differences between the dark and light layers are mainly due to groundwater and organic acid activity and not changes in cultural activities or differences in taxonomic content. As Stein and Teltser (1989) determined with the analyisis of other materials, the data and technique reported here demonstrate that the analysis of grain size distributions of artifacts is a powerful tool to use in the investigation of site-formation processes. This tool has provided insights into the history of archaeological facies at British Camp.

Acknowledgments

The excavated materials that form the basis of this analysis were curated by C. Beck, E. Dennis, K. Kornbacher, A. Linse, and M. L. Parr. K. Kornbacher and J. Tyler prepared the Harris Matrix for each of the units. L. K. Alford provided the photograph for Figure 1. D. K. Grayson provided very helpful comments on earlier versions of this paper. J. K. Stein contributed substantially to my understanding of site-formation processes, and her persistent and effective editing skills have significantly improved the earlier versions of this chapter.

References

Abbott, R. T.

1974 American seashells. New York: Van Nostrand Reinhold.

Bernick, K.

1983 A site catchment analysis of the Little Qualicum River Site, DiSc 1: a wet site on the east coast of Vancouver Island, B.C. Archaeological Survey of Canada Paper No. 118. National Museum of Man, Ottawa.

Blatt, H. G., G. Middleton, and R. Murray

1980 *Origin of sedimentary rocks.* Englewood Cliffs, New Jersey: Prentice-Hall.

Carlson, C.

1979 The early component at Bear Cove. *Canadian Journal of Archaeology* **3**, 177–194.

Carlson, R. L.

1979 The early period on the central coast of British Columbia. *Canadian Journal of Archaeology* **3**, 211–227.

Chave, K. E.

1964 Skeletal durability and preservation. In *Approaches to paleoecology*, edited by J. Imbrie and N. Newell. New York: John Wiley & Sons. Pp.377–387.

Conover, K.

1978 Matrix analysis. In *Studies in Bella Bella Prehistory*, edited by J. J. Hester and S. Nelson. Simon Fraser University, Department of Archaeology Publication 5, 67–100.

Driscoll, E. G., and T. P. Weltin

1973 Sedimentary parameters as factors in abrasive shell reduction. *Palaeogeography, Palaeoclimatology, Palaeoecology* **13**, 275–288.

Dunnell, R. C., and J. K. Stein

1989 Theoretical issues in the interpretation of microartifacts. *Geoarchaeology: An International Journal* **4**, 31–42.

Ford, P. J.

1989 Molluscan assemblages from archaeological deposits. *Geoarchaeology: An International Journal* **4**, 157–173.

Ham, L. C.

1976 Analysis of shell samples from Glenrose. In *The Glenrose Cannery sites*, edited by R. G. Matson. National Museum of Man, Archaeological Survey of Canada No. 52. Pp.42–78.

1982 Seasonality, shell midden layers, and Coast Salish subsistence activities at the Crescent Beach Site, DgRr1. Ph.D. dissertation, Department of Anthropology and Sociology, University of British Columbia, Vancouver, B.C.

Hamblin, N. L.

1984 *Animal use by the Cozumel Maya.* Tucson: University of Arizona Press.

Kornbacher, K. D.

1989 A methodological stride in shell midden archaeology. Paper presented at the 54th Annual Meeting of the Society for American Archaeology, Atlanta.

Luebbers, R.

1978 Excavations: stratigraphy and artifacts. In *Studies in Bella Bella prehistory*, edited by J. J. Hester and S. M. Nelson. Department of Archaeology, Simon Fraser University Publication No. 5,11–66.

Muckle, R. J.

1985 Bivalve mollusk shells as archaeological sediments. Master's thesis, Department of Archaeology, Simon Fraser University, Burnaby, B.C.

Nelson, M. A., P. J. Ford, and J. K. Stein

1986 Turning a midden into mush: evidence of acidic conditions in a shell midden. Paper presented at the 51st Annual Meeting of the Society for American Archaeology, New Orleans.

Norman, L. K.

1985 Sea urchins: an analysis from an archaeological perspective. Unpublished manuscript.

Schwartz, M. L., and G. Grabert
1973 Coastal processes and prehistoric maritime cultures. Coastal Geomorphology, Publications in Geomorphology, State University of New York. Pp.303–320.
Smith, A.
1984 *Echinoid paleobiology*. Boston: George Allen and Unwin.
Stein, J. K.
1984 Interpreting the stratigraphy of Northwest shell middens. *Tebiwa* **21**, 26–34.
1987 Deposits for archaeologists. *Advances in archaeological method and theory, vol. 10*. Edited by M. B. Schiffer. Orlando: Academic Press. Pp.337–395.
1989 Prehistoric occupation of coastal islands in Washington: view from a shell midden. Paper presented at the Circum-Pacific Prehistory Conference, Seattle.
Stein, J. K., and P. A. Teltser
1989 Size distributions of artifact classes: combining macro- and micro-fractions. *Geoarchaeology: An International Journal* **4**, 1–30.
Visher, G. S.
1969 Grain size distributions and depositional processes. *Journal of Sedimentary Petrology* **39**, 1077–1106.
Wessen, G. C.
1988 The use of shellfish resources on the Northwest Coast: the view from Ozette. In *Prehistoric economies of the Northwest Coast*, edited by B. L. Isaac. Research in Economic Anthropology, suppl. 3. Pp.179–210.
Zar, J. H.
1974 *Biostatistical analysis*. Englewood Cliffs, New Jersey: Prentice-Hall.

14

Is Bone Safe in a Shell Midden?

Angela R. Linse

I. Introduction

Bone preservation has long been correlated with the acidity of archaeological sediments. Traditionally, archaeologists have simply considered that bones decompose under acidic conditions and are preserved under alkaline conditions (e.g., Gilbert 1977). Bone is a composite of organic materials (collagen, proteins, and fats) and inorganic materials (hydroxyapatite) which are differentially affected by weathering processes. Both the organic and inorganic constituents are highly susceptible to decomposition in acidic environments. Archaeologists have not recognized that the inorganic component is also soluble in alkaline environments. The diagenesis of the bone mineral hydroxyapatite can be linked to bone preservation because it is the primary constituent of bone. Therefore, alkaline as well as acidic conditions influence the preservation of buried bone.

Archaeologists have not reached a consensus, however, on the mobility of ionic calcium and phosphorus in buried bone (Whitmer *et al.* 1989) and the relationship such mobility has to the preservation of bone. The movement of elements and compounds during pedogenesis has been studied extensively by soil scientists. Soil science research on the solubility of mineral compounds has provided the theoretical basis for the research and results described in this paper. Studies have shown that the solubility of calcium phosphates varies along the pH continuum (Lindsay 1979). Hydroxyapatite, the primary bone mineral, is a calcium phosphate that is highly soluble in acidic solutions, decreasing in solubility with increasing pH up to pH 7.88. As the pH becomes increasingly alkaline, hydroxyapatite solubility increases (Figure 1).

Bone and sediments from the British Camp shell midden, located in the San Juan Island National Historic Park, Washington (see map, Stein, Ch. 1, this volume) are the subject of this study. They are examined in order to address the theoretically determined relationship between bone preservation and the increased solubility of hydroxyapatite in alkaline depositional conditions. The hypothesis that archaeological

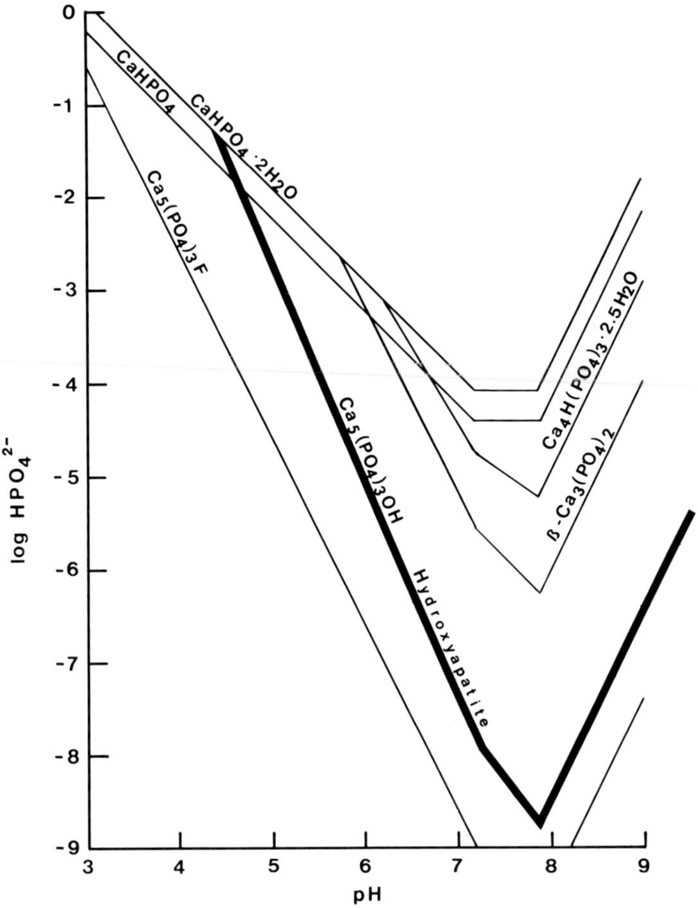

Figure 1 Calcium phosphate solubility as a function of pH. Modified from Lindsay (1979, Figure 12.8, p. 181).

bone decomposes in alkaline sediments is tested by examining the mineral fraction of bone from alkaline deposits. The bones from alkaline deposits should reflect the effects of mineral dissolution. Alternatively, bone from deposits whose pH coincides with the lowest solubility of hydroxyapatite (7.8–7.9) should be well preserved.

A. *Previous Research*

Interest in the chemistry of archaeological sites and preservation of archaeological bone has a long history (Cook 1951; Cook and Heizer 1965; Graf 1949; Watanabe 1950). Physical anthropologists and researchers in the medical professions have provided most of the information utilized by archaeologists on the morphology and internal structure of bone. During the past 25 years, however, increased attention has been paid to the chemical and microscopic study of archaeological bone, particularly human bone (Hedges and Wallace 1978; Parker and Toots 1970; Price 1989; Race *et al.* 1968, 1972; Solomon and Hasse 1967; Stout and Simmons 1979; Szpunar *et al.* 1978). Archaeologists interested in bone chemistry and preservation have also begun to study the effects of the depositional environment on bone diagenetic processes (e.g., Keeley *et al.* 1977; Lambert *et al.* 1984; Pate and Brown 1985; Pate and Hutton 1988; Whitmer *et al.* 1989).

Archaeologists who study the preservation or destruction of artifacts, including bone, are necessarily concerned with the chemical environment of deposition and its postdepositional alteration. Three primary factors that affect the preservation or destruction of buried bone are pH, temperature, and moisture content (Whitmer *et al.* 1989). The latter two factors influence the pH of a deposit. Water is required for chemical activity to take place, and temperature affects the amount of available moisture in a deposit. A single deposit subject to spatially different temperature or moisture conditions after deposition will exhibit more than one pH value.

A number of researchers have investigated the relationship between the pH of deposits and the bone contained therein. White and Hannus (1983) have examined the chemical weathering of the inorganic component of bone. They establish, using changes in the Ca/P ratio of archaeological bone, that the weathering of bone is governed by chemical reactions influenced by water, pH, oxygen, calcium, and phosphorus in bone and the sediment in which the bone is deposited. Gordon and Buikstra (1981) establish a correlation between the preservation of buried human bone and pH of the surrounding sediment based upon an ordinal scale of decomposition. They assert that pH can be a strong predictor of the state of bone preservation. Others have considered this relationship as a part of studies on elemental diffusion (e.g., Whitmer *et al.* 1989) and exchangeable ion activity (e.g., Pate and Hutton 1988) between buried bone and the surrounding sediment. Few, however, have considered the alkaline portion of the pH spectrum.

B. *Bone Chemistry and Decomposition*

Both the composition and structure of bone are important in its diagenesis. In general, bone is composed of 60% calcium phosphate (hydroxyapatite), 25% collagen, and 5 to 10% bone fat. The remaining 5 to 10% includes a combination of other minerals, muco-polysaccharides (complex sugars), calcium fluoride, magnesium phosphate, sodium salts, and elements such as iron and manganese (Berger *et al.* 1964). Structurally, bone consists of collagen fibers embedded with hydroxyapatite crystals. The structural interdependence between collagen and bone mineral plays an important role in the resistance of bone to decay.

1. Organic Fraction

The major organic constituent of bone is collagen [approximately 90% of the organic fraction (Price *et al.* 1989)]. Archaeologists sometimes assume that this organic matrix decomposes soon after deposition, as does the soft tissue adhering to bone.

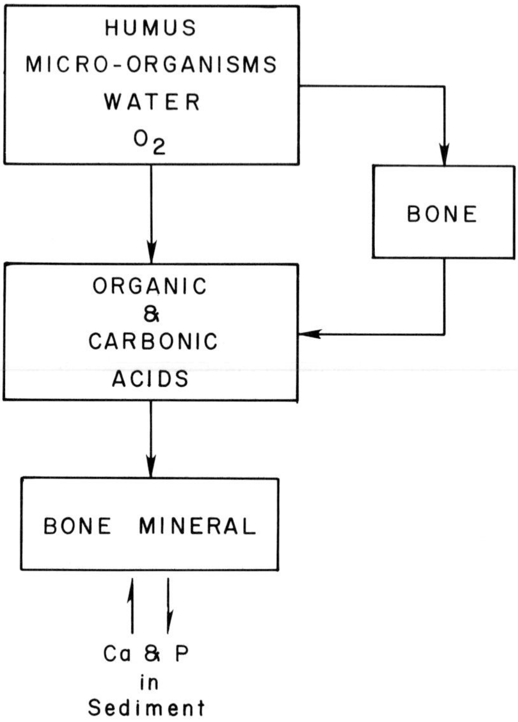

Figure 2 Bone decomposition.

In fact, the collagen in buried bone is highly stable chemically. Collagen stability is a result of three primary factors: its intergrowth with the inorganic matrix of bone, its fibrous structure, and its relative insolubility. Collagen is somewhat less stable structurally than it is chemically, but the processes involved in the loss of collagen from bone are not well known (Wyckoff 1972). Denatured collagen, collagen which has had its initial form or structure disturbed, is susceptible to the effects of proteolytic enzymes that reduce the collagen to relatively soluble substances (Garlick, 1979). The enzyme most commonly involved in the breakdown is collagenase, produced by only a few microorganisms and active within a fixed pH range. Hedges and Wallace (1978:379) suggest that collagen protein can survive "under most conditions in macroscopic amounts in bone buried for up to a few millennia." The factors of collagen preservation however, can be significantly altered with the destruction of the inorganic structure of bone.

2. Inorganic Fraction

The primary inorganic component of bone is hydroxyapatite. The initial weathering of hydroxyapatite in buried bone is caused by a reaction between the bone mineral, soil humus, and acids produced during organic tissue and collagen decomposition.

Subsequent diagenesis of hydroxyapatite is dependent upon the abundance of calcium and phosphorus in a deposit because the mineral loses or gains calcium and/or phosphate ions (as well as other elements) as the surface of the bone comes into chemical equilibrium with the deposit (White and Hannus 1983).

In acidic sediments, hydroxyapatite reacts with carbon dioxide (CO_2), bicarbonate (HCO_3^-), and hydrogen ions (H^+). In calcareous sediments, calcium from carbonates in the soil may replace other ions that are leached from hydroxyapatite in bone, and thereby impede the dissolving reactions (Lambert *et al.* 1985; Nelson and Sauer 1984).

Bone diagenesis is not a straightforward progression; it is a process wherein the partial or complete decomposition of any one component affects every other component. The general sequence of buried bone decomposition can be outlined as follows: (1) humus, microorganisms, water, and oxygen work to break down tissue and collagen; (2) the decay of the tissue and collagen produces more organic and carbonic acids in the buried sediment; (3) these acids react further with exterior bone where hydroxyapatite is most dense; and (4) the soluble bone mineral then exchanges calcium and phosphate ions with the surrounding sediments to attain equilibrium (White and Hannus 1983) (Figure 2). The chemical changes occurring in bone during the equilibrium process are governed by the availability of ions in the surrounding sedimentary solution which, in turn, is affected by the hydrogen ion activity (pH) in the solution near the bone. The chemical processes of bone decomposition may, in part, be predicted by measuring the pH of the solution in a deposit.

C. *Theoretical Premise for the Research*

As the major constituent of bone, the solubility of hydroxyapatite is potentially significant in bone diagenesis. Though archaeologists know that bone decomposition is accelerated in an acidic depositional environment, few have considered the effects of the entire pH continuum on bone preservation. The preservation potential for bone may be lower than previously thought because of the increased solubility of hydroxyapatite in alkaline environments. The amount of this mineral destroyed in bone should increase with both increased acidity and increased alkalinity, thus, bone decomposition should occur at pH values both above and below the critical pH of 7.9 (see Figure 1).

As bone is exposed to acidic conditions hydroxyapatite will dissolve. The ratio of calcium to phosphorus remains stable because both calcium and phosphorus ions are released from bone mineral (in exchangeable forms). In alkaline deposits, the stability of calcium and the instability of phosphorus will result in an increased calcium/phosphorus ratio. The ratio will change within and adjacent to the bone because phosphorus is leaching from the bone into the surrounding sediment, while the calcium remains at a constant level. In chemically unaltered hydroxyapatite, the theoretical mass ratio of calcium to phosphorus is 2.16 (Kyle 1986; see also Price et al. 1989; White and Hannus 1983), but in bone buried in alkaline deposits the calcium/phosphorus ratio should be higher.

II. Methods

A. *Sediment Sampling*

The British Camp shell midden offers a unique opportunity to test the hypothesis that bone decomposes in alkaline deposits. Deposits, called "facies," were selected from the midden based upon their pH values. Sediment samples of known provenience and volume were available from each facies. The pH of the sediment was determined by measuring a sample of the silt and clay fraction (< 63 microns). Two groups of bone samples were obtained based on the pH values of the facies in which they were found. One group includes bones from facies with the highest recorded pH (8.4–8.8). The second group comprises bones from facies with pH measurements near the point where hydroxyapatite is least soluble (pH = 7.8 or 7.9). These groups have been labeled the "high" and "near-neutral" pH groups (Table 1).

B. *Measuring pH*

The value obtained during pH analysis is a measurement of the hydrogen ion activity in the solution within a deposit rather than a assessment of the quantity of an element. The accuracy and precision of pH measurements are affected by a number of factors, including moisture content and the amount of soluble salts and carbonates (Jackson 1958). Measurements of pH are used to predict how elements will behave in a solution. The procedure for pH measurement used in this study is the water saturation percentage method presented by Jackson (1973) with a 20-g sample of sediment.

C. *Bone Selection*

The amount and kind of mineral in bone varies between taxa, elements, and sizes of bone (Jowsey 1968; MASCA 1970). The same element from a single taxon, within a limited size range, was examined. Restricting the analyses in this manner controls for differential mineralization and ensures that any significant variation observed between the two pH groups is not a result of variation across taxa or element.

Table 1 Facies Selected for Analysis

Near-neutral pH deposits			High pH deposits		
Unit	Facies/level	pH	Unit	Facies/level	pH
310E/304S	1D/1	7.8	310E/304S	1I/1	8.5
310E/304S	1E/1	7.9	310E/304S	2G/1	8.4
310E/304S	1G/1	7.9	310E/304S	2K/1	8.5
310E/304S	1P/1	7.8	310E/304S	2K/2	8.6
310E/302S	1U/1	7.9	310E/304S	2M/1	8.6
310E/302S	1Z/1	7.9	310E/304S	2M/4	8.5
308E/304S	1T/1	7.8	310E/304S	2L/1	8.6
			310E/302S	1Z/2	8.4

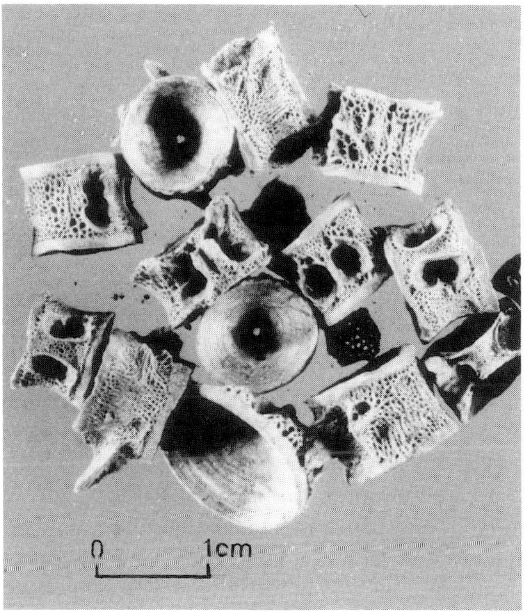

Figure 3 Salmonidae vertebrae.

Salmon vertebrae were chosen as the unit of analysis because they are one of the most common elements recovered from the site. They are usually recovered whole (minus the vertebral spines) and are easily identified to the family level, even when partially decomposed or fragmented. The bones examined in this study were identified to the family level Salmonidae. All of the vertebrae were passed through nested 1/2-inch and 1/4-inch screen; the bones were selected from the sample retained on the 1/4-inch mesh (Figure 3). Variation in mineralization or the loss of bone mineral during life (osteoporosis) is presumed to be random within the samples.

The presence of annual growth rings, or *annuli*, on the centra of fish vertebrae that are the product of differential mineralization during distinct periods of the year is generally acknowledged (Casteel 1976; Monks 1981). The seasonal variation in mineralization, however, is not a significant problem because the entire bone was analyzed in the primary method used in this study.

In addition, differences due to harvest during different seasons of the year are likely to be visible only at the species level. Although individual species may have been harvested at different times, the time of death of the animal is not recoverable, nor is it necessary, at the current level of analysis. Salmon have a regular spawning cycle (Hasler 1966; Johnsen 1978) and thus, in terms of cost effectiveness, harvest can be assumed to have taken place during times of abundance (e.g., during spawning season). If harvest correlates with regular periods of abundance, and these periods are a function of the life cycle of salmon (Hasler 1966; Hourston *et al.* 1965), then there is a high probability that harvests of large quantity are obtained during specific times of the year, and the season of harvest is assumed to be a constant. If a number of different species are represented in the deposits, the assumption can be made that all have been harvested during approximately the same point in the mineralization cycle of salmon, during spawning season. Thus, the effects of differences in season of harvest are assumed to be negligible in the study of mineral preservation.

III. Analysis

A. *Scanning Electron Microscope Microanalysis*

The initial investigation of bone diagenesis in alkaline depositional conditions was attempted through elemental microanalysis. Analysis was performed on a scanning electron microscope fitted with an energy-dispersive spectrometer (EDS), or microprobe. Thirty salmon vertebrae were examined, 19 from alkaline deposits (pH ≥ 8.4) and 11 from near-neutral deposits (pH 7.8–7.9) (Linse 1989). The samples were not modified (i.e., impregnated, cut, or polished) prior to the analysis, to ensure that exterior bone, where hydroxyapatite diagenesis occurs first, was the subject of examination (Figure 4). The microanalysis produced a ratio of the relative percent of calcium to phosphorus for a point on the sample, usually a high ridge of compact bone.

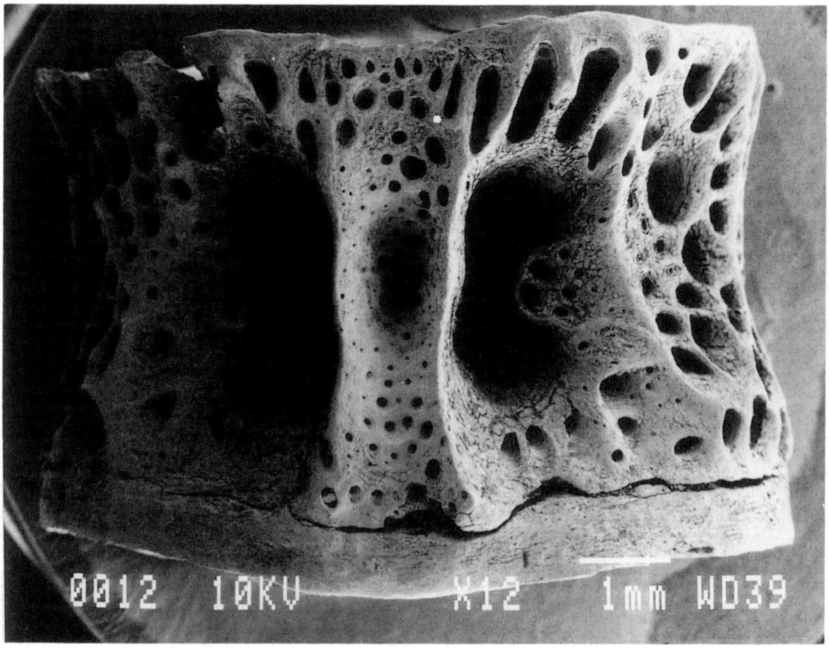

Figure 4 Salmonidae vertebra at 12× magnification.

Table 2 Microanalytic Results

Near-neutral			High		
Facies	Sample	Ca/P	Facies	Sample	Ca/P
1	a	3.10	1	a	2.09
2	a	2.23		b	3.08
	b	3.78		c	2.16
	c	2.27		d	2.75
	d	3.21	2	a	2.53
	e	5.27		b	2.38
	f	2.66		c	3.08
3	a	5.01		d	3.60
	b	2.21	3	a	4.00
	c	2.84		b	4.69
4	a	2.08		c	2.40
$\overline{X} = 3.14 \pm 1.11$				d	2.48
			4	a	10.55
				b	3.79
				c	2.33
				d	4.59
				e	4.03
				f	4.56
			5	a	2.48
				$\overline{X} = 3.55 \pm 1.11$	

The results for individual samples vary widely, and the Ca/P ratios of the two groups overlap, thus there is no significant difference between the bones from the alkaline and near-neutral deposits (Table 2). The variability of the results was due primarily to the structural heterogeneity of bone mineral. Each reading produced a measure of the Ca/P ratio at a particular point, and thus the measure is not necessarily representative of the bone as a whole. Furthermore, the technique is dependent on sample surface homogeneity. The EDS requires a flat, smooth surface to produce precise and accurate readings. Figure 5 shows that the topography of untreated bone varies dramatically over a few microns. The microanalytic technique was abandoned in favor of an analytic method that was insensitive to variation within individual bones, and able to provide a relative Ca/P for the entire bone.

B. *Inductively Coupled Plasma Spectrometry Analysis*

Inductively coupled plasma (ICP) spectrometry provides a means of simultaneously determining the concentration of a suite of elements in a single analysis of each specimen. The ability to obtain comparable data simultaneously for both calcium and phosphorus makes it preferable to standard techniques, such as electrodes for calcium, and titration, colorimetry, or spectrographic techniques for phosphorus. Although the use of ICP spectrometry as a technique of multi-element analysis is routine in the geological sciences (Walsh and Howie 1986), its application to organic

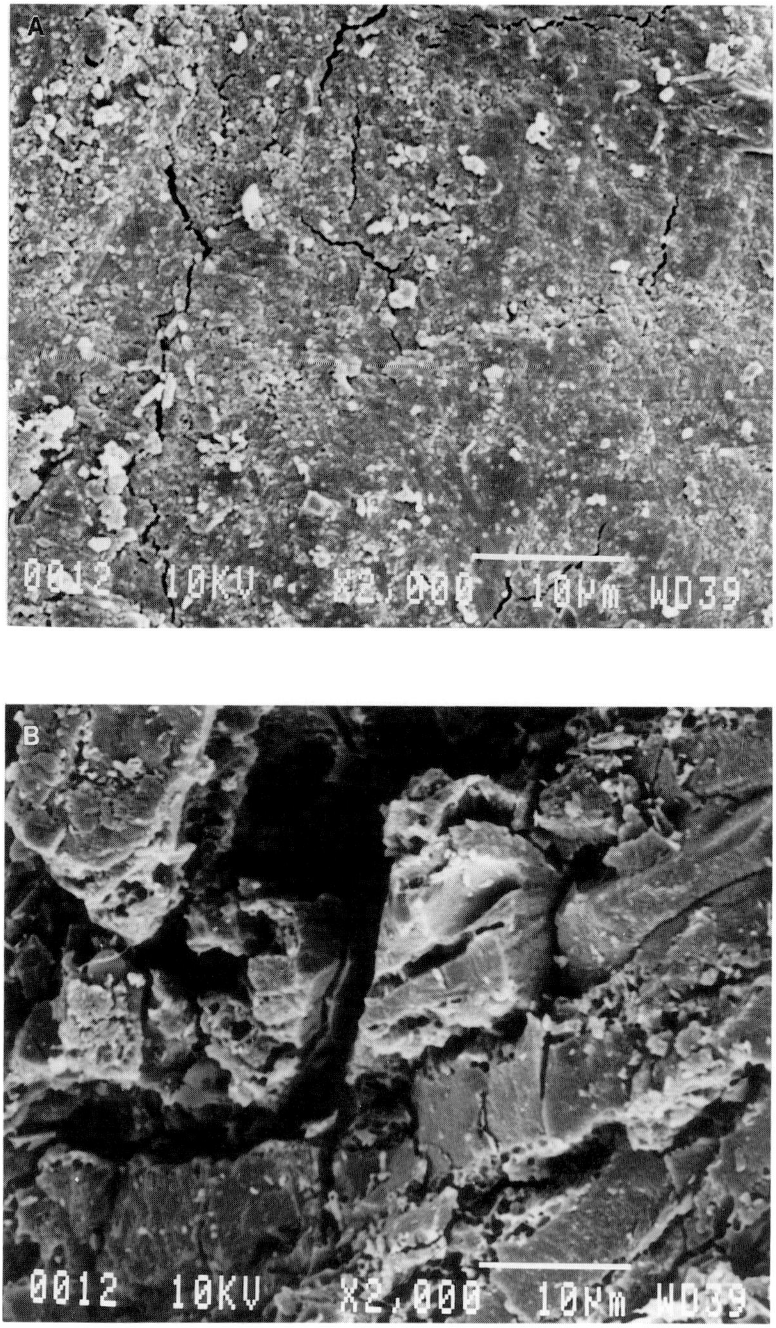

Figure 5 Salmonidae vertebra surface at 2000× magnification. A and B are microns apart.

materials has been less common and relatively recent (Burton and Price 1990a, 1990b; Price *et al.* 1989). ICP analysis resolves the problem of intrasample mineral variability encountered in the application of the microprobe method (Linse 1988, 1989) because the ICP uses a solution method of analysis that provides quantitative measures for the sample as a whole.

The ICP solution method of analysis involves the conversion of a dissolved sample into an aerosol using argon gas. The aerosol is introduced to the high temperatures of the ICP torch and atomization of the analyte solution is complete. The atomized solution emits light at various wavelengths for the corresponding atoms and ions in the solution. Each element has a number of wavelengths at which it can be measured and analyzed; the choice of wavelength depends on the sensitivity of the wavelength and the analytical system. The light emitted by a sample solution is resolved into its component wavelengths using a diffraction grading, and measured with light-sensitive photomultiplier tubes.

Standard solutions (either natural or synthetic), with known elemental concentrations, are run prior to the samples of unknown concentrations. The light intensities emitted by the standards are used to construct calibration curves. The intensity of the light at each wavelength is a linear function of the quantity of an element. Thus, intensities can be measured to quantitatively determine elemental concentrations. The light intensities emitted by the sample solution at various wavelengths are compared to the calibration intensities to determine the composition of the solution. ICP analysis can determine both the presence and concentration (ppm or ppb) of a number of elements. For a detailed discussion of ICP analysis see Walsh and Howie (1986) or Thompson and Walsh (1983).

Analyses were conducted at the Department of Geological Sciences ICP Laboratory, University of Washington. The samples were analyzed using a Baird PS-1 polychromator ICP and the data analyzed with a D.E.C. computer system. In this study, three samples of modern salmon, and six archaeological Salmonidae vertebrae were examined. The relatively high cost of ICP analysis precluded examination of a large number of specimens.

Each bone sample underwent similar treatment in initial preparation. All were ground with ethyl alcohol and dried in a 40° C (105° F) drying oven. The archaeological bones were selected based on the pH of the deposits from which they were recovered and include three vertebrae from sediments with a high pH (8.5–8.6) and three from near-neutral pH (7.8–7.9) deposits. The samples were washed in distilled water to remove any sediment clinging to the specimen and dried in a low-temperature drying oven.

Fresh salmon skeletons, obtained from a local fish market, were cooked in distilled water until the meat was easily removed from the bone. The vertebral spines were removed after drying in a low-temperature oven. The modern bones used in these analyses were selected for their size (as noted above) and low visible oil content. Oil can interfere with ICP analyses and oil-free bones are more comparable to the archaeological bone.

Eight modern vertebrae were crushed, mixed together, and subsequently split into three samples. This procedure ensured that the modern sample was homogeneous, and that any resultant variation can be interpreted as a product of sample treatment rather than intersample variability. One sample of modern bone was analyzed while the other two were placed in buffer solutions after which the solutions were analyzed.

ICP analysis requires that samples be dissolved prior to analysis. All of the archaeological bone and one of the modern samples were pretreated using a lithium metaborate ($LiBO_2$) fusion technique (Linse 1989) that is traditionally used to dissolve geological specimens [simple dissolution in acid is preferable for organic specimens, and is used in subsequent research (Linse and Burton 1990)].

In order to examine the theoretical basis for the hypothesis that bone will decompose in alkaline deposits, the two remaining modern samples were subject to a slightly different treatment. Two samples of modern bone were added to buffer solutions prepared at pH values of 7.9 and 9.0 and left for 2 weeks (Linse 1989). If the hypothesis proposed above is correct, bone immersed for an extended period of time in an alkaline solution will lose phosphorus while calcium concentrations will remain stable. Alternatively, bone immersed in a solution with a pH that corresponds to the point at which hydroxyapatite is least soluble (7.9) will have relatively little ionic exchange and thus, only small amounts of phosphorus should be leached into the solution. The Ca/P ratios of the solutions, rather than the bone, are examined after the bone samples have soaked in the buffer solutions for an extended period of time. The buffer solutions are not initially in equilibrium with the bone, thus ionic diffusion should occur, and phosphorus ion mobility should be greater in the alkaline solution. A lower Ca/P ratio for the alkaline solution relative to that obtained for the near-neutral solution reflects the greater phosphorus mobility in alkaline environments.

1. ICP Results

The ICP analysis provided concentrations of calcium and phosphorus in the form of percent by weight of oxides in the sample. The oxide data have been converted to elemental parts per million. Table 3 shows the results of all ICP analyses and their calculated ratios.

a. Archaeological Bone The average ratio of calcium to phosphorus for the near-neutral group (2.34) is slightly lower than the average Ca/P ratio for the high pH group (2.42). The results are inconclusive, however, because no significant difference exists between the average Ca/P ratios of the two groups. Although considerable overlap exists in the values obtained for the archaeological specimens, a noticeable qualitative difference is observed between the bones from the highly alkaline deposits and those from the near-neutral pH deposits. They differ both in color and apparent state of preservation. The bones from the sediments with a high pH are fragile and chalky, and more friable than both the modern bone and the archaeological bone from near-neutral depositional conditions.

b. Modern Bone The analyses of the modern bone provide the most intriguing results. The Ca/P ratio of the modern bone (2.10) is slightly less than the 2.16 mass

Table 3 ICP Results. Treatment 1: Modern salmon vertebrae soaked in buffer solutions; buffer solutions were analyzed after the bone was removed. Treatment 2: Modern salmon vertebrae fused with $LiBO_4$, dissolved, and analyzed. Treatment 3: Archaeological salmon vertebrae fused with $LiBO_4$, dissolved, and analyzed.

Sample	pH	Treatment	ppm Ca[a]	ppm P[a]	Ca/P Ratio
1	7.9	1	16.2	42.2	0.38
2	9.0	1	2.5	59.0	0.04
3	—	2	18.6×10^4	8.8×10^4	2.10
4	7.9	3	26.7×10^4	11.6×10^4	2.29
5	7.8	3	27.2×10^4	11.7×10^4	2.33
6	7.8	3	26.1×10^4	10.8×10^4	2.42
7	8.6	3	26.6×10^4	10.3×10^4	2.57
8	8.6	3	27.5×10^4	11.7×10^4	2.34
9	8.5	3	29.9×10^4	12.8×10^4	2.34

[a]All data are in parts per million by weight.

ratio derived from theoretical calculations (Kyle 1986; White and Hannus 1983). The ratio is an average of three measurements (2.03, 2.08, 2.19) whose range includes the theoretical mass ratio. In addition, there is a substantial difference between ratios of the archaeological samples and the modern salmon vertebrae. All of the archaeological vertebrae have considerably higher Ca/P ratios and concentrations of both calcium and phosphorus than the modern bone.

The results of the buffer solution experiment are also very informative (Table 3). As expected, the Ca/P ratio of the highly alkaline solution is less than in the ratio of the near-neutral solution. The analysis shows that more phosphorus did dissolve in the solution with a pH of 9.0 than in that with a pH of 7.9, lending experimental support to the hypothesis.

2. Discussion

The examination of two groups of archaeological salmon bone is the primary concern of this study. However, experiments with modern bone have also provided important information about bone diagenesis and the theoretical basis for the research. Although the results for the archaeological analyses are inconclusive, the study has provided both qualitative and quantitative information on bone diagenesis in alkaline deposits.

a. Technical Accuracy There are two factors that may have affected the results of all of the ICP analyses. They involve a possible systematic error in the precision and accuracy of the measurements. First, calcium is the element to which the ICP is most sensitive, leading to an approximate error rate of 2%. Alternatively, ICP phosphorus analysis is less precise, with error rates as high as 10%. The ICP used in this study is normally used to analyze rocks, some of which are high in calcium and few of which are high in phosphorus. Thus, the precision of measuring phosphorus is usually not a problem in geological analyses, and extensive measures have not been undertaken to correct the situation. Second, the calcium and phosphorus standards used for geological analyses have lower concentrations than bone. Therefore, calibration lines must be extrapolated for the higher concentrations of phosphorus and calcium found in bone. The extrapolated number may be incorrect and thereby affect the accuracy of phosphorus measurements.

The above factors can cause a systematic measurement error leading to ratios that are inaccurate. Given that errors in accuracy are systematic, however, all samples have been affected similarly and are comparable to each other. In this analysis, the ICP results are assumed to be correct because the Ca/P ratio of 2.10 for the modern bone is within 0.06 of the theoretical ratio of 2.16 and well within the experimental range obtained by other analysts. If the ICP readings for calcium and phosphorus can be considered both accurate and precise, then the reason for the insignificant results must lie elsewhere.

b. Archaeological Bone Two possible explanations exist for the results obtained for the archaeological bone. First, the sample size is too small. Second, pH measurements are inherently unstable. Any measure of the alkalinity or acidity of a deposit is a measure of the sample in laboratory conditions, and not necessarily an accurate measure of the conditions in the deposit. The water table in the midden area does fluctuate (Stein 1984) and assuredly affects the availability of the elements in the solution and, thereby, the pH. Sediments, however, do not commonly change from neutral to highly alkaline over a short period of time.

c. Modern Bone The Ca/P ratios derived from the modern bone, as well as the inflated ratios of the archaeological samples, serve to strengthen the hypothesis that bone mineral is unstable in alkaline depositional conditions. The results obtained from modern bones soaked in the high pH solution were very different from those soaked in the near-neutral solution. There is an increase in the quantity of phosphorus in the solution, which suggests that the initial proposition is correct, that in alkaline conditions, phosphorus in bone mineral is more soluble than calcium. The data also show that there is considerably less dissolved calcium in the high pH solution. The low concentration of calcium is more likely to be responsible for the low Ca/P ratio than is a high concentration of phosphorus. A cloudy substance (perhaps a precipitate of calcium or hydroxyapatite) was present in the highly alkaline solution and may account for the low concentration of calcium. When substances similar in appearance were analyzed by Burton (personal communication, 1989) using X-ray diffraction and chemical analysis, they were found to be virtually pure hydroxyapatite. Whether or not the cloudy substance was precipitate hydroxyapatite, the solution experiment shows that ion exchange was undoubtedly occurring in the solutions.

The disparity between the ratios of the modern bone (Ca/P = 2.10) and the archaeological bone (Ca/P = 2.29–2.57) also indicates that there has been considerable ionic exchange between the archaeological bone and their associated sediments. The Ca/P ratio for the modern salmon vertebrae is within the range of variation found in modern bone of all kinds. The uniformly high Ca/P ratios of the archaeological samples, relative to the theoretical Ca/P ratio of 2.16, suggest that the archaeological bone mineral has been altered since deposition.

IV. Summary and Conclusions

The purpose of this study was to examine the preservation of buried bone in alkaline environments. The primary constituent of bone is the mineral hydroxyapatite [$Ca_{10}(PO_4)_6OH_2$]. Changes in the preservation of archaeological bone are assumed to be a function of changes in the chemistry of hydroxyapatite. Soil science research (Lindsay 1979) has determined that hydroxyapatite is soluble in acidic and alkaline conditions, and least soluble in solutions with a pH value of 7.88. Calcium and phosphorus are the primary elements of bone mineral and calcium is stable in alkaline solutions. Therefore, changes in the solubility of bone should be identifiable as differences in the ratio of calcium to phosphorus in archaeological bone. Use of the ratio of the two elements insures comparability of results between bones of different size and presumably different initial concentrations of calcium and phosphorus.

The research in this chapter tests the hypothesis that bones from alkaline depositional conditions have a higher Ca/P ratio than bones from deposits with a near-neutral pH, because phosphorus has been leached from the bone while calcium has not. Two matters need to be considered when evaluating the analytic results. First, why was the difference between the bones from the alkaline deposits and the bones from the near-neutral deposits not significant? Second, what are the implications of the results of the modern bone experiments which appear to support the hypothesis?

The difference between the Ca/P ratios of archaeological and modern bone and the ratios of the solution experiment suggest that the calcium/phosphorus ratio may not be the optimal factor to differentiate bones from alkaline and near-neutral deposits. It is, however, one of the best quantitative measures of bone preservation available at this time (Price *et al.* 1989). Thus, the insignificance may reflect a sample-size problem rather than an inappropriate measure of bone preservation.

The disparity in the Ca/P ratios of the dissolved bone, as opposed to the ratios of the solutions in which bone has been soaked, may be a function of the initial concentrations of calcium and phosphorus in the bone. The concentrations of these elements are very high in bone, in contrast to their absence in the buffer solutions. Although removal of a small amount of phosphorus from bone alters the structural integrity of bone, it does not appear to produce a significant shift in the Ca/P ratio in the residual bone. The amount of bone dissolved in the alkaline solution is small relative to the total amounts of calcium and phosphorus in the bone. A very small increase in the amount of phosphorus in the solution relative to the increase in the concentration of calcium, however, produces a substantial difference in the Ca/P ratios of the solutions. The ratios obtained for all of the archaeological bones differ significantly from those obtained from the modern bone. The ratios reflect differential ionic exchange between the buried archaeological bone and the surrounding sediment in alkaline and neutral deposits.

The results of the solution analyses support the proposition that hydroxyapatite, and thereby bone, is chemically altered in alkaline environments. The difference between the ratios of the alkaline and near-neutral solutions indicates that while calcium and phosphorus appear to be stable in the near-neutral solution, phosphorus is considerably more mobile than calcium in the alkaline solution.

Additional research (Linse and Burton 1990) expands the above study, incorporating the immersion of modern fish and mammal bone in alkaline buffer solutions with pH values from 7.0–13.0. The latter modern bone experiments confirm the results of this study, and greatly expand the pH range of the above interpretations. Linse and Burton's (1990) study examines changes in the concentration of calcium and phosphorus in untreated and burned bone, as well as the interrelationship between the elements. The effect of time on solubility is also investigated by sampling the solutions at regular intervals. The amount of phosphorus leached from untreated bone increases through time as the pH value increases above pH 7.8. The study also documents the insolubility of burned bone and the extreme susceptibility of bone to decomposition in highly alkaline (>11 pH) conditions. Bone mineral solubility increases dramatically at values above pH 11, because collagen begins to break down. The destruction of collagen exposes more surface area of the mineral to alkaline solutions.

The relationship between alkaline environments and bone decomposition has previously been overlooked in archaeological analyses. Negative effects on the potential for bone preservation are likely to be produced by extremes in alkalinity as well as extremes in acidity. Bones may be relatively "safe" in a shell midden in terms of traditional morphological classification, however, the chemical integrity of bone in alkaline environments is suspect.

Acknowledgments

I would like to thank J. Burton for his insights and comments on earlier drafts of this paper, and V. Holliday for his time and comments. J. Stein deserves special thanks for her advice and support. The archaeological samples were obtained from excavations at the San Juan Island National Historic Park-British Camp, funded by the National Park Service and the University of Washington. This research was, in part, supported by a National Science Foundation Graduate Fellowship.

References

Berger, R., A. G. Horney, and W. F. Libby

1964 Radiocarbon dating of bone and shell from their organic components. *Science* **144**, 999–1001.

Burton, J. H., and T. D. Price

1990a Paleodietary applications of barium values in bone. In *Proceedings of the 27th international symposium on archaeometry, Heidelberg, 1990*, edited by E. Pernicka and G. A. Wagner. Basel: Berkhauser Verlag. Pp.787–795.

1990b The ratio of barium to strontium as a paleodietary indicator of consumption of marine resources. *Journal of Archaeological Science* **17**, 547–557.

Casteel, R. W.

1976 *Fish remains from archaeological sites*. New York: Academic Press.

Cook, S. F.
1951 The fossilization of human bone: calcium, phosphate, and carbonate. *University of California publications in American archaeology and ethnology* **40**(6) 263–280.
Cook, S. F., and R. F. Heizer
1965 Studies on the chemical analysis of archaeological sites. *University of California publications in anthropology* Vol. 2, Berkeley, California: University of California Press.
Garlick, J. D.
1979 Buried bone: the experimental approach in the study of nitrogen content and blood group activity. In *Science in archaeology, a survey of progress and research*, edited by D. Brothwell and E. Higgs. New York: Basic Books. Pp.503–512.
Gilbert, R.
1977 Applications of trace element research to problems in archaeology. In *Biocultural adaptation in prehistoric America*, edited by R. L. Blakely. Athens: University of Georgia Press. Pp.85–100.
Gordon, C. C., and J. E. Buikstra
1981 Soil pH, bone preservation, and sampling bias at mortuary sites. *American Antiquity* **46**, 566–571.
Graf, W.
1949 Preserved histological structures in Egyptian mummy tissue and ancient Swedish skeletons. *Acta Anatomy* **8**, 236–350.
Hasler, A. D.
1966 *Underwater guideposts: Homing of salmon.* Madison: University of Wisconsin Press.
Hedges, R. E. M., and C. J. A. Wallace
1978 The survival of biochemical information in archaeological bone. *Journal of Archaeological Science* **5**, 377–386.
Hourston, A. S., E. H. Vernon, and G. A. Holland
1965 *The migration, composition, exploitation and abundance of odd-year pink salmon runs in and adjacent to the Fraser River convention area.* International Pacific Salmon Fisheries Commission, New Westminster, British Columbia, Canada.
Jackson, M. L.
1958 *Soil chemical analysis*. Englewood Cliffs, New Jersey.
1973 *Soil chemical analysis: Advanced course*. Englewood Cliffs, New Jersey.
Johnsen, P. B.
1978 Contributions on the movements of fish: behavioral mechanisms of upstream migration and homestream selection in Coho salmon (*Oncorhynchus kisutch*): winter aggregations of carp (*Cyprinus carpio*) as revealed by ultrasonic tracking. Ph.D. dissertation, University of Wisconsin, Madison.
Jowsey, J.
1968 Age and species differences in bone. *Cornell Veterinarian* **58**, 74–94.
Keeley, H. C. M., G. E. Hudson, and J. Evans
1977 Trace element contents of human bones in various states of preservation, 1. the soil silhouette. *Journal of Archaeological Science* **4**, 19–24.
Kyle, J. H.
1986 Effect of post-burial contamination on the concentrations of major and minor elements in human bones and teeth—the implications for palaeodietary research. *Journal of Archaeological Science* **13**, 403–416.
Lambert, J. B., S. V. Simpson, J. E. Buikstra, and D. E. Charles
1984 Analysis of soil associated with Woodland burials. In *Archaeological chemistry III*, edited by J. B. Lambert. Washington, D.C.: American Chemical Society. Pp.97–113.
Lambert, J. B., S. V. Simpson, S. G. Weiner, and J. E. Buikstra
1985 Induced metal ion exchange in excavated bone. *Journal of Archaeological Science* **12**, 85–92.
Lindsay, W. L.
1979 *Chemical equilibria in soils*, New York: John Wiley & Sons. Pp.181–187.

Linse, A. R.
1988 Is bone safe in shell middens? Paper presented at the 53rd annual meeting of the Society for American Archaeology, Phoenix, Arizona.
1989 Bone preservation and diagenesis in alkaline archaeological deposits. Master's thesis, Department of Anthropology, University of Washington, Seattle.
Linse, A. R., and J. H. Burton
1990 Bone solubility and preservation in alkaline depositional conditions. Paper presented at the 55th Annual Meeting of the Society for American Archaeology, Las Vegas.
MASCA
1970 Bone from domestic and wild animals: crystalographic differences. *MASCA Newsletter* **6**, 1–2.
Monks, G. G.
1981 Seasonality studies. *Advances in Archaeological Method and Theory* **4**, 177–240.
Nelson, D. A., and N. J. Sauer
1984 An evaluation of postdepositional changes in the trace element content of human bone. *American Antiquity* **49**, 141–147.
Parker, R. B., and H. Toots
1970 Minor elements in fossil bone. *Geological Society of America Bulletin* **81**, 925–932.
Pate, D., and K. A. Brown
1985 The stability of bone strontium in the geochemical environment. *Journal of Human Evolution* **14**, 483–491.
Pate, D. F., and J. T. Hutton
1988 The use of soil chemistry data to address postmortem diagenesis in bone mineral. Unpublished manuscript, Department of Anthropology Brown University, Providence, Rhode Island.
Price, T. D.
1989 *Chemistry of prehistoric bone*. Cambridge: Cambridge University Press.
Price, T. D., J. H. Burton, and J. Blitz
1989 Bone composition studies in archaeology. Paper presented at the 54th annual meeting of the Society for American Archaeology, Atlanta.
Race, G. J., E. I. Fry, J. L. Matthews, J. H. Martin, and J. A. Lynn
1968 The characteristics of ancient Nubian human bone by collagen content, light and electron microscopy. *Clinical Pathology* **45**, 704–713.
Race, G. J., F. Wendorf, F. B. Humphreys, and E. I. Fry
1972 Paleopathology of ancient Nubian human bone studied by chemical and electron microscope methods. *Journal of Human Evolution* **1**, 263–279.
Solomon, C. D., and N. Hasse
1967 Histological and histochemical observations of undecalcified sections of ancient bones from excavations in Israel. *Israel Journal of Medical Science* **3**, 747–754.
Stein, J. K.
1984 Interpreting the stratigraphy of Northwest shell middens. *Tebiwa* **21**, 26–34.
Stout, S., and D. J. Simmons
1979 Use of histology in ancient bone research. *Yearbook of Physical Anthropology* **22**, 228–249.
Szpunar, C. B., J. B. Lambert, and J. E. Buikstra
1978 Analysis of excavated bone by atomic absorption. *American Journal of Physical Anthropology* **48**, 199–202.
Thompson, M., and J. N. Walsh
1983 *A handbook of inductively coupled plasma spectrometry*. New York: Blackie.
Walsh, J. N., and R. A. Howie
1986 Recent developments in analytical methods: uses of inductively coupled plasma source spectrometry in applied geology and geochemistry. *Applied Geochemistry* **1**(1), 161–171.
Watanabe, N.
1950 The preservation of bony substances in the soil of prehistoric sites. *Zinriugaku Zassi* (*Journal of the Anthropological Society of Japan*) **61**(2), 1–8.
White, E. M., and L. A. Hannus
1983 Chemical weathering of bone in archaeological soils. *American Antiquity* **48**, 316–322.

Whitmer, A. M., A. F. Ramenofsky, J. Thomas, L. J. Thibodeaux, S. D. Field, and B. J. Miller
1989 Stability or instability: the role of diffusion in trace element studies. *Archaeological Method and Theory* **1**, 205–273.
Wyckoff, Ralph W. G.
1972 *The biochemistry of animal fossils*. Baltimore: Williams and Wilkens.

15

Burned Archaeological Bone

Patrick T. McCutcheon

I. Introduction

Bone, particularly burned bone, is a common constituent of archaeological deposits and can contribute in a variety of archaeological research questions, varying from subsistence and taphonomic studies, to formation processes, and dating. Traditionally, bone has not been treated as an important archaeological constituent; however, consequent realization that archaeological bone is a source of information about the past has firmly established its significance in the archaeological record. One aspect that has received only casual attention until recently is the effect of heating or burning on bone. The degree of thermal alteration carries direct implications for the condition of bone when it enters the archaeological record as well as when it is recovered. In order to undertake analyses of subsistence, taphonomy, dating, and formation processes with burned bone, it is necessary to have an explicit understanding of the physical and chemical alterations that bone undergoes when heated. This research addresses formation processes in a shell midden; additional research on taphonomy and dating are in preparation.

Recent studies (Armstrong and Singer 1965; Bonucci and Graziani 1975; Knubovyets *et al.* 1979; Legros *et al.* 1978; Shipman *et al.* 1984) suggest that heating causes both chemical and physical alterations in bone, which are related to specific temperature thresholds. Because the changes are irreversible, the maximum temperature to which a bone has been heated can be determined. However, scanning electron microscopy (SEM) and X-ray diffraction (XRD) analyses, involved in determining maximum heating temperature, are expensive and destructive, precluding the direct determination of heating temperature in archaeological analysis on a routine basis. Because the changes induced by heating are chemical and physical, the appearance of the bone is necessarily altered in potentially predictable ways. If these chemical and physical states in bone are irreversible and can be associated unambiguously with temperature classes based on less expensive observations (e.g., color), then ordinal-level determinations of the heating temperatures to which bone was exposed can be made routinely.

The present study includes a repetition of previously published experiments, in order to calibrate my techniques with the earlier results, and refine the thresholds at which specific physical and chemical changes take place. The initial objective of this study is to identify both macro- and microscopic surrogates in the appearance of thermally altered bone. An ordinal classification of burned bone constructed from the experimental results of SEM, XRD, and Munsell color analyses, is used to identify thermally altered bone from the archaeological deposit at British Camp, San Juan Island, Washington. Specific attention is directed to the dimension of color. The results are used to assess postdepositional alterations of bone from two lithostratigraphic layers of the British Camp shell midden.

II. Previous Research

A. *Bone Chemistry*

Previous studies have demonstrated that bone material is made up of both mineral and organic phases. Approximately 35% of the dry weight of bone is organic (Posner and Belts 1975). Of this, 90% is collagen (Glimcher and Krane 1968), the remainder being made up of various compounds, such as glycosaminoglycans and proteoglycans (Herring 1972). The mineral phase predominantly consists of hydroxyapatite, $Ca_{10}(PO_4)_6(OH)_2$, whose crystal size is 0.2–0.3 μm (cf Posner 1969). In addition to calcium phosphate, bone mineral contains approximately 4 to 6% carbonate, 0.9% citrate, 0.5% magnesium, 0.7% sodium, trace amounts of Cl^-, F^-, K^+, Sr^{+2}, and other metal ions (Armstrong and Singer 1965).

B. *Burned Bone*

Several branches of anthropology share an interest in burned bone: archaeology, physical anthropology, and forensic anthropology. The number of specific studies of burned bone is large (Table 1). Some of the characteristics of burned bone have received more attention than have others in the previous studies (e.g., color and macroscopic characteristics).

Traditionally, color has been used from a commonsensical perspective. Identification of bone as burned, as well as the degree of thermal alteration, is traditionally measured by difference in color (Table 1). A standardized color system, such as the Munsell color charts, has been used only recently (Shipman *et al.* 1984). Color has had great success as an attribute for identifying thermally altered bones in the archaeological record because change in color of bone due to heating is drastic, and thereby easily observed. Color is observed macroscopically and is distinguished from macroscopic morphological attributes.

Macroscopic morphological attributes are those that are observed without the aid of magnification and were used consistently in all the previous studies (Table 1). These attributes were of interest for those archaeologists undertaking measurements

Table 1 Summary of Prior Work

Citation	Color	Macroscopic	Microscopic	Experimental	Provenience studies
Franchet 1934	X	X		X	
Krogman 1939	X	X	X	X	
Haury 1945		X			
Angel and Coon 1954		X		X	
Baby 1954	X	X		X	X
Wells 1960	X	X			X
Anderson 1963		X			
Binford 1963	X	X		X	X
Merbs 1967	X	X			X
Spence 1967		X			
Buikstra and Goldstein 1973	X	X			X
Bonucci and Graziani 1975	X	X	X	X	
Herrmann 1977	X	X	X	X	
Pfeiffer 1977	X	X			X
Stewart 1979	X	X	X	X	
Therman and Willmore 1981	X	X		X	
Shipman *et al.* 1984	X	X	X	X	
Ubelaker 1984	X	X			
Knight 1985	X	X	X	X	
Buikstra and Swegle 1989	X	X		X	

of cremated skeletons. Understanding skeletal dimensions of aboriginal populations which practiced cremation of their dead is undertaken successfully only when the amount of bone shrinkage as a result of heating is taken into account. Baby (1954) made descriptions of cremated bone based on macroscopic morphological attributes so that determination of precrematory condition of the individuals could be made; that is, if the skeletons were disarticulated at the time of cremation. Traditionally, as with color, macroscopic observations have been measured in a qualitative fashion with some exception (Shipman *et al.* 1984).

To a lesser extent, microscopic morphological attributes were noted during experimental studies (Table 1). When investigators want to address questions which deal with the crystalline phase of bone, instrumentation for microscopic analysis is needed. As noted previously, the size of an individual hydroxyapatite crystallite is approximately 0.2–0.3 μm in size, clearly beyond macroscopic observation. Frequently, microscopic information is used to assess specific changes in bone structure resulting from heating, for example, shrinkage.

Provenience studies are those in which the condition or position of the burned bones in an archaeological site is given meaning by assessing the degree of thermal alteration, for example, how close the bones were to the fire (Table 1). This information has also been used to address physical postdepositional alterations of the archaeological record (Buikstra and Goldstein 1973).

Although color and macro- and microscopic attributes have been used with some consistency in burned bone studies, a rather casual treatment of the chemical and physical changes induced by heating has prevailed, with some exception. As a result, our understanding of bone material and the changes it undergoes when heated is just beginning to become clear.

In an early experimental study undertaken by Franchet (1934, cited in Pfeiffer 1977:124), eleven stages of transition were noted for bone exposed to heat: (1) yellow color; (2) light brown; (3) dark brown; (4) black (completely burnt); (5) indigo-blue; (6) blue-gray; (7) white (600° C); (8) deformation; (9) torsion splits; (10) vitrification begins (causing an appearance like porcelain); and (11) bone melts (1200° C). It is interesting that Franchet's study relies heavily on color to describe the degree of thermal alteration up to 600° C. Macroscopic morphological attributes are used to characterize changes in bone material burned at temperatures above 600° C. The notation of "vitrification begins, causing an appearance like porcelain" suggests that there may have been some microscopic analysis as well.

As archaeologists began to investigate deposits containing burned bone more frequently (e.g., Baby 1954; Haury 1945), a need for a more explicit understanding of burned bone material developed. The need probably originated out of concern that traditional skeletal measurements of populations or individuals might be erroneous due to the alteration in size of bone when burned, as well as genuine archaeological interest in the condition of skeletal material before cremation. Baby's (1954) study has had the greatest impact in North American archaeology (e.g., Binford 1963; Buikstra and Swegle 1989; Knight 1985; Merbs 1967; Ubelaker 1984). In his study, he burned both fleshed and unfleshed bones, and then observed the changes that had taken place, presumably as results of heating. Baby's classification consists of three general classes: (1) completely incinerated (fragments range from light to blue-gray buff and show deep checking, diagonal transverse fracturing, and warping); (2) incomplete incineration (fragments are blackened, through the incomplete combustion of organic material present in bone); and (3) normal bone or nonincinerated bone (not affected by heat, but show some smoking along the edge). Baby was investigating Hopewell cremations. His primary goal was to delineate the preburning status of the bodies, that is, if they were fleshed or defleshed. In a subsequent study, Binford (1963) replicated Baby's experiments with similar results. Specific assessments of temperature were not noted by Baby or Binford.

Others have inferred or actually measured the temperature at which bone has been burned (Bonucci and Graziani 1975; Buikstra and Swegle 1989; Franchet 1934; Herrmann 1977; Shipman *et al.* 1984; Stewart 1979; Ubelaker 1984). For instance, Stewart (1979:66) notes: "Prehistoric burnt bones of a blue-gray color appeared to agree in macro- and microscopic structure with test specimens exposed to temperatures of 500–800° C." Ubelaker (1984:34) tells us that combustion of the organic matter in bone remains incomplete below 800° C. Herrmann (1977:101) notes that at the "critical level" (> 700–800° C), complete cremation occurs, that is, the organic matter is cremated completely and crystals of bone mineral fuse. Below the critical level (< 700–800° C), incomplete cremation occurs, resulting in extensive carbon coloration due to "non-quantitative" cremation of organic matter.

Ascertaining the degree of thermal alteration by observing bone as it is burned in either experimental (e.g., Bonucci and Graziani 1975; Shipman *et al.* 1984) or actualistic (e.g., Knight 1985) studies, has served to render provenance studies possible. In two studies, Pfeiffer (1977) and Buikstra and Goldstein (1973) used the degree of thermal alteration of archaeological bone to infer the position of the bodies when burned. Buikstra and Goldstein find this inference useful in distinguishing the postdepositional from the depositional processes, that is, the postcremation movement of skeletal parts. Pfeiffer (1977:124–125) notes that the color of any given bone is closely tied to the shape of the bone. She suggests combining observations of shape and color to make inferences about the condition of the body prior to cremation. In doing so, color may suggest which parts of skeletons were in articulated positions when burned (cf Wells 1960). She also suggests using color to establish the position of the body when burned. However, one must note that because color may be dependent on shape of bone, conclusions about the location of a bone in relation to heat source may not be accurate.

Most recent studies suggest that bone, when heated, goes through specific chemical and physical alterations and that these alterations can be related to specific temperature thresholds. In Bonucci and Graziani (1975) and Shipman *et al.* (1984), changes in bone material induced by heating were characterized using macroscopic, microscopic, and color analyses. In addition, both of these studies used X-ray diffraction to characterize crystal morphology change. In the 1975 study, thermal gravimetric (TG) and differential thermal analysis (DTA) were employed. Based on these studies, and the work of others (Armstrong and Singer 1965; Glimcher and Krane 1968; Knubovyets *et al.* 1979; Legros *et al.* 1978; McConnell 1973; Posner 1969), the following stages of thermal alteration in bone material are summarized as

1. 20° C (room temperature) to 300–350° C: loss of unbound water and some initial carbonization of the organic phase (Civjan *et al.* 1970; Holager 1970).
2. 350° C to 500–600° C: complete combustion of the organic phase, loss of bound crystal water, and possible formation of pyrophosphates (cf Dean 1974; Glimcher and Krane 1968; Knubovyets *et al.* 1979; Posner 1969).
3. 600° C to 950° C: increase in crystal size of hydroxyapatite, based on XRD; from microscopic analysis, a fusing of the crystals themselves (Franchet 1934; Herrmann 1977).

These stages represent a compilation of numerous studies of bone material, both archaeological and biomedical. As outlined above, these stages will be used for comparing results of the present study.

III. Materials and Techniques

A. *Experimental Bone*

Two sets of fresh artiodactyla bone were prepared for the experimental analysis. One set was used in scanning electron microscopy and the other for X-ray diffraction analysis. In addition to these analyses, both sets of bones were measured with Munsell color charts. Compact bone (as opposed to cancellous or subchondral bone) was chosen for analysis. Each bone specimen was cut from the diaphysis of a tibia. The average thickness of the bones was 0.5 cm, the width was always 1.0 cm, and the length varied from 1.0 to 2.25 cm. At certain temperatures the bone, when burned, exhibited a "core." The core of a particular bone specimen is analogous to that of a ceramic sherd, that is, a core is an interior area of the bone which was not exposed to an amount of oxygen necessary for complete combustion of the organic phase (cf Rice 1987:334).

The reason for choosing only artiodactyla bone is threefold: (1) it occurs frequently in archaeological deposits; (2) no evidence exists to suggest any major structural differences in either mineral or organic phases in mammalian taxa (cf Jowsey 1968; Armstrong and Singer 1965); and (3) variability in the surficial texture of subchondral and cancellous bone types is complex, especially when compared to the surface of compact bone.

B. *Burning*

In previous studies the exact temperatures at which specific thermally induced chemical or physical alterations occur, as well as the kind of bone analyzed (i.e., compact, subchondral, or cancellous), varies. In the present study, the experimental sets of bone were exposed to a range of temperatures (20, 130, 185, 240, 285, 340, 440, 500, 550, 600, 650, 750, 800, and 950° C) in an effort to pinpoint the temperatures at which alterations in bone chemistry and structure occur, in addition to replicating previous results.

The bones were exposed to the various temperatures for a duration of 2 hours. Buikstra and Swegle (1989) found that when bone (unburned, green, or fleshed) was exposed to heat it did not reach a calcined state in less than 70 minutes in some cases. In Therman and Willmore (1981:280) a humerus was observed to have blackened at first and then "calcined white" in 15 minutes in an oak fire. In the present research, the bones were placed in a muffle furnace alongside a thermocouple attached to a programmer. The programmer raised the temperature of the oven over a period of 2 hours until the desired temperature was reached, at which point the bone was allowed to dwell in the oven for 2 hours at the specified thermal increment. The bone was cooled for 4 hours inside the furnace. An identical heating program was used by Shipman *et al.* (1984).

C. *Scanning Electron Microscopy*

General macroscopic observations of bone structure can be made routinely with the unaided eye or low-powered hand lens. However, investigations into crystallographic structure of bone, particularly burned bone, require a high-powered analytical technique, such as scanning electron microscopy (SEM). As noted previously, the average size of an individual hydroxyapatite crystal is 0.2–0.3 μm in size, clearly beyond the range of low-powered imaging techniques.

The SEM is particularly useful for inspecting small-scale topographic relief. Under the scanning electron microscope, three specimens for each temperature were observed. Each specimen was obtained from an independent burning episode. Each specimen was mounted on a stub and coated with gold-palladium (AuPb) in order to make it conductive (Hayat 1978). In Shipman *et al.* (1984), magnifications from 15× to 15,000× were employed. After some exploration of contrast evident at various magnifications, 500× and 2500× magnification were used. A split-image dual-zoom mode allowed both levels of magnification to be compiled on one SEM image. Higher magnifications were informative on a few of the bones. During analysis, a representative surficial pattern was identified by making systematic transects across the surface of the bone.

D. *X-ray Diffraction*

X-ray diffraction (XRD) analysis was used to characterize changes in hydroxyapatite (HA) crystal morphology over a range of temperatures to which the bones were exposed. Bone specimens were prepared for XRD by grinding them into a fine powder with mortar and pestle. The powder was mixed with acetone to produce a dilute aqueous solution. The solution was spread over a glass slide in a smooth and even layer. This procedure assures minimal preferred orientation of the grains reducing variations in diffractogram patterns (Klug and Alexander 1974). A Picker diffractometer, using copper Kα radiation, was used. The machine parameters were: rate = 300, 2 Θ/minute, voltage = 36. Samples were prepared for bone burned at 20, 185, 285, 440, 500, 550, 600, 650, 800, and 950° C.

E. *Color*

Color, as a descriptive attribute, has been used consistently in burned bone studies (Table 1). The usefulness in measuring color of bone, as it is burned, comes from an implicit *a priori* assumption that when bone is burned, changes in color can be correlated with temperature. The dominant color exhibited by a bone was assumed to be the representative color of that particular temperature. Given that the shape and size of a bone may influence its color, in this study a standard size and shape of bone was used in all cases, in order to limit variation in color due to structural difference. By using a standardized size of bone, variation in color is assumed to be related only to variation in heating threshold. Munsell color charts were used to record the predominant hue, value, and chroma. All bone specimens were measured for color in a dry condition under incandescent lights. When a core (center of a bone specimen that was not completely oxidized) was present, its color was recorded. The Munsell measurements are defined as follows: (1) the *hue* notation of a color indicates its relation to red, yellow, green, blue, and purple; (2) the *value* notation indicates its lightness; and (3) the *chroma* notation indicates its strength (Munsell Color Company, Inc. 1954). In each munsell measurement, the hue is noted first, followed by value and chroma [e.g., 10YR (hue) 7 (value)/2 (chroma), or 10YR 7/2].

IV. Results

A. *Scanning Electron Microscopy*

Scanning electron microscopy examination of the surface of burned artiodactyla compact bone resulted in qualitative groupings, based on three trends or patterns in the surface morphology observed from the SEM images (Figure 1). Bones heated up to 340° C (130, 185, 240, 285, and 340° C) exhibit a slight qualitatively derived increase in porosity on the bone surface when compared to the surface of unheated bone. In the SEM images of bone burned at 440, 500, and 550° C, the surface exhibits a noticeable increase in surficial porosity. The surface appears considerably less homogeneous and smooth than that of unheated bone. Individual aggregates of bone material can be observed in the SEM images. At temperatures ranging from 600 to 650° C, the aggregates become poorly defined and appear globular (Figure 2). From 650 to 950° C little change in surficial porosity was observed. However, as the temperature approaches 950° C, the surface appears increasingly globular (cf Shipman *et al.* 1984).

B. *X-ray Diffraction*

The results of the X-ray diffraction analysis suggest a drastic change in the hydroxyapatite crystal morphology (Figure 3). Diffractograms for bone burned at 20 and 185° C show small, poorly defined peaks, but the diffractograms for bones burned at 285, 440, and 500° C exhibit an increase in area under the peaks and a slight narrowing of the 500° C peak. An increase in area under the diffractogram peak is seen again at 550 and 600° C, as well as additional narrowing of these two peaks. Finally, at 650, 800, and 950° C, the peaks become very narrow and well defined. It is interesting that the peak for bone burned at 650° C appears to have a greater area contained under the diffractogram than that of either of the two burned bone specimens heated at higher temperatures.

C. *Color*

In bones burned at temperatures ranging from 20 to 600° C, the color varies with each thermal increment (Figure 4). Unheated bone is pale yellow (2.5Y 8/4) to white (2.5Y 8/2) in color. It begins to turn very pale brown (10YR 8/4) to reddish yellow (7.5YR 6/6) as temperatures of 130 to 240° C are reached. From 240 to 340° C the bone goes from dark reddish brown (5YR 2.5/2) to black (5YR 2.5/1) to a very pale brown (10YR 7/3), and at 440° C a light brownish gray (10YR 6/2) color is exhibited. Cores (center of a bone specimen that was not completely oxidized), were detected in specimens burned at 440, 500, 550, and 600° C. From 600° C on, the sole color observed was a neutral white (N 10/0).

V. Interpretation of the Results

A. *Scanning Electron Microscopy*

The organic phase of bone makes up approximately 1/3 of the bone structure, and its progressive combustion from an unheated state (20° C) to 550° C accounts for the large increase in porosity represented in the SEM images. Individual aggre-

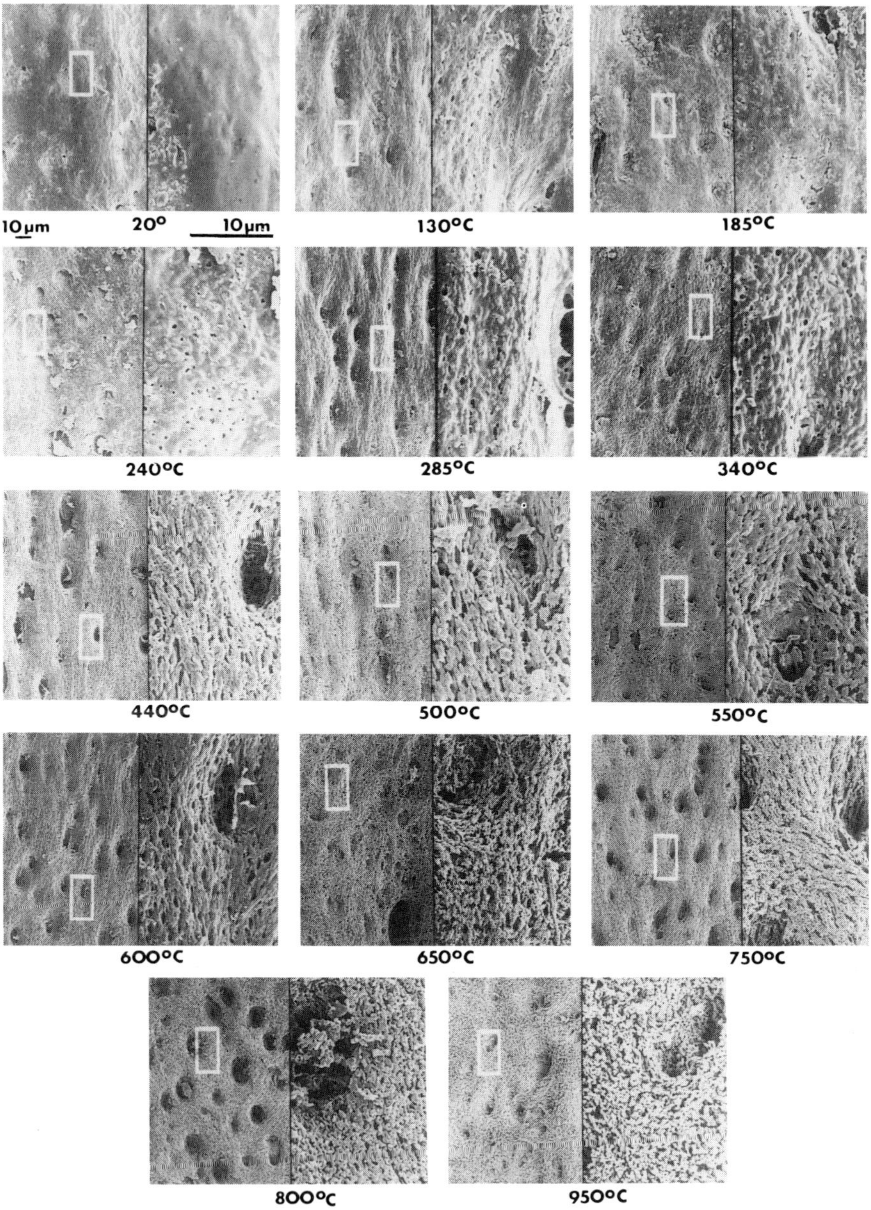

Figure 1 Scanning electron microscopy images of experimentally burned bone. A "split-image dual-zoom mode" was used in these and in the following SEM images. In the left side of each SEM image, the area encompassed by the white box is enlarged on the right side of the image. Note the increasing surficial porosity, especially after the temperature of 440° C is reached. At 650° C the bone surfaces become globular in appearance.

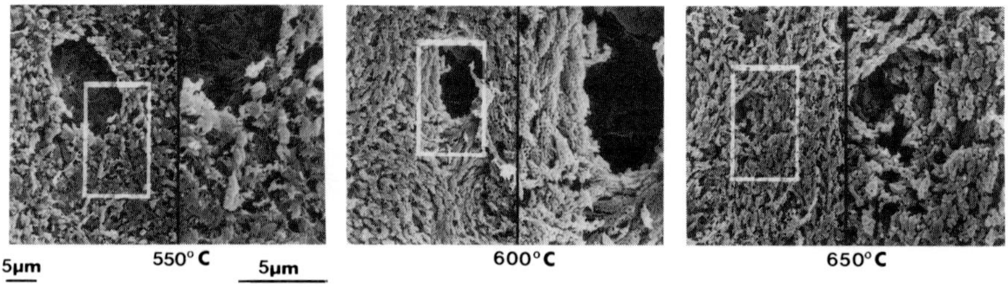

Figure 2 High magnification SEM images of experimentally burned bone. These three images illustrate the increasing globular appearance of the surface of experimentally burned bone. A magnification of 2500× was used on the left side of each image. The area encompassed in the white box is enlarged to a magnification of 6500× on the right side of the image.

gates of bone mineral can be observed in the high-magnification SEM images (Figure 2); the hydroxyapatite crystals are located in their loci of deposition, as they were arranged around the organic collagen fibrils before that phase was combusted. Scanning electron microscopy images of bone burned at 500° C are very similar to SEM images of chemically deproteinized bone (cf Jones and Boyde 1970). In the images of bone burned at 600° C, the crystals become somewhat fused, that is, the crystals on the surface of the bone look less distinct, giving the crystal surface a globular appearance. The surfaces of bones burned at 650, 750, 800, and 950° C appear increasingly fused in the SEM images. As exhibited by the SEM images, the trend is best described as an increase in porosity, largely as a function of the combustion of the organic phase, followed by a fusion of the individual hydroxyapatite crystals.

B. *X-ray Diffraction*

An examination of the X-ray diffractograms suggests two trends in the burned bone hydroxyapatite crystal morphology: (1) an increase in crystal size, as exhibited by the sharpening of the diffraction peak; and (2) a change in volume fraction of the crystallites themselves, as indicated by the gradual increase in area under each X-ray diffraction pattern (Figure 3). As the bone specimens are heated, the small hydroxyapatite crystals begin to grow at the expense of other crystals (i.e., the crystals get larger). Evidence for this lies in the sharpening of the diffraction peaks. Posner (1969) has done a detailed crystallographic study of hydroxyapatites, both biological and synthetic types. He notes that well-crystallized synthetic hydroxyapatite exhibits a better defined X-ray diffraction peak than the hydroxyapatite in bone, because in synthetic hydroxyapatite the crystals are larger. The difference is due to the preparation technique used when forming synthetic hydroxyapatite; well-crystallized synthetic hydroxyapatite is created by heating amorphous calcium phosphate. The formation of biological hydroxyapatites occurs continuously during life as hydroxyapatite crystals precipitate from amorphous calcium phosphate. The formation of the former includes heating at temperatures much greater than the latter.

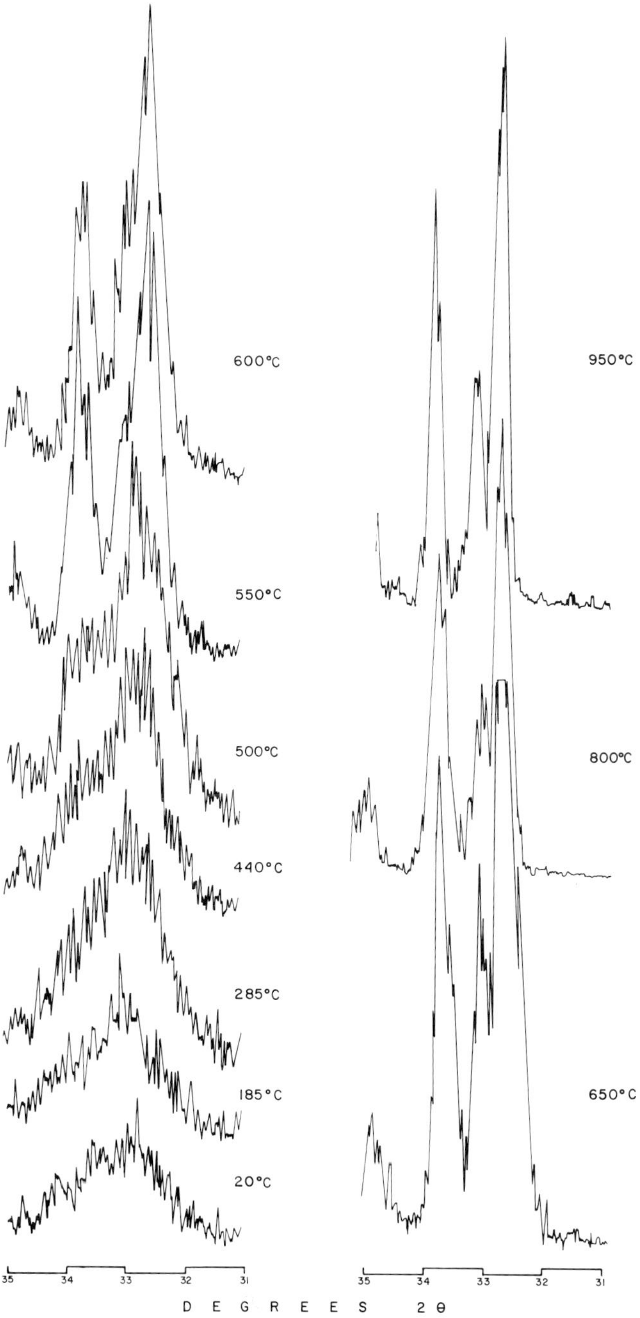

Figure 3 X-ray diffractograms of experimentally burned bone. Diffractograms of powdered specimens of heated bone. The °C on the right of each diffractogram denotes the temperature to which that particular bone was exposed. Note the drastic increase in peak intensity at 550° C and again at 650° C. The 650, 800, and 950° C peaks are identical to synthetic well-crystallized hydroxyapatite.

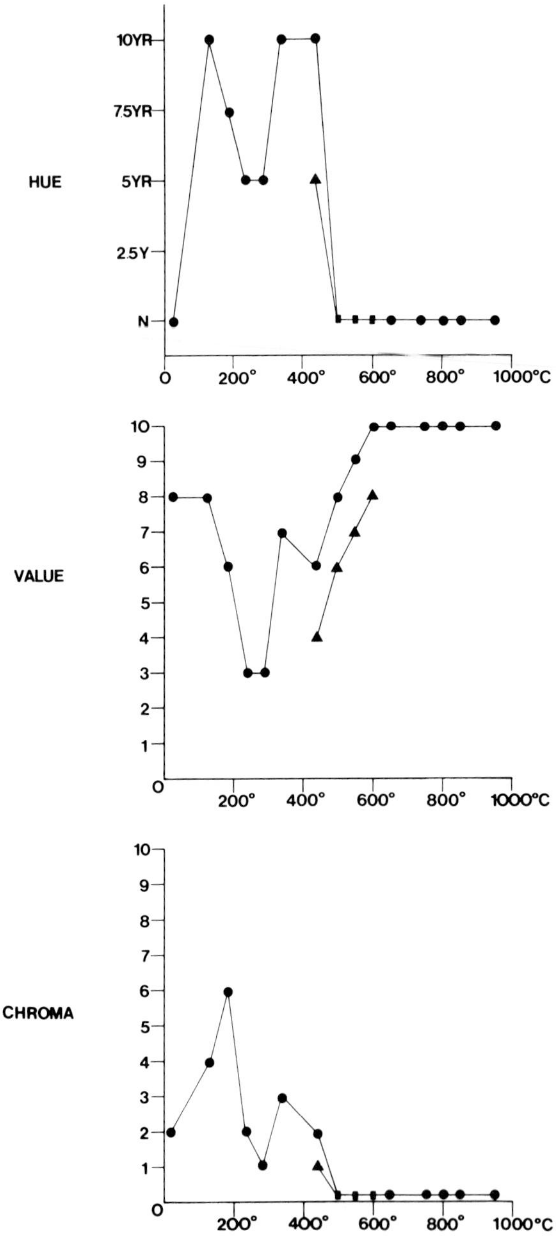

Figure 4 Munsell measurements of experimentally burned bone. In each of the graphs, the dark circles represent predominant hue, value, and chroma; the dark triangles represent the core hue, value, and chroma; and the dark squares represent Munsell measurements on specimens with the same predominant and core hue and chroma. Note that hue and chroma do not change after 440° C, but value, as a measurement of light and dark, documents the final combustion of any remaining organic matter in the bone structure.

As noted above, based on the increase in area under each diffraction peak, the volume fraction increased (i.e., the volume of the unit cell in the crystal lattice increased, an expansion of one thing at the expense of another). A detailed study of this phenomenon is beyond the scope of this study; however, previous studies suggest that the change in volume fraction is the result of an octocalcium phosphate phase converting to a beta-tricalcium phosphate phase (cf Armstrong and Singer 1965; Glimcher and Krane 1968; Knubovyets *et al.* 1979; Legros *et al.* 1978; Posner 1969), which results in the production of pyrophosphates and an elimination of crystalline water. A phase change on such a microscopic scale is not of any particular value in determining the maximum heating temperature to which a bone has been exposed, especially in an ordinal-level analysis.

C. *Color*

As noted previously, color of burned bone specimens changes with each thermal increment, up to 600° C, at which point the same hue, value, and chroma are recorded for each of the remaining thermal increments (Figure 4). The consistent change in color below 600° C demonstrates the differential combustion of the organic phase in bone at 130, 185, 240, 285, 340, 440, 500, and 550° C. Large variation in core color is documented only in the measurement of value (Figure 3). With the exception of the bone specimen burned at 440° C, the hue and chroma measurements are the same as the predominant color measured on all other burned bone specimens with cores present. Value as a measurement of lightness or darkness documents the progressive and complete combustion of organic material in the bones. As such, a measurement of value may be the best indication of bone material with a completely combusted organic phase.

The organic phase appears to be the main constituent of bone structure contributing to thermally induced color change. At thermal increments of 600, 650, 750, 800, 850, and 950° C a solitary color is present. All of these temperatures are well above the temperature at which organic matter will combust (i.e., 500° C), when exposed to an ample amount of oxygen (Dean 1974).

D. *Burned Bone Classification*

When the results of the SEM, X-ray diffraction, and Munsell color analyses are considered together, three classes of burned bone can be distinguished (Table 2). For a bone to be included in Class I it must yield a poorly defined diffraction pattern, an SEM image which exhibits minimal increase in surficial porosity, and a color measurement within the range of colors observed for experimental bones burned from 185 to 340° C. Admittedly, a single Munsell hue, value, or chroma measurement does not correlate with this entire range, and as such, I would suggest that color be used to further divide this first class into different levels (e.g., slightly smoked, smoked, charred, etc.).

Class II (440 to 600° C) consists of those bones that have had much, if not all, of their organic phase combusted. The SEM images show a marked increase in porosity; a function of the removal of the organic phase, leaving only hydroxyapatite

Table 2 Summary of Burned Bone Classes

Class	Temperature (°C)	Munsell	XRD	SEM	Interpretations
I	20	2.5YR 8/2 white	Slight increase in peak intensity	Surface becomes increasingly porous	Loss of free water, some carbonization of organics
	130	10YR 8/4 very pale brown			
	185	7.5YR 6/6 reddish yellow			
	240	5YR 2.5/2 dark reddish brown			
	285	5YR 2.5/1 black			
	340	10YR 7/3 very pale brown			
II	440	10YR 6/2 light brownish gray	Large increase in peak intensity; crystals grow bigger	Surface becomes very porous	Complete loss of organic phase, loss of crystal-bound water
	500	N 8/0 white			
	550	N 9/0 white			
	600	N 10/0 white			
III	650	N 10/0 white	Peaks become defined; resemble synthetic HA diffractograms	Crystal surface becomes fused; slight increase in surface porosity	Change in crystal size
	750	N 10/0 white			
	850	N 10/0 white			
	950	N 10/0 white			

crystals on the surface. The diffractograms are distinct from those that precede them with the exception of the peak at 440° C. The Munsell hue and chroma correlate well with each other in this range of temperatures as well. Value, as a Munsell measurement of lightness or darkness, documents the transitional loss of the organic phase from 440 to 600° C.

Finally, Class III consists of all those bones burned at 650 to 950° C. From the SEM images the texture of the hydroxyapatite mineral appears to remain more or less the same, but differs significantly from the previous class in that the size of the hydroxyapatite crystals changes as the crystals fuse together. The X-ray diffraction patterns are indicative of a larger crystal size than that found for bones heated at lower temperatures. All Munsell measurements are the same: White (N 10/0).

VI. Comparing the Experimental Results with Archaeological Data

To apply the experimentally derived classification to archaeological burned bone, bones from the British Camp shell midden were analyzed in the same manner as outlined above. Bone specimens were taken from facies that exhibited other evidence of burning (e.g., high ash content, burned shell, no humus), called tan facies. Tan facies 1T and 2C from excavation unit 310/302 were analyzed (Table 3). A "facies" is a deposit representative of a single depositional event in prehistory (Stein 1987).

Table 3 Summary of Archaeological Bone Specimens

Specimen number	Facies	Color	XRD	SEM
1	1T	10YR 8/4 very pale brown (I)	Slight increase in peak intensity (II)	Surface exhibits minimal porosity (I)
2	1T	5YR 4/1 dark gray (I)	Slight increase in peak intensity (II)	Surface exhibits minimal porosity (I)
3	2C	N 10/0 white (III)	Peak is well defined (III)	Porous surface, appears fused and globular (III)
4	2C	7.5YR 7/2 pinkish gray (I)	Peak is unusual (?)	Surface exhibits minimal porosity (I)
5	2C	N 9/10 white (II)	Peak is well defined (III)	Porous surface, but not fused (II)

The facies is the principal unit of excavation at the British Camp shell midden. Tan facies 1T is situated in the "light," upper stratigraphic layer and tan facies 2C is situated in the "dark," lower stratigraphic layer. The dark layer is described as a greasy, dark brown to black unit containing less shell than the light stratigraphic layer, which is described as ". . . shell-rich, lighter in color, and dry rather than greasy" (Nelson *et al.* 1986:1). "Burned" bone was intuitively identified previously from both stratigraphic units.

Bones identified as compact mammal bone were chosen for the classification. A total of five bone specimens were examined for evidence of burning; two from facies 1T and three from facies 2C (Table 3). Initially, Munsell color measurements were made on all five archaeological specimens and then SEM and X-ray diffraction analyses were undertaken in an effort to assess the accuracy with which color can be used to assign the five archaeological bone specimens to the burned bone classification. In Table 3, the results of Munsell, SEM, and XRD analyses are summarized. A Roman numeral is located next to each analytical summary. Each Roman numeral represents the Burned Bone Class (I, II, or III) each archaeological bone specimen would be assigned to based on the results of that particular analytical technique (color, SEM, or XRD).

A. *Results*

Bone specimens 1, 2, and 3 from unit 310/302, tan facies 1T and 2C have Munsell colors of 10YR 8/4, 5YR 4/1, and N 10/0 (Table 3). Similar Munsell measurements were given to bones burned at 130, 285, and 650° C, members of classes I and III (Table 2).

The XRD patterns for facies 1T, bone specimens 1 and 2, are similar to bone heated experimentally at less than 500° C (Figure 5). The diffraction pattern for bone specimen 3 from facies 2C is indicative of bone heated experimentally to 650° C or greater.

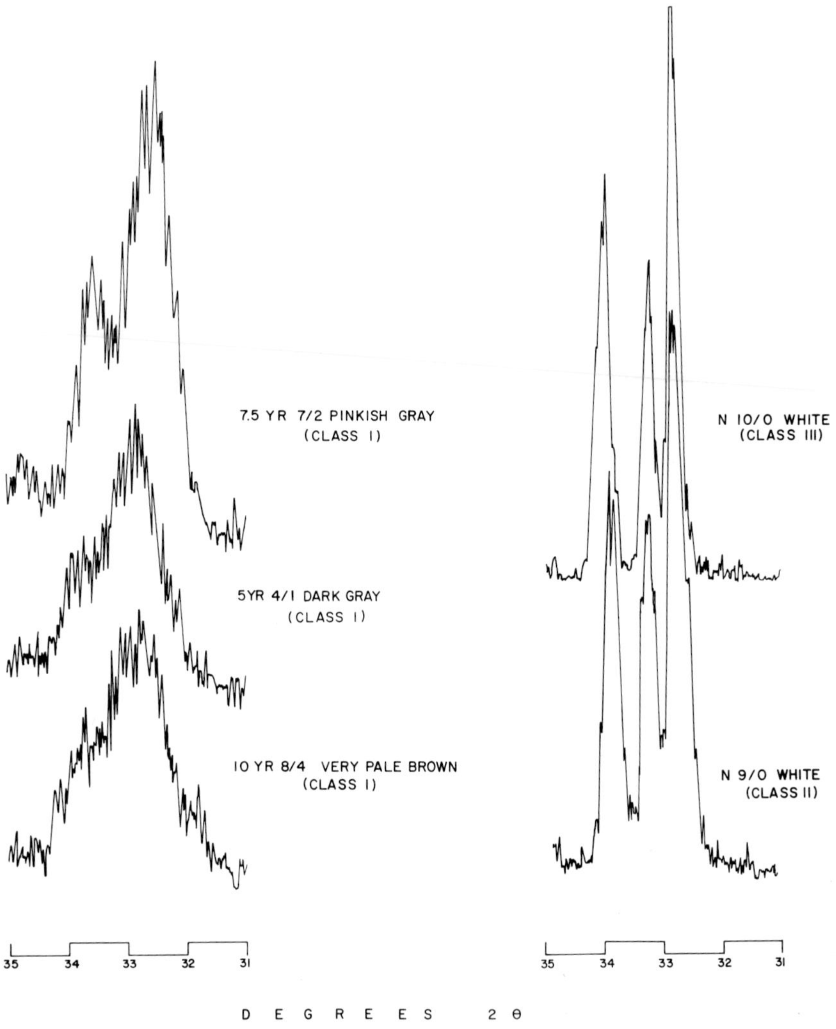

Figure 5 X-ray diffractograms of burned archaeological bone. The archeological specimen's Munsell measurement and the burned bone class to which that color belongs are located to the right of its diffractogram.

In comparing the SEM images of the bone specimens from facies 1T, the surfaces are similar in porosity and texture to fresh bone burned between 130 and 185° C, differing primarily in the amount of particles on the surface of the archaeological specimens (Figure 6, A and B).

B. *Interpretations*

The XRD and SEM analyses of bone specimens 1 and 2 from facies 1T indicate they belong to Class I, the same class previously assigned by color analysis alone (Table 3).

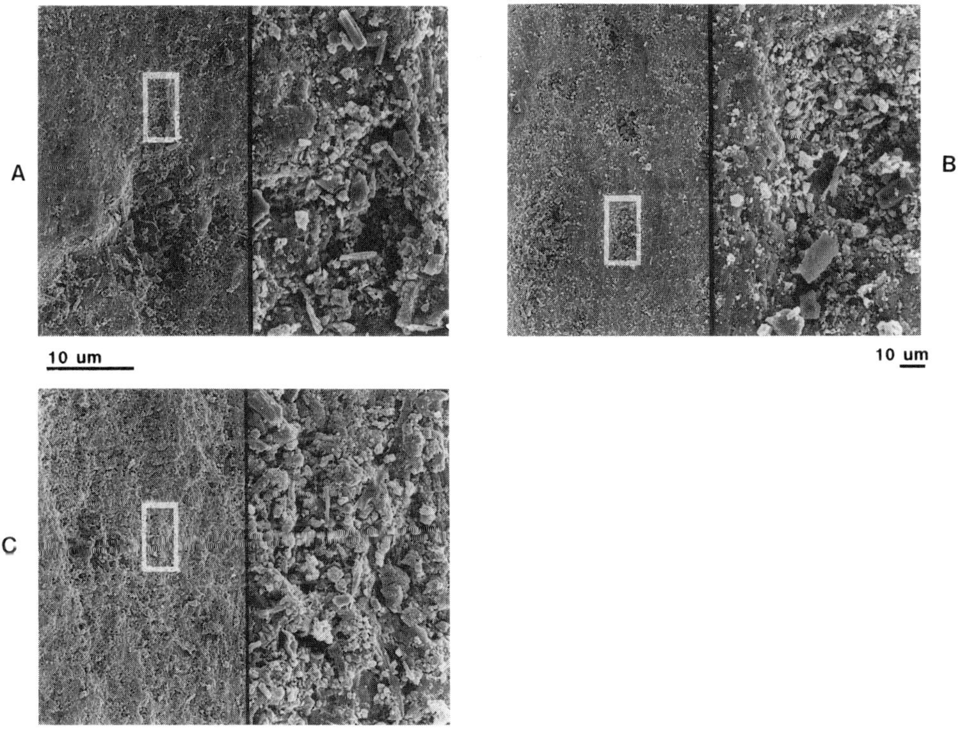

Figure 6 Scanning electron microscopy images of archaeological burned bone. A, Bone from upper light layer of shell midden, facies 1T, 10YR 8/4 very pale brown, Class I; B, bone from upper light layer of the shell midden, facies 1T, 5YR 4/1 dark gray, Class I; and C, bone from dark layer of shell midden, facies 2C, N 10/0 white, Class III.

Bone specimen 3, with a Munsell color of N 10/0, from facies 2C is assigned to Class III when SEM and XRD analyses are considered (Figures 5 and 6C; Table 3). The SEM image is very similar to images of bone burned experimentally at 650° C. Again, color assignment proves to be as reliable as SEM or XRD.

C. *Discrepancies*

Bone specimen 4 from facies 2C with a Munsell color of 7.5YR 7/2 is assigned to Class I based on color analysis alone (Table 3). The SEM image looks very similar to bone burned at 185° C (Figure 7a); however, the X-ray diffraction pattern is unusual (Figure 5). Based on color alone, bone specimen 5 with a Munsell of N 9/0 is assigned to Class II (Table 3). The SEM images are similar to those of the experimental bones from Class II (Figure 7B), but the XRD pattern suggests the archaeological bone should be a member of Class III (Figure 5).

Based on the above, spurious conclusions can result. Two technical issues warrant consideration. First, the technique used for mounting XRD specimens may

A

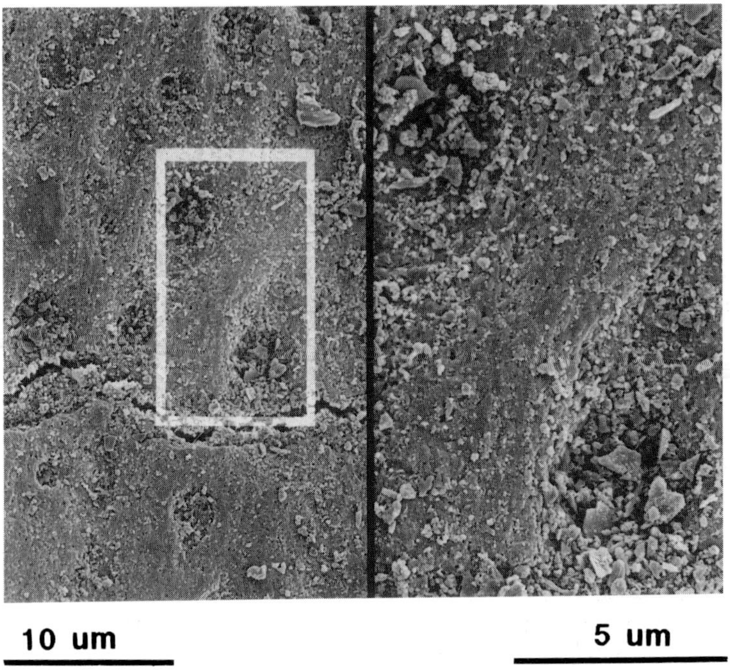

B

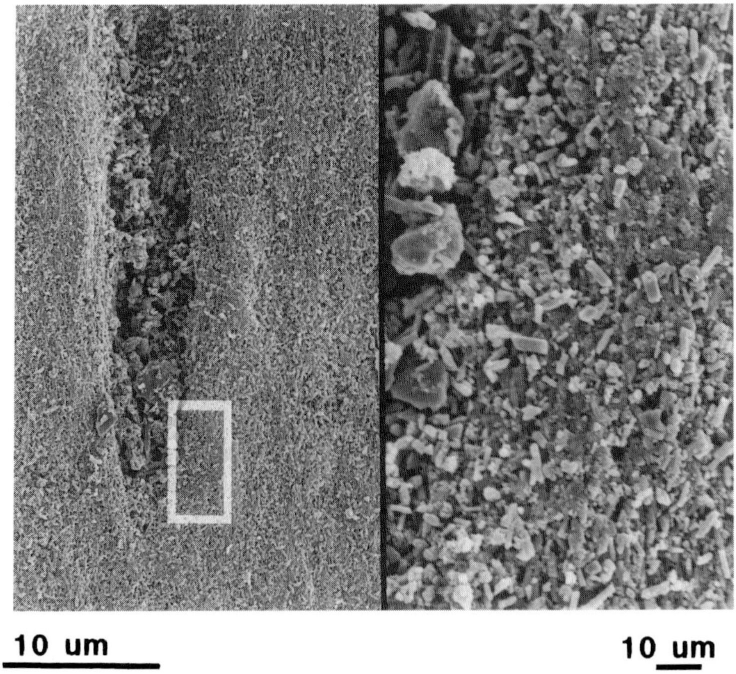

affect the diffractograms (Klug and Alexander 1974:374). As the specimen is bombarded by X rays, irregularities in the surface of the specimen or insufficient bone material can give rise to inconsistent diffractograms. Bones, when burnt, are very fragile and break easily. As such, the size of the archaeological specimen is small and normally does not provide an adequate sample size to ensure consistent results in XRD analysis. I suggest that heavier emphasis be placed on SEM imagery than on XRD analysis unless large samples are available and thereby more precise mounting techniques are available.

Second, color may not be a precise measurement of degree of thermal alteration (cf Shipman *et al.* 1984). Color of a bone is a function of the amount of organic phase left in the bone and the chemical and physical state of the remaining organic matter. Stimulated by paleodiet reconstructions, investigations into the diagenesis of buried bone have yielded important information regarding organic and mineral interactions in bone, particularly the role of diffusion (Parker and Toots 1980; Whitmer *et al.* 1989). Diffusion is the movement of minerals and/or individual elements between sediment and bone, and in the case of buried bone, ". . . occurs whenever elements are mobile, it is necessarily a basic component of diagenesis" (Whitmer *et al.* 1989). Minerals that may alter the color of a bone, burned or not, through diagenetic processes (diffusion) can move into the buried bone in two manners: (1) those that fill surface voids and holes; or (2) those that become chemically combined with the crystal structure of hydroxyapatite (Parker and Toots 1980:199). The role of diffusion (and its effect on color) in the diagenesis of buried burned bone in the British Camp shell midden has not been examined adequately, but the SEM images of bones from the deposit clearly show that minerals have moved into bone material voids (Figures 6 and 7).

VII. Postdepositional Processes in the Shell Midden

In a recent chemical investigation of the British Camp shell midden (Nelson *et al.* 1986; Stein in press; Stein, Ch. 7, this volume) stratification of light and dark layers in the midden are thought to be primary results of postdepositional alterations of the midden, and not depositional, as has been traditionally assumed. One of the differences between dark and light layers lies in the presence of water in the lower dark layer, in the form of the water table. Because there is groundwater in the dark layer, decomposition of organic matter and the concomitant production of organic acids has occurred. When analyzing the flora, fauna, and chemical attributes of the shell midden, the major differences between the dark and light layers are: (1) shell surfaces from the dark layer are etched making it difficult to identify the shells to species (Ford, Ch. 13, this volume); (2) carbonates have been leached from the dark layer (Stein, Ch. 7, this volume); and (3) artifact frequencies change (Kornbacher, Ch. 8, this volume).

Figure 7 Scanning electron microscopy images of archaeological bones. A, bone from facies 2C, 7.5YR 7/2 pinkish gray, Class I; B, bone from facies 2C, N 9/0 white, Class II. The SEM images resemble bones with similar Munsell measurements. Results of XRD analysis suggests that either color alone is not sufficient to accurately place these two archaeological specimens correctly into the burned bone classification, or misleading diffractograms can result from sample preparation.

A. *Problem*

Considering these results, I hypothesize that bones identified as having minor thermal alteration (i.e., bones which would be members of Class I) will show significant surficial differences (due to decomposition of organic phase) under the scanning electron microscope if they are from the dark layer as opposed to the light layer. In Jones and Boyde's (1970) "Experimental Studies on the Interpretation of Bone Surfaces with the SEM," bone material was deproteinized chemically to interpret the mineral state of bone at any given point in its life history. They made numerous SEM images of hydroxyapatite mineral as it appears on the surfaces of bone, which can be used to compare archaeological specimens from the dark layer of the shell midden with specimens from the light layer. If groundwater deproteinized bone in the lower dark layer, then the two sets of bone should not resemble each other.

B. *Results*

From both the light and dark layers, three bones were chosen on the basis of their Munsell color. Based solely on color, they are members of Class I. The bones from the light layer exhibit an SEM image in which there is a general smooth surface and only minimal porosity noted. Individual aggregates of hydroxyapatite crystal are not detectable (Figure 6, A and B).

Bones from the dark layer exhibit differences from those in the light layer with the same Munsell hue. The similarity of the surficial texture of the archaeological bone's SEM images, the degree of porosity, and the arrangement of hydroxyapatite crystals, to the deproteinized images is substantial, indicating some if not all of the organic constituents have decomposed (Figure 8).

C. *Discussion*

Based on the condition of bone as observed under the SEM, the bone in the dark layer has been affected at a greater rate than that in the light layer. Although the sample size of bone is small, the results agree with theoretical predictions. As Hare (1980:218) has noted, few data are available concerning the effects of groundwater and microorganisms on the organic phase and preservation of bone.

The effects of the groundwater in the dark layer, at the British Camp shell midden, on the color of burned bones, seems to be an important consideration. In the present study, SEM and XRD analyses were used to assess the accuracy with which bones can be correctly assigned to the burned bone classification. In two instances, spurious results were observed from bone specimens 4 and 5 from the dark layer. The chemistry of the dark layer may have affected the color of these bones. Future applications of the burned bone classifications should consider the chemical properties of the archaeological deposit and any postdepositional alterations that act differentially on one stratum as opposed to another. In addition, consideration should

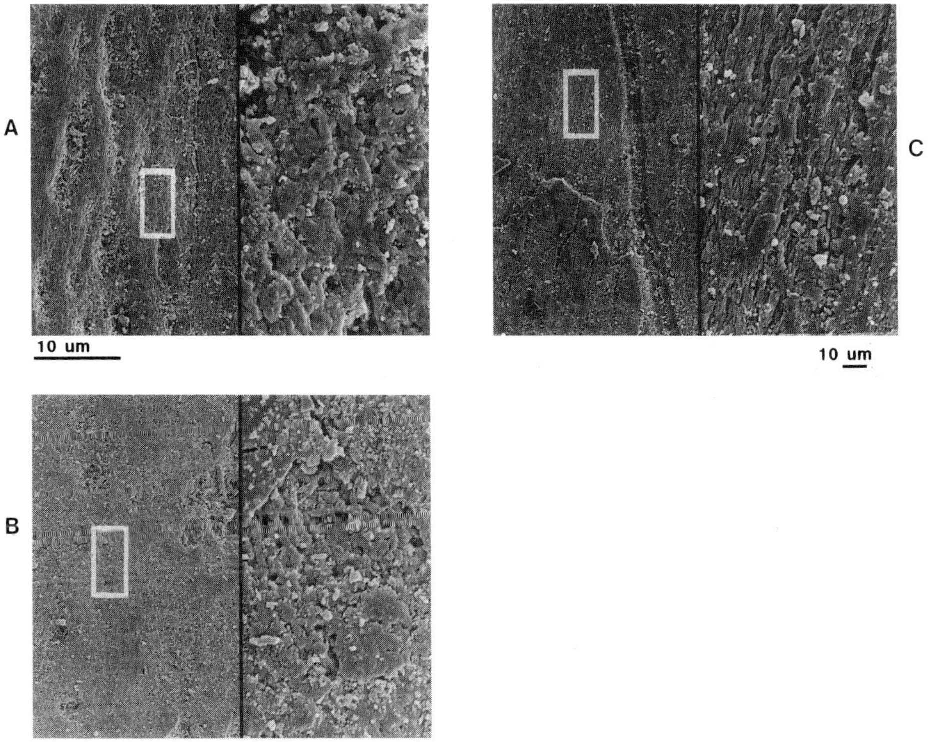

Figure 8 Scanning electron microscopy images of Class I burned archaeological bone. All three bones are from the lower dark layer of the shell midden facies 2C. Note the degree of porosity of the bone surfaces and their similarity (in porosity) to the experimentally burned bones at 440, 500, and 550° C (Figure 1).

be given to the preburning condition of bone. Although it was not considered here, the degree to which color changes proposed above can be generalized to small and large mammals, to nonmammalian species, and to bone burned in fleshed or dry conditions should be considered.

VIII. Conclusions

Inferences of the specific temperature at which a bone buried in an archaeological deposit has been burned are problematic, but ordinal-level assignments of burned bone into classes constructed from my experimentation can be made with a fair degree of accuracy based on color alone. My analysis of archaeological bone from "tan facies" at the British Camp site suggests that the bone was thermally altered. As an extension of this study I suggest that color be used to ordinally assign bone to the burned bone classification. From the resulting classes weights could be recorded and used to compare amounts of burned bone in tan facies, and other facies as well. Spatial data from such a study could yield structural information on the British Camp shell midden. Grayson (1991, and in preparation) is using a version of the burned bone classification to investigate "intrasite" structure in high-elevation archaeological deposits in the White Mountains, California.

While inferences of exact temperature were not suggested, ordinal assignments of degree of thermal alteration were made using color, SEM, and XRD analyses. The discrepancies that occurred in the results suggest that XRD analysis can yield inconsistent results and that a better understanding of the affect of groundwater on bone color is needed. Clearly, studies which render burned bone a more sensitive indicator of degrees of thermal alteration are useful in expanding our archaeological knowledge of not just shell middens but other prehistoric deposits containing burned bone as well.

Acknowledgments

My sincere appreciation is extended to those people and academic departments who aided in the completion of this research. The Material Science and Engineering, Botany, Geology, and Anthropology departments at the University of Washington permitted me access to the necessary equipment to undertake this project. Timothy Canaday provided expertise in identifying faunal archaeological specimens for analysis. Sarah C. Sherwood and Jane E. Buikstra provided editorial assistance in the preparation of the final draft. Finally, a special thanks is extended to Julie Stein for her insights into the geochemical aspects of the study and her editorial assistance, and Robert Dunnell for the guidance and encouragement to undertake and complete this project.

References

Anderson, J. E.

1963 The people of Fairty: An osteological analysis of an Iroquois ossuary. National Museum of Canada Bulletin 193, Contributions to Anthropology Part I, pp.28–129. Ottawa, Canada.

Angel, J. L., and C. S. Coon

1954 La Cotte de St. Brelade II: present status. *Man (London)* **54**, 53–55.

Armstrong, W. D., and L. Singer

1965 The composition and constitution of the mineral phase of bone. *Clinical Orthopaedics* **38**, 179–194.

Baby, R. S.

1954 Hopewell cremation practices. Ohio Historic Society, Papers in Archaeology 1.

Binford, L. R.

1963 An analysis of cremations from three Michigan sites. *Wisconsin Archaeologist* **44**, 98–110.

Bonucci, E., and G. Graziani

1975 Comparative thermogravimetric, x-ray diffraction and electron microscope investigations of burnt bones from recent, ancient and prehistoric age. *Atte della Accademia Nazionale dei Lincei, Rendiconti Sci. Ris. Mat. Nat. Series 8* **59**, 517–532.

Buikstra, J. E., and L. Goldstein

1973 The Perrins ledge crematory. Illinois State Museum Report of Investigations 28.

Buikstra, J. E., and M. Swegle

1989 Cremated bone: experimental evidence. In *Bone modification*, edited by M. Sorb and R. Bonnichsen. Pp.248–258. Orono: University of Maine Press.

Civjan, W. J., W. J. Selting, L. B. De Simon, G. C. Battistone, and M. F. Grower
1970 Characterization of osseous tissues by themogravimetric and physical techniques. *Journal of Dental Research* **51**(2), 539–542.

Dean, W. E., Jr.
1974 Determination of carbonate and organic matter in calcareous sediments and sedimentary rocks by loss-on-ignition: comparison with other methods. *Journal of Sedimentary Petrology* **44**, 242–248.

Franchet, L.
1934 [in Morel, Ch. Le Tumulus N de Freyssinel (causse de Sauverterre)]. *Bulletin Societe Prehistorique Francaise* **4**, 77–194.

Glimcher, M. J., and S. M. Krane
1968 The organization and structure of bone, and the mechanism of calcification. In *Treatise on Collagen*, edited by B. S. Gould. London: Academic Press. Pp.68–252.

Grayson, D. K.
n.d. Alpine faunas from the White Mountains, Califronia: adaptive change in the late prehistoric Great Basin. *Journal of Archaeological Science* (in press).

Hare, P. E.
1980 Organic geochemistry of bone and its relation to survival of bone in the natural environment. In *Fossils in the making: vertebrate taphonomy and paleoecology*, edited by A. Behrensmeyer and A. Hill. Chicago: University of Chicago Press. Pp.208–219.

Haury, E. W
1945 The excavation of Los Muertos and neighboring ruins in the Salt river valley, Southern Arizona. Papers of the Peabody Museum of American archaeology and ethnology, Harvard University, volume 24, number 1. Cambridge.

Hayat, M. A.
1978 Introduction to biological scanning electron microscopy. Baltimore: University Park Press.

Herring, F.
1972 The organic matrix of bone. In *Biochemistry and physiology of bone*, edited by O. H. Bourne. New York: Academic Press. Pp.127–190.

Herrmann, B.
1977 On histological investigations of cremated human remains. *Journal of Human Evolution* **6**, 101–103.

Holager, J.
1970 Thermogravimetric examination of enamel and dentine. *Journal of Dental Research* **49**, 546–548.

Jones, S. J., and A. Boyde
1970 Experimental studies on the interpretation of bone surfaces studied with the SEM. In *Scanning electron microscopy/1970*, proceedings of the third annual scanning electron microscope symposium. Chicago: ITT Research Institute. Pp.195–200.

Jowsey, J.
1968 Age and species differences in bone. *Cornell Veterinarian* **LVIII**, 74–94.

Knight, J.
1985 Differential preservation of calcined bone at the Hirundo site, Alton, Maine. Manuscript on file, Department of Quaternary Studies, University of Maine, Orono.

Klug, H. P., and L. E. Alexander
1974 *X-ray diffraction procedures*. New York: John Wiley & Sons.

Knubovyets, R. G., T. V. Zubkova, G. I. Cherenkova, and V. T. Gnedenkova
1979 Synthesis and study of carbonate-hydroxyl apatite. *Zhurnal Neorganicheskoi Khimii* **24**, 2072–2076.

Krogman, W. M.
1939 A guide to the identification of human skeletal material. Federal Bureau of Investigation, the law enforcement bulletin, volume 8, number 8. Washington D.C.

Legros, R., G. Bonel, N. Balmain, and M. Juster
1978 Precise identification of inorganic constituents of a bone by studying its thermal behavior. *Journal of Chim. Phys. Phys.-Chim. Biol.* **75**, 761–766.

McConnell, D.
1973 *Apatite: Its crystal chemistry, mineralogy, utilization, and geologic and biologic occurrences.* New York: Springer-Verlag.
Merbs, C. F.
1967 Cremated human remains from Point of Pines, Arizona: A new approach. *American Antiquity* **32**, 498–506.
Munsell Color Company, Inc.
1954 *Munsell Soil Color Charts.* Baltimore, Maryland.
Nelson, M., P. J. Ford, and J. K. Stein
1986 Turning midden into mush: evidence of acidic conditions in a shell midden. Paper given at the 51st annual meeting of the Society for American Archaeology. New Orleans, Louisiana.
Parker, R. B., and H. Toots
1980 Trace elements in bones as paleobiological indicators. In *Fossils in the making: vertebrate taphonomy and paleoecology*, edited by A. Behrensmeyer and A. Hill. Chicago: University of Chicago Press. Pp.197–207.
Pfeiffer, Susan
1977 The skeletal biology of archaic populations of the Great Lakes region. National Museum of Man Mercury Series. Archaeological Survey of Canada, Paper No. 64. National Museum of Canada, Ottawa.
Posner, A. S.
1969 Crystal chemistry of bone mineral. *Physiological Review* **49**(4), 760–793.
Posner, A. S., and D. Belts
1975 Synthetic amorphous calcium phosphate and its relation to bone mineral structure. *Accounts of Chemical Research* **8**, 273–281.
Rice, P.
1987 *Pottery analysis: a sourcebook.* Chicago: The University of Chicago Press.
Shipman, P., G. Foster, and M. Schoeninger
1984 Burnt bones and teeth: an experimental study of color, morphology, crystal structure and shrinkage. *Journal of Archaeological Science* **11**, 307–325.
Spence, T. F.
1967 The anatomical study of cremated bone fragments from archaeological sites. *Proceedings of the Prehistoric Society* **33**, 70–83.
Stein, J. K.
1987 Deposits for Archaeologists. *Advances in Archaeological Method and Theory* **11**, 319–396.
n.d. Formation processes of coastal sites: view from a Northwest Coast shell midden. In *Circum-Pacific prehistory conference: Western North America maritime prehistory*, edited by A. R. Blukis-Onat. Washington State University, Pullman (in press).
Stewart, T. D.
1979 *Essentials of forensic anthropology*. Springfield, Illinois: Charles C. Thomas.
Therman, M. D., and L. James Willmore
1981 A replicative cremation experiment. *North American Anthropologist* **2**(4), 275–283.
Ubelaker, D. H.
1984 Human skeletal remains: excavation, analysis, and interpretation. Manuals in Archaeology No. 2. Smithsonian Institution, Taraxacum Washington.
Wells, C.
1960 A study of cremation. *Antiquity* **34**(33), 29–37.
Whitmer, A. M., A. F. Ramenofshy, J. Thomas, L. J. Thibodeaux, S. D. Field, and B. J. Miller
1989 Stability or instability: the role of diffusion in trace element studies. *Archaeological Method and Theory* **1**, 205–273.

Index